TEXACO
AND THE
$10 BILLION JURY

To Tom Zaal
with best wishes
from Pat's brother
and a seeker after truth
(or something like that)

James Shannon

20-August-88

THE
PRENTICE HALL
CORPORATE LIBRARY

SHANNON, Texaco and the $10 Billion Jury

SLATER, This . . . is CBS

TEXACO
AND THE
$10 BILLION JURY

James Shannon

Prentice Hall
Englewood Cliffs, New Jersey 07632

Library of Congress Cataloging-in-Publication Data

SHANNON, JAMES, (date)
  Texaco and the $10 billion jury.

  Bibliography: p.
  Includes indexes.
  1. Pennzoil Company—Trials, litigation, etc.
2. Texaco, inc.—Trials, litigation, etc. 3. Getty
Oil Company. 4. Petroleum industry and trade—United
States—Consolidation. 5. Consolidation and merger of
corporations—United States. I. Title. II. Title:
Texaco and the ten billion dollar jury.
KF1866.P46S5   1988   345.73'0268   88–9885
ISBN 0–13–911959–0   347.305268

Editorial production supervision and interior design: *Denise Gannon*
Acquisitions editor: *Jeffrey A. Krames*
Copy editor: *Sally Ann Bailey*
Cover design: *Lundgren Graphics, Ltd.*
Manufacturing buyer: *Mary Ann Gloriande*

The publisher offers discounts on this book when ordered
in bulk quantities. For more information, write:

Special Sales/College Marketing
College Technical and Reference Division
Prentice-Hall, Inc.
Englewood Cliffs, New Jersey 07632

Printed in the United States of America

10 9 8 7 6 5 4 3 2 1

ISBN 0-13-911959-0

PRENTICE-HALL INTERNATIONAL (UK) LIMITED, *London*
PRENTICE-HALL OF AUSTRALIA PTY. LIMITED, *Sydney*
PRENTICE-HALL CANADA INC., *Toronto*
PRENTICE-HALL HISPANOAMERICANA, S.A., *Mexico*
PRENTICE-HALL OF INDIA PRIVATE LIMITED, *New Delhi*
PRENTICE-HALL OF JAPAN, INC., *Tokyo*
SIMON & SCHUSTER ASIA PTE. LTD., *Singapore*
EDITORA PRENTICE-HALL DO BRAZIL, LTDA., *Rio de Janeiro*

FOR MY PARENTS,
James J. and Shirley Kilpatrick Shannon,
who taught me the difference between right and wrong
and gave me the courage to do something about it.

CONTENTS

AUTHOR'S NOTE

This book is something of a hybrid—in part a personal memoir of my service on the jury in the case of *Pennzoil Co.* v. *Texaco, Inc.*, but also a presentation of actual documentary exhibits and testimony from the trial itself. The form in which it appears here did not immediately suggest itself; it evolved to meet the demands of the narrative I had constructed. The book is not the product of extensive interviews (although a large number of interviews were conducted) but rather the experiences of the author supplemented by a sheaf of depositions; appeal briefs; trial transcript; and newspaper, book, and magazine accounts.

The author wishes to acknowledge the encouragement and support of a number of people who made contributions to my consideration of this difficult case:

First, the lead attorneys from the trial, Richard B. "Dick" Miller for Texaco and Joseph D. Jamail for Pennzoil. After spending five months in the courtroom with these two gentlemen, I felt I knew them well enough to ask the tough questions, which they answered almost without exception. The generosity they expressed in sharing their time and thoughts with me greatly aided in my research and writing.

Similarly, input from Pennzoil attorneys John Jeffers and Irv Terrell of the Houston law firm of Baker & Botts and Texaco senior attorney David Luttinger was also of value.

My sincere thanks to Sarah McGill of Baker & Botts and Denise Davidson of Jamail & Kolius who did yeoman's work helping me to navigate the tens of thousands of pages of paper generated by the case.

This book would not have been published in anything resembling its present form without the help and guidance of my editor, Jeffrey A. Krames III of Prentice Hall. My debt to Jeff is significant. He believed in this project when it seemed few others did, and his enthusiasm was a source of light on those inevitable dark days when doubt crept into my studio. In a very real sense, this is his book too.

Production editor Denise Gannon's material contribution to the finished book is apparent on every page of the text. The value of her work in wrestling a bulky manuscript into a readable format cannot be overestimated.

Several reporters I came to know in the weeks and months after the trial made contributions to this book, though in some cases it was unwitting on their part. In particular, though, Patricia Manson of the *Houston Post*, Barbara Shook of the *Houston Chronicle*, and Thomas Petzinger, Jr., of the Houston bureau of *The Wall Street Journal* all displayed grace under pressure when the tide turned and *my* questions began to probe *them*.

Finally, the support of my friends and family who (almost) never tired of hearing me talk about the case was very important to me in the two years consumed by the preparation of the manuscript. A list like this is necessarily incomplete but must include my thanks to Bill and Cynthia Davis, Charles Shannon, Richard Glen Smith, Jack and Ann Falloure, Malcolm and Elizabeth Bales, T. Sean Shannon, Al, Ruth, Randy, and Melissa LeBlanc, John C. Bailey, John and Edith Shannon, and Edward Mallett.

Special thanks to my brother, Patrick Shannon, who contributed to the success of the project in ways both material and spiritual, and to my wife, Susie, an ever-present source of inspiration who was finally convinced that a writer can be working even when he is not sitting down at the keyboard.

J.S.

DRAMATIS PERSONAE

The Six Day War

THE GETTY ENTITIES:

1. GETTY OIL COMPANY
Management
  SIDNEY PETERSEN, Chairman and CEO
  R. DAVID COPLEY, General Counsel
  CHAUNCEY J. MEDBERRY III, Directory
  HENRY WENDT, Director
  LAURENCE TISCH, Director

Investment Banker: GOLDMAN, SACHS,
represented by GEOFFREY T. BOISI

Attorneys: BARTON J. WINOKUR
of Dechert, Price & Rhoads
HERBERT GALANT
of Fried, Frank, Harris, Shriver & Jacobson

2. SARAH C. GETTY TRUST (THE TRUST)
  GORDON P. GETTY, Trustee

Investment Banker: KIDDER, PEABODY & CO.
represented by MARTIN SEIGEL

Attorneys: MOSES LASKY, CHARLES COHLER,
THOMAS WOODHOUSE of Lasky, Haas, Cohler & Munter

3. J. PAUL GETTY TRUST (THE MUSEUM)
  HAROLD WILLIAMS, President

Investment Banker: SALOMON BROTHERS,
represented by JAY HIGGINS

Attorneys: MARTIN LIPTON and PATRICIA VLAHAKIS
of Wachtell, Lipton, Rosen & Katz

PENNZOIL

Management:
    J. HUGH LIEDTKE, Chairman and CEO
    BAINE P. KERR, President
    PERRY BARBER, General Counsel

Investment Banker: LAZARD FRERES & CO.
represented by JAMES GLANVILLE

Attorneys: ARTHUR LIMAN and SEYMOUR HERTZ
of Paul, Weiss, Rifkind, Wharton & Garrison

MOULTON GOODRUM and JOSEPH CIALONE
of Baker & Botts

TEXACO

Management:
    JOHN McKINLEY, Chairman and CEO
    JAMES KINNEAR, Vice Chairman
    ALFRED C. DeCRANE, President
    WILLIAM WEITZEL, General Counsel

Investment Banker: FIRST BOSTON, INC.
represented by JOSEPH PERELLA, BRUCE
WASSERSTEIN, and TOM PETRIE

Attorney: MORRIS KRAMER of Skadden, Arps, Slate,
Meagher & Flom

The Trial

PENNZOIL

Lead Attorney: JOSEPH D. JAMAIL of Jamail & Kolius

Attorneys: G. IRVIN TERRELL, JOHN JEFFERS,
RANDALL HOPKINS, SUSAN ROEHM, PAUL YETTER,
and BRIAN WUNDER
of Baker & Botts

JAMES KRONZER, Abraham & Kronzer

Of Counsel: MARK YUDOF, Dean
University of Texas Law School

TEXACO

Lead Attorney: RICHARD B. MILLER
of Miller, Keeton, Bristow & Brown

Attorneys: RICHARD KEETON, ROBERT BROWN,
R. MICHAEL PETERSON, and J. C. NICKENS
of Miller, Keeton, Bristow & Brown

Of Counsel: W. PAGE KEETON, Former Dean
University of Texas Law School

151st DISTRICT COURT

Judges: ANTHONY J. P. FARRIS (excused due to illness)
SOLOMON C. CASSEB

Bailiff: CARL SHAW

Clerk: CLIFTON BENNETT

Court Reporter: JACQUELINE MILES

THE JURY

FRED D. DANIELS	ISRAEL JACKSON
DIANA STEINMAN*	OLA SPERLING GUY
JUANITA SUAREZ	THERESA LADIG
RICHARD LAWLER	SHIRLEY J. WALL
SUSAN FLEMING	JAMES SHANNON
LILLIE FUTCH	DOUGLAS B. SIDEY†
LAURA JOHNSON	LINDA J. SONNIER†
VELINDA ALLEN	GILBERT STARKWEATHER†

---

*Excused.

†Alternate jurors.

INTRODUCTION

There are more pompous, arrogant, self-centered, mediocre-type people running corporate America who should be out on some postal route delivering mail.

JOE JAMAIL
PENNZOIL ATTORNEY

The free-enterprise system has gone to hell.

LEE A. IACOCCA

Poor Texaco. This hidebound corporate giant had stood on the sidelines as hundreds of millions of barrels of oil changed hands in the megamergers of the early 1980s.

Despite a reputation in the industry for hardball tactics, Texaco's avowed policy to participate only in "friendly" takeovers had caused the firm to walk away from a bid for Conoco, whose lucrative reserves were later acquired by DuPont.

Texaco's record for finding new oil reserves had been one of the worst in the industry. And when the oil crunch of 1974 confronted the industry, Texaco entered a period of steep decline. It could no longer rely on a cheap, plentiful supply of foreign oil, and its older fields in this country had begun to run dry. Oil reserves are the lifeblood of any oil company, especially one with the large, thirsty refining and marketing capacity of Texaco.

With exploration and drilling costs soaring (and given the firm's miserable record in the field), Texaco was desperate to replenish its dwindling oil reserves, which had plummeted dramatically in recent years. Currently, most of the "new" reserves were being found on Wall Street.

On January 2, 1984, Texaco's chairman, John McKinley, was vacationing on his Alabama ranch. At Texaco headquarters in White Plains, New York, President Alfred DeCrane was monitoring the rapidly changing situation over at the Getty Oil Company.

Pennzoil, the Houston-based oil company that had made a tender offer for 20 percent of the stock of Getty the week before, and was currently in New York negotiating with the various Getty factions.

Trading in Getty stock was suspended on the New York Stock Exchange at the company's request. Although there had been no public announcement, something was clearly happening.

Investment bankers are crucial players in the mergers and acquisitions game, where takeovers with billion-dollar price tags are not uncommon. The only way investment bankers can lose on a major acquisition is if they don't represent one of the parties involved—the raiders, the targets, or the so-called "white knights."

Joseph Perella and Bruce Wasserstein of the investment banking firm First Boston had identified Getty Oil as a possible takeover target months before. They had conducted an in-depth study of the reserves and assets of Getty, assembling the information an acquiring firm would need to evaluate a potential bid. Their only problem was that even as the Getty endgame was being played out, First Boston still didn't have a client. A last-ditch round of phone calls ensued.

Back at Texaco, Al DeCrane took a frantic phone call from Joe Perella. First Boston had a pipeline into the Getty situation, and Perella provided DeCrane with some details from this inside source. Handwritten notes taken by DeCrane during this phone call and subsequent calls from Wasserstein would show the evolution of Texaco's thinking about Getty—and the role that inside information from First Boston played in that thinking.

Even as Chairman McKinley hurried back to New York on January 4, DeCrane was defining the strategy that would lead to Texaco's move on Getty, while still maintaining that Texaco only participated in "friendly" mergers. That morning, however, the Getty Oil Company and the two Getty trusts that were the majority shareholders of the company had announced a merger agreement—with Pennzoil.

Texaco was about to give a whole new definition to the word "friendly."

Nearly two years later, I find myself sitting in a cramped back room on the fifth floor of the old Harris County Courthouse in Houston, Texas. I can't help reflecting on the chain of events just outlined and how those events and the participants involved have dominated my life for the past five months.

Gathered around a table piled high with hundreds of pages of documents admitted as evidence during the trial sit twelve average citizens from all walks of life: eight women and four men, white, black, and Hispanic.

In the trial that will conclude this day, the actions of the "friendly" Texaco will invoke the wrath of the members of the jury—ordinary people with no ax to grind, who have served in this case at great personal sacrifice. This has been no day at the beach.

The conversation is subdued as the morning wears on; we have finally reached our verdict. We are unanimous on all counts and damage awards except one. A member of the jury is unwilling to accept the reduction in punitive damages thrashed out by the other jurors. We have been instructed that punitive damages are also called exemplary damages, and are intended to punish a guilty party for illegal conduct and to serve as an example to others not to engage in this type of conduct.

I fully identify with the view held by the recalcitrant juror. A persuasive argument has been made that Texaco's actions have merited the full measure of damages allowed by law. However, I had recognized that after awarding the full measure of actual damages the previous afternoon, several of the jurors might require some rationale to award punitive damages in a manner commensurate with the actions of Texaco. A final review of the critical evidence and a fairly well-thought-out formula on punitive damages I had articulated to the panel had finally won over the others, many of whom had initially shared the feelings of our last holdout.

She sighed, shook her head, and looked sadly from face to face around the table: "After what they did . . . I don't like it, but all right." Weary resignation hung heavy in the room. There was no joy in the task we had been called upon to perform.

Our presiding juror knocked on the door and told the bailiff we had a message for the judge. It was a message that would send shock waves throughout the country in ways we could never have imagined.

The trial of *Pennzoil Co.* v. *Texaco, Inc.* was a struggle of titanic proportion, a real battle of the corporate giants. Some have likened this case to the biblical tale of David and Goliath—but in fact, although Texaco is one of the largest, most powerful corporations in the world and a genuine Goliath, Pennzoil is no David. A large, diverse company firmly rooted in the upper echelon of the *Fortune* 500, Pennzoil had sufficient resources and the will to use them in pursuing this lawsuit. Otherwise, it very likely would not have succeeded in bringing Texaco to the bar of justice.

The events that made up the basic facts of this case occurred during a period I call the Six Day War—from January 1, when Pennzoil and Gordon Getty reached their initial accord, to January 6, when Texaco announced that it, in fact, would acquire all of Getty Oil.

With some important exceptions, these facts were not generally in dispute. What was bitterly contested at every turn was the interpreta-tion of these facts. That's what the jury was there to decide.

The list of witnesses who gave testimony during the trial reads like a *Who's Who* in American business and law:

Investor *Laurence Tisch* (now president and chief executive officer of CBS) was a Getty director for less than thirty days and pocketed over $4 million on the Texaco deal after exercising Getty stock op-tions.

Arthur Liman (later Senate counsel in the Iran-Contra hearings) was Pennzoil's New York attorney and emissary to the critical Getty board meeting. Equally adept as a trial lawyer and corporate repre-sentative, Liman was a savvy witness who greatly advanced Pennz-oil's case.

Thomas Barrow (former chairman of Kennecott Copper and vice-chairman of Sohio) was an impressive expert witness for Pennzoil. He appeared reluctantly, and was not paid for his testimony.

Chauncey J. Medberry III (feisty former chairman of the Bank of America) was a long-time Getty director. Despite his best efforts, he was not a very good witness for Texaco.

Martin Lipton (inventor of the "poison pill" defense and the top take-over lawyer in the country by most accounts) was a critical Texaco witness who's testimony cinched the case—for Pennzoil.

Gordon Getty (who inherited control of the massive family fortune by default) had previously spent his time pursuing his twin passions of anthropology and music. Then he gained control of 40 percent of the Getty Oil Company. Obviously, things changed.

Martin Siegel (with Kidder, Peabody & Co., Gordon's investment banker) was a key player throughout the Six Day War and the subse-quent lawsuits. What no one could have known was that less than a year later he would plead guilty to criminal charges for passing in-side information to Ivan Boesky, who used the information to make huge profits speculating in a number of takeover stocks, including Getty Oil.

John McKinley (Texaco's chairman) was called as a hostile witness by Pennzoil. The soft-spoken CEO didn't serve his own cause particu-larly well.

J. Hugh Liedtke (Pennzoil's chairman) was an impressive witness in his own behalf. He survived a rigorous cross-examination and Tex-aco's best efforts to portray him as the villain of the story.

Despite the magnitude of the stars involved, the trial was virtually ignored by the business community and the news media. With the reams of copy that have been generated since the verdict, it is remarkable that *The Wall Street Journal* didn't send over a reporter from its Houston bureau three blocks from the courthouse. After a preliminary story filed during jury selection, the *Journal* ignored the parade of Lipton, Tisch, McKinley, et al., to the witness stand.

The post trial aspects of the case are more important than they might first seem, and have greatly colored public perception of what had actually transpired. With no national coverage of the trial or the issues at stake in the case, the huge judgment seemed to have come out of nowhere.

Traditionally, the common tactic for the losing party in a trial is to decry the verdict, proclaim its innocence, and express confidence that the verdict will be overturned on appeal.

After the trial, Texaco did all this and a whole lot more. To fill the sudden demand for information about the case, the normally inaccessible Texaco retained Hill & Knowlton, the world's largest public relations firm, which within a matter of days, painted a picture of Texaco as the victim of a "legal lynch mob" in a kangaroo court that perpetrated the "Texaco Common Law Massacre." This last quote was the title of not one but two editorials in *The Wall Street Journal.*

This attitude was adopted to a lesser extent by many reporters and editors; Texaco's version of the case was given wide credence. The expertise of Hill & Knowlton maximized the effectiveness of what I felt was largely a disinformation campaign. Texaco was able to have gross distortions and outright lies printed as fact. It widely circulated "evidence" it didn't dare bring to court, where it would be subject to cross-examination and verification.

The jurors were not left out of all the fun, however. We were portrayed as "ignorant," "unsophisticated," "confused," "regionally biased," "anti-Semitic," " a jury run amok," and on and on.

No doubt the first accounts you read contained some or all of the elements described here. Later accounts would balance the coverage somewhat, but the "Texaco as innocent victim" theme was played early and often.

In fact, without that vituperative campaign, this book might never have been written. Having involuntarily surrendered five months of my life to this case, my inclination was to walk away and not look back.

Stung by the attacks on the intellect and character of the jury, however, I wrote a lengthy, strongly worded opinion for the *Houston Chronicle* that was published five days after the end of the trial. (It is reproduced elsewhere in this book.)

My naive hope was to balance the coverage of the case with a view from the jury box and, at the same time, achieve a degree of personal catharsis to get this bizarre interlude out of my system and return to my own life. Instead, the opinion piece provoked immediate and intense reaction from all sides. It was widely excerpted in newspaper and magazine accounts, with some reporters making it seem I had given the statements quoted in response to some probing questions. I have since learned that this practice is not at all uncommon.

Invitations to defend the verdict on radio and television followed, and I accepted a few of these, as had Richard Lawler, the jury foreman.

After the *Chronicle* piece broke, several reporters asked me if I planned to write a book. During the trial, I had often thought that the case would make compelling reading but that it was not my book to write. I had never read or even heard of a book by a juror (I have since read at least half a dozen). The sharp reaction to the opinion piece generated the first serious consideration of what ultimately became this book.

The issue of jurors writing about their courtoom experiences has recently been called into question. It was raised by the actions of some jurors in the trial of the so-called "subway vigilante" Bernhard Goetz, who sold accounts of their experiences to rival tabloids, the *New York Post* and *the Daily News*. This prompted cries of outrage from Goetz's lawyer, Barry Slotnick, who urged that jurors be prohibited from writing about the cases they hear by enacting laws similar to those that prohibit criminals from profiting from accounts of their crimes.

The other example of "juror journalism" cited in the ensuing news coverage was the book you are reading now, whose imminent publication had become known by the time the Goetz jury story broke.

Although somewhat chagrined at being lumped together with the Goetz jury, I fully support their right to freedom of expression. At the same time I harbor grave doubts about the sort of law proposed by Mr. Slotnick. Besides being constitutionally dubious in the extreme, such a law ignores the fact that, for decades, prosecutors, defense lawyers, and judges have written books about the cases they tried. Should citizens who have fulfilled their jury duty be stripped of their First Amendment rights just because some lawyer didn't like what they said about him?

The current laws and rules of procedure offer adequate safeguards against a trial being tainted by someone getting on the jury with the idea of commercially exploiting it. The manner in which jury pannels are selected makes such a move mathematically unlikely, if not impossible. Under existing law, the examination of prospective jurors can easily include questions on this particular issue if an attorney feels it is that important.

The draconian measures Slotnick proposed are far more chilling than any documented abuses.

Texaco wasn't thrilled by a juror who turned out to be a fly in the ointment of its massive public relations campaign. It filed two counts of Jury Misconduct against me in its motion for a new trial; for good measure, it filed one count of Jury Misconduct against Lawler, the only other member of the jury to defend the verdict openly. Since the firm couldn't silence us and our posttrial statements had little legal significance, Texaco fabricated charges in an attempt to discredit us. Fortunately, the trial judge dismissed these baseless allegations out of hand, but not before they had been given wide circulation in the national press—no doubt the reason they were filed in the first place.

The charges of jury misconduct were made with a bang; they died with a whimper. Even though Texaco had offered no evidence to support its claims, this alleged jury misconduct has remained part its continuing lament.

When a prominent national political figure was acquitted in a criminal case last year, I could identify with his plea to the horde of reporters who had sensationalized the serious charges: "Where do I go to get my good name back?"

Thankfully, personal vindication is not the driving force behind this book. Among the many individuals Texaco has attempted to tar with the brush of its own wrongdoing, I count myself in pretty good company.

The Texas judicial system doesn't require my defense; neither does Pennzoil. It is my firm belief that the evidence in this case is so overwhelming that any twelve fair-minded Americans in any state would have had the courage to say "No" to Texaco.

This jury didn't award Pennzoil the largest damage award in legal history on the basis of a handshake, as has been widely reported. There was a signed merger agreement between Pennzoil and the majority stockholders of Getty Oil. You might not have known that; it's not very sexy, it doesn't make for "hot copy," but it's true.

Rather than merely describe that agreement or tell you what it says, it is reproduced in this book for your inspection, along with other critical pieces of evidence. As documents admitted into evidence during the trial, they are a matter of public record. Their use didn't require permission from Pennzoil, Texaco, or what's left of Getty Oil. The decision to include them here is strictly my own.

Although hundreds of documents totaling thousands of pages were admitted into evidence, I have focused on those that played a major role in the jury's decision. These include exhibits Texaco used to bolster its case. Evidence that preponderates both ways is included; this deck doesn't need to be stacked.

Reading these documents is not much fun, but in an attempt to place the reader in the jury box, you will see what we saw. The highlighted sections and explanatory notes were prepared for this book and are intended to help guide you through this paper maze. They are clearly marked, so the original document can be easily discerned.

Rather than utilize the complicated document numbering system of the trial (where the same document introduced by both sides would be assigned two different exhibit numbers), I will call the documents by the name commonly used in the trial; hence, the Memorandum of Agreement, the "Dear Hugh" letter, the "DeCrane notes," and so on. In some instances, I have included only the critical page(s) of a multipage document: page 63 of the "Copley notes," pages 8 and 13 of the "Lynch notes," and so on.

Similarly, much of the trial testimony was repetitive. In an attempt to make the case come alive for the reader, the crucial sections are included, as are exchanges that capture the flavor of the testimony. The actual transcript of the trial is 25,445 pages long, with many more volumes of pretrial depositions, appellate briefs, and opinions. It takes up five times more space on the bookshelf than the *World Book Encyclopedia*, so even lengthy excerpts are necessarily incomplete.

If your inclination at this point is to throw up your hands, take heart. What this book covers in several hours of reading required five often-tedious months of involuntary servitude for the jury.

I don't believe that the story of *Pennzoil* v. *Texaco* is a metaphor for a crumbling society, but it is absolutely as American as apple pie. Our system of justice is unique in all the world. It requires twelve citizens, with no ax to grind in the matter at hand, to wield the awesome power of the state. It is a deadly serious responsibility, no matter what the nature of the case.

Although jury verdicts are sometimes unpopular, they are infrequently overturned. Appeals courts have to determine if the law was correctly applied, but the words on the pages of the transcript don't equal the experience of the jury seeing and hearing the witnesses.

Neither are appeals courts allowed to consider extraneous issues not introduced in the original trial. One example of this is what Pennzoil attorneys have called the "widows and orphans" argument, part of Texaco's postverdict legal strategy that sought to excuse the firm's conduct because the judgment will adversely affect employees and stockholders. The latter group includes numerous retirees and pension plans.

As the District Court of Appeals said in its unanimous decision upholding the verdict, although "we are sympathetic with those who might be affected by the verdict through no fault of their own, we are not authorized by law to substitute our judgment for that of the jury."

Every member of the jury has genuine compassion for those employees and shareholders who have been hurt by this verdict. Their grievance is a legitimate one, but it is with Texaco management, not Pennzoil, the courts, or the jury.

Here, for you to decide, is the case of *Pennzoil* v. *Texaco*. Take your seat in the jury box.

PART I

BLOOD IN THE WATER

The Getty Saga

("Blood in the water") . . . refers to a situation
where a target company has been wounded, in
effect has bled, and as we all know in that
situation, when sharks are around, as soon as
there is blood in the water, they immediately
close in on the victim.

BARTON J. WINOKUR
GETTY OIL ATTORNEY

My father always told me that all business men
were sons-of-bitches, but I never believed him
until now.

JOHN F. KENNEDY

1

The beginnings of the imbroglio that would lead to the landmark legal decision in a Texas courtroom in 1985 did not involve Texaco or Pennzoil, but a family and an oil company named Getty.

The year 1983 was one of discontent for the Los Angeles–based Getty Oil Company. When news of this great internal strife was made public in December of that year, the company was identified as a possible takeover target. In other words, the blood was in the water.

The roots of the dispute could be traced back more than fifty years, when J. Paul Getty was locked in a struggle for control of the family business with his 78-year-old mother, Sarah C. Getty.

When his father, George Getty, died in 1930, he willed the bulk of his estate to his wife. By that time, J. Paul Getty had already been married three times, divorced twice (eventually five times), and spawned the first two of his five children. Fearful he would bring his profligate ways to the family business, George left him only 5 percent of his $10 million estate.

A Getty family tradition was born as mother and son battled over the Getty Oil Company, which had been formed in 1916. Generations of lawyers have been profiting from Getty family feuds ever since.

By 1934, ill health and constant pressure from her son led the strong-willed Sarah to agree to a compromise solution that would extend benefits to a generation of Gettys yet unborn. J. Paul Getty would indeed exercise control over the company and its resources, but the bulk of the money and stock would be placed in a trust that would benefit her grandchildren and great grandchildren.

The terms of the Sarah C. Getty Trust, as it was called, were spelled out in an eighteen-page instrument that was executed on December 31, 1934.

Although it named J. Paul Getty as trustee, it limited his ability to dispose of the assets of the trust (primarily stock in the Getty Oil Company) except "to save the trust estate from a substantial loss." He would receive 5 percent of the Trust's annual income as trustee, with the rest distributed to his sons at his discretion. After the last of his sons died, the remainder of the Trust would be divided among their children.

Many years after Sarah Getty and her son had gone to their respective rewards, copies of the Trust instrument would be distributed to members of the jury in a Texas courtroom to puzzle over as attorneys and witnesses argued over its meaning.

J. Paul Getty accepted the terms of the Trust and spent the next forty-two years pursuing his greatest passion—making money. By acquiring other companies and discovering new oilfields overseas, he built Getty Oil into one of the nation's largest integrated oil companies.

An integrated oil company takes its oil from the ground through the refinery and delivers it to the consumer at the gas pump. Getty Oil bought leases, drilled oil wells, operated refineries, and sold gasoline at its own filling stations under various brand names.

Essentially a loner, J. Paul Getty ran an extremely profitable empire that nonetheless was overshadowed by the giant oil companies, the so-called "Seven Sisters." As he told *The New York Times* in 1974, "Actually, I've never felt very rich, because I've always been in a business where I was a moderate-sized fellow compared to Exxon, Shell, Gulf, Socal. I'm a small-sized fellow, a small-sized outfit, so I've never had delusions of grandeur."

The "small-sized fellow" broke new ground in 1949, when he negotiated with King Ibn Saud for oil concessions in a strip of land between Saudi Arabia and Kuwait called the Neutral Zone. After several dry holes, the gamble paid off. By 1955, Getty Oil had fifty-five producing wells in the Neutral Zone.

Despite his business success, J. Paul Getty remained a relatively unknown figure to the public. Then, in 1957, *Fortune* magazine published a list of the wealthiest people in the United States. In first place, above the Rockefellers, the Fords, and the Mellons was J. Paul Getty.

The notoriety that would follow Getty for the rest of his life can in large part be traced to the sensation created by the *Fortune* article. Other magazines and newspapers followed suit, with the result that the sobriquet "legendary billionaire" frequently preceded his name in print. For his part, Getty did little to discourage the attention lavished on him; if anything, the opposite would seem to be true.

The exploits of J. Paul Getty have been well documented in a number of books over the years, including his rather fanciful autobiography that he had revised at least once. His dynamic success in business did not extend into his personal life. As has been noted, he was married five times and had five sons.

Jeanette Dilmont was the first Mrs. J. Paul Getty. The marriage took place in 1923. His first son, George Franklin, was born in 1924. He divorced Jeanette a year later.

His second marriage was to Allene Ashby in 1926 and produced no children. They were divorced in 1928.

A few months later in Havana, Cuba, he married Adolphine ("Fini") Helme, the daughter of a German industrialist. Their union lasted long enough to produce his second son, Ronald. When divorce terms were being negotiated in 1932, his wife's father demanded and won a large financial settlement for his daughter. Getty, who knew how to drive a hard bargain, could not abide being squeezed. He was also a man who knew how to carry a grudge. When he died forty-four years later, his son Ronald was still excluded as a beneficiary of the family trust.

In 1932, when he was 40 years old, he married his fourth wife, Ann Rork, an actress half his age. She bore him two sons, Paul, Jr., and Gordon, before they were divorced in 1935.

Louise ("Teddy") Lynch was a nightclub singer in New York who became the last Mrs. Getty in 1939. His fifth son, Timothy, was born prematurely in California in 1946. Getty was in Tulsa on business at the time, but he quickly returned to Los Angeles. The child was frail and anemic at birth, and would be plagued with health problems throughout his brief life.

After the birth of Timothy, Getty's focus returned, as always, to business. It was in this period that his business interests in Europe and the Middle East were rapidly escalating, and he relocated to England. He would never return to America again for any extended period. He ran his far-flung business interests by letters, cablegrams, and phone calls, with the assistance of one harried secretary.

His marriage to Teddy was essentially over, although they would not be divorced until 1958, when Timothy died at age 12. In his short, unhappy life, he had been frequently hospitalized, operated on for a brain tumor, and gone blind. Although J. Paul Getty often professed his love for his youngest son, he was thousands of miles away when the boy died.

J. Paul Getty was not the first father who, while he may have loved his children, had a hard time expressing it. Nor would he be the last. Obviously, this cannot excuse the lack of attention he gave his children. Following the death of his youngest son in 1958, Getty was even more determined to build a family dynasty that would endure for generations.

Even though his contact with the offspring of his previous marriages was sporadic, at best, it is revealing that his four remaining sons all chose to follow their distant father into the family business.

George Getty, his oldest son, was the first. Despite an uneasy relationship with his father, he achieved a great degree of business success, working his way up the ladder. By age 34, he was already president of Tidewater Oil Company, a key Getty acquisition. Later, he would become chief operating officer of Getty Oil Company. It is questionable how much independence George enjoyed in these positions, however. His father was a stern taskmaster whose reach easily extended from England to California.

Ronald, the second son, was in charge of marketing for Getty's European operations. He was given his big chance to head Getty operations in France, but didn't last long. He hired a number of executives from a rival oil company, violating a French law that prohibits a company from seducing the employees of another. He had to leave the country immediately to avoid prosecution. Chastened, he returned to California to try his hand in the movie business, with little success.

Paul was sent to Italy to take over Getty Oil Italiana, and managed to stick it out until the mid-1960s, when he dropped out, grew a beard and long hair, quit his job, and left his wife Gail and their four children, including Paul III. This is no different from what happened to a lot of those of his generation at that time, but the Italian paparazzi had a field day with the son and namesake of J. Paul Getty. The resulting photos soon began appearing in scandal sheets around the world.

Needless to say, the old man was less than thrilled. In 1966, Paul married Talitha Pol, a Dutch actress, and hung out with the Rolling Stones. He also became involved in drugs. His second marriage produced a son, whom they whimsically named Tara Gabriel Galaxy Gramaphone Getty. Paul's psychedelic sojourn came to an abrupt and tragic end in 1971, when Talitha died of a massive heroin overdose. She was 30 years old.

Gordon, the youngest surviving son, had also entered the family business after graduating from college in 1956. He was sent to the Neutral Zone in 1958 to learn the nuts and bolts of one of Getty Oil's most lucrative operations, the oil concessions J. Paul Getty had obtained from King Saud in 1949.

Many accounts have depicted Gordon during this period as not the most worldly of young men. At any rate, in the Neutral Zone a tangle of seemingly minor events escalated into a major incident that placed him at odds with the sensitive laws of Saudi Arabia.

As recounted in Russell Miller's *The House of Getty*, Gordon had inadvertently managed to offend the local emir almost immediately upon his arrival. This proved most inconvenient several months later when a Getty Oil truck crashed into a pipeline, causing serious damage. When the driver and his immediate superior fled the country, the hapless Gordon found himself the victim of the perverse "order of succession" of Saudi law that assigns blame for the actions of subordinates to their superiors.

This situation was greatly exacerbated by his new enemy, the emir. Gordon was arrested and held under house arrest for two weeks. He later recalled using his two weeks of confinement to read the complete works of Keats as well as "Measure for Measure" and other Shakespeare plays.

In defense of Gordon, a witness at a trial twenty-five years later about another matter entirely would recall the incident and testify that he, too, had been to Saudi Arabia and "it's not very hard to get into trouble over there."

J. Paul Getty was livid, however, no doubt fearing the potential economic impact on his business interests in the region. Upon his release, Gordon was recalled from the Neutral Zone and safely exiled back to the states as a consultant on such family business interests as a mobile home factory in Tulsa and the Pierre Hotel in New York, where it was felt he could do no harm. There is no indication that his recommendations received much serious consideration, and the financial rewards were commensurate with the low level of the position and his father's tightfistedness.

Gordon's early withdrawal from the family business may in retrospect seem inevitable, given the events just recounted. He devoted much of the next twenty years to the study of music and anthropology. He began to compose music seriously and worked with the prestigious J. S. Leakey Foundation, in funding anthropological exploration. Gordon's inability to please his famous father, however, matched that of his brothers.

All except for George. J. Paul Getty's oldest son was like his father in many ways: he loved business and was notoriously frugal. It was George who came to represent the old man's great hope for a family dynasty. Although they would argue bitterly over business matters and there was never any doubt about who was in charge, George strove mightily to please the chairman of the board.

"Getty Oil expanded its operations and made significant strides during the years George was its chief operating officer," his father would record. "I took great pride in his accomplishments and derived confidence from them. I felt certain that the business founded by my father and built further by me would eventually pass into the most capable hands."

These hopes were cruelly dashed in early June 1973. George Getty, the chosen son, died suddenly under mysterious circumstances. He was 48 years old. The official coroner's report listed the cause of death as an overdose of alcohol and barbiturates.

A month later, J. Paul Getty III disappeared in Rome. As a teenager, he had reacted to the desertion of his father Paul by emulating him, letting his hair grow long, running wild, and taking drugs. The Italian scandal sheets described him as "the Golden Hippie" and breathlessly regaled readers with accounts of his motorcycle and car crashes and affairs with older women before he was 16 years old.

Nine days after his disappearance his mother received a note demanding payment of several million dollars in ransom or the boy would be killed.

At Sutton Place, his Tudor mansion outside London, J. Paul Getty was still grieving over George's death. Executives of Getty Oil Italiana had kept him informed of Paul III's escapades. He was not at all convinced that the kidnapping of his grandson was genuine and initially declined to pay.

In a provocative demonstration of their intent, the kidnappers cut off Paul III's ear and mailed it to an Italian newspaper. The incident made headlines around the world. At that point, the ransom was quickly paid, but not before the old man did a little bargaining. He loaned the ransom money to Paul at 4 percent interest, to be repaid from his share of the proceeds of the Trust. By making it appear to the kidnappers that his son would have to pay the ransom, it was reduced to $850,000. The old man's parsimony was an integral part of the legend; this was not the first time he had used it to save himself some money.

But at what cost? Despite persistent rumors that Paul III had engineered his own abduction, there was little doubt that he had been deeply traumatized by the experience. The trauma lingered for years; he became dependent on tranquilizers and at age 24 suffered a severe stroke—aggravated by alcohol—that left him blind and paralyzed.

Paul was no more responsive to his son than his father had been to him. Living in London as a virtual recluse, he flatly refused to pay his son's medical bills. His only public comment about the matter was in a written statement to the *London Times*, where he said: "Anyone who believes I am unmoved by my son's tragedy, or willing to see him become a public charge, simply does not know me. I have never failed to meet my obligations toward my children under the legal settlements as agreed and my paternal responsibilities as I see them."

Ultimately, Paul III's mother had to file suit in California to force Paul to assume responsibility for his son's medical bills. The suit was settled out of court.

*   *   *

J. Paul Getty was 81 when George died, and never really recovered from the shock of his death. In 1976, after an illness of several months, he died at Sutton Place at the age of 83.

His last will and testment left the bulk of his personal fortune to the museum in California where he had shipped the many pieces of art he had acquired over the years. Overnight, the J. Paul Getty Museum in Malibu became by far the most richly endowed museum in the world. Its principal asset was 4 million shares of stock in the Getty Oil Company, valued then at around $700 million.

The lion's share of the Getty wealth, however, remained concentrated in the Sarah C. Getty Trust. Its value had grown exponentially since that 1934 compromise between a mother and a son.

The death of George, the dissipation of Paul, and the disqualification of Ronald had constituted a process of elimination that left Gordon as the successor trustee. To ensure the survival of the Trust, J. Paul Getty had provided that his long-time lawyer C. Lansing Hays serve as co-trustee, with a California bank named as a second trustee.

The continuity of the Trust was essential for the stability of the Getty Oil Company. The Sarah C. Getty Trust owned over 40 percent of the company's stock.

2

In 1976, Gordon Getty took the seat on the board of the Getty Oil Company vacated by the death of his famous father.

C. Lansing Hays, the old man's long-time friend and trusted legal advisor, also sat on the Getty board. His status was enhanced by his position as co-trustee of the Sarah C. Getty Trust (which owned 40 percent of the company stock). J. Paul Getty had designated the Security Pacific Bank as a corporate co-trustee. The bank declined the appointment, fearing unprofitable entanglements with the litigious family. This left Gordon as the only other co-trustee, a fact that would later prove to be of great significance.

J. Paul Getty had run his empire from abroad for over two decades of his life, so his death did not create in an immediate vacuum in the company's management. In addition, the domineering Hays routinely continued the old man's practice of subjecting Getty Oil's plans and policies to sharp questioning and generally behaved as if he believed himself to be "J. Paul's messenger on earth."

Hays also became something of a mentor to Gordon, who almost always followed his lead in matters brought to a vote before the Getty board. One vote of particular importance was the elevation of Sidney Petersen to chairman and chief executive officer. The dapper, urbane Petersen had been with Getty Oil virtually all of his adult life, working his way up through a variety of positions on the financial side of the company. Although he now ran the eighth largest oil company in the world, he was not an oilman.

Only three years older than Gordon Getty, Petersen had been with the company during Gordon's generally disastrous attempts at becoming a productive company executive through a series of low-level positions he had been placed in by his father.

In 1980, Lansing Hays merged his small law firm with the New York office of Dechert, Price & Rhoads, a 109-year-old firm with over two

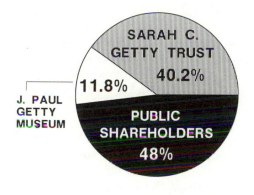

The GETTY Oil Company

Ownership in December, 1983

SARAH C. GETTY TRUST 40.2%

11.8%

J. PAUL GETTY MUSEUM

PUBLIC SHAREHOLDERS 48%

hundred lawyers. A large amount of the Getty Oil legal work went to Barton Winokur, a hard-nosed 40-year-old partner described as "strong-willed, outgoing, and articulate" and "a rising star" within the firm. Winokur quickly established a close working relationship with Getty CEO Sidney Petersen.

Gordon's confidence was not handed to Winokur along with the company's legal work; it remained exclusively with Lansing Hays.

In May 1982, there was a crucial turn of events. Lansing Hays died. Gordon lost his mentor and at the same time become sole trustee of the Sarah C. Getty Trust. Needless to say, things changed.

Under the terms of the Trust, the annual income was distributed to the sons of J. Paul Getty. Since virtually all the Trust's assets consisted of those 32 million shares of Getty Oil stock, the annual income was restricted to the dividend paid on those shares. Gordon's share of that income had exceeded $28 million in the most recent year, but he became convinced that the company could do much better.

The ownership situation at Getty Oil was unique among large, publicly held companies (such as those whose shares are traded on the New York Stock Exchange). The majority of the shares were held by two trusts: the Sarah C. Getty Trust, with 32 million shares (40.2 percent), and the J. Paul Getty Museum in Malibu, with 9 million shares (11.8 percent). The remaining 38 million shares (48.0 percent) were held by public stockholders, none of whom owned a significant percentage.

In 1981, Harold Williams had become president of the Museum. The former chairman of Norton Simon had also been chairman of the Securities & Exchange Commission under President Carter. Despite that background, Williams was a logical choice to head the J. Paul Getty Museum. Although it housed one of the world's most valuable private art collections, its huge endowment consisted almost exclusively of shares of stock in a *Fortune* 500 company. The pivotal position of the Museum's block of shares resulted from its role as the swing vote in any power struggle within the Getty Oil Company.

In early 1983, Gordon launched a campaign to increase the value of the company's stock, which would in turn increase the value of the Trust.

There was no doubt the stock was undervalued. A respected oil industry analyst had calculated that the stock (price in the $50 range) reflected less than one-third the actual value of the assets of Getty Oil. To exacerbate this fact, company's profits were also down, performing at a level below the industry standard.

Under Petersen's stewardship, the company had launched a large-scale diversification program, investing in areas other than oil and gas, such as insurance and cable TV. Getty Oil was the majority owner of ESPN, the cable TV sports channel that had yet to show a profit. The insurance venture had proven more lucrative, however. In truth, the verdict was still out on the Petersen diversification program.

Gordon's desire to improve the company's performance undoubtedly stemmed from a genuine desire to improve shareholder earnings (including his own). But it ran deeper than that. As one of his close advisors described the situation, "with Hays dead and with Petersen acting like he disdained him, and with Winokur having no relationship with him because he was a Petersen man, Gordon felt left out and like he wanted in."

It seems likely that Petersen did, in fact, disdain Gordon. Accounts of every misstep of Gordon's earlier career were well known among company insiders. After working his way up the ladder to the pinnacle of Getty Oil, Petersen may have felt that *he* was the true lineal descendant of J. Paul Getty, who had held the title Petersen now so proudly clutched to his chest: Chairman of the Board.

The clash between Sidney Petersen and Gordon Getty began in earnest in early 1983. Convinced that major changes were needed, Gordon presented a number of proposals to the company.

Included among these was an ambitious equity restructuring program. Under this plan, Getty Oil would buy back enough of its outstanding shares to cause the stock price to rise. By reducing the total number of shares, it would also have the effect of raising the Trust's percentage from 40 percent to a majority.

Another proposal recommended backing off from Petersen's diversification program to concentrate on finding new oil and gas reserves, an area where the company had historically enjoyed its greatest success.

All Gordon's suggestions were rejected out of hand by the princely Petersen, who had the support of the Getty board. A large majority of the board members shared Petersen's view of Gordon as an unqualified upstart.

When Gordon announced that the Trust would retain investment bankers Goldman, Sachs & Co. to "evaluate what their options were," he finally got Petersen's attention. From his long experience in the world of finance, Petersen knew that a 40 percent shareholder retaining an investment banker would be tantamount to hanging a "For Sale" sign on the entire company. With merger activity increasingly dominating the financial pages, such a move could have fatal implications for Petersen's vision of a strong, diverse industry giant.

Instead, he convinced Gordon to allow the company to retain Goldman, Sachs. Mollified by the fact that one of his proposals had finally found a receptive ear (and mindful of saving the Trust astronomical fees charged by investment bankers), Gordon acquiesced.

It would prove a hollow victory for Petersen, however. In July, Goldman, Sachs presented its analysis of Getty Oil's options. Prominently mentioned was a stock repurchase plan not unlike the one proposed by Gordon, which was again rejected by Petersen and the board. Feelings were running high on both sides.

Petersen apparently realized his grounds for opposing the plan were eroding, so he offered a counterproposal: the stock repurchase plan would be undertaken to increase the value of the shares, but although this would increase the Trust's position to over 50 percent, Gordon would be restricted to his current 40 percent level of shareholder votes.

A dismayed Gordon vehemently rejected the plan, which he could hardly interpret as anything other than a slap in the face. Petersen may have possessed considerable financial prowess, but communication skill was not apparent in his dealings with Gordon Getty. Although Gordon had voted for Petersen's appointment as chairman, his respect for him had to have been eroded by the events of the past few months.

This scenario is not offered as a psychological profile of either man, but if Petersen could have found some way to accommodate Gordon's desire to play a meaningful role in the affairs of the company that bore his family name, there is a strong likelihood that Getty Oil would still be intact today. Having at last attained the whip hand, however, Petersen was not inclined to relinquish it to any stockholder, even one with 40 percent.

The Museum, with almost 12 percent of the company stock, had played no direct role in these events. By remaining neutral, it in fact sided with management. Its paramount interest was in preserving the value of the Getty shares that represented the bulk of the fabulous endowment.

To achieve that goal for the Museum, Williams had hired his old friend Martin Lipton, perhaps the top mergers and acquisitions (M & A) lawyer in the country. His law firm, Wachtell, Lipton, Rosen & Katz of New York, had gotten rich from the its lucrative M & A work. Lipton, who had invented the "poison pill"* defense for companies that were takeover targets, had been involved in some capacity in practically all the big corporate mergers of recent years.

Gordon Getty also served on the Museum's board of directors. As at Getty Oil, his influence on the Museum board was minimal.

When Petersen heard rumblings that Gordon was trying to convince the Museum to join him in forming a majority that would take control of Getty Oil, he decided that preventive action had to be taken.

These actions commenced on October 1 and would ultimately trigger a war ninety days later, a war in which Getty Oil would not be a combatant but the prize sought.

Corporate protocol (and indeed, the bylaws of most corporations) requires advance notice to all directors of impending board meetings. Aware that Gordon Getty was in London for a meeting of the Museum board, Petersen decided to utilize a little finesse. He called a board meeting for Sunday, October 2. The other members of the board were notified by telephone, but a telegram announcing the meeting was delivered to Gordon's San Francisco home on the afternoon of Saturday, October 1.

The meeting was held as scheduled; Gordon was still in London. Although telephonic hookups are frequently employed to bring a far-flung board together, no such effort was made to ensure Gordon's participation.

---

*A "poison pill" is an issue of securities designed to discourage a hostile merger bid. Upon completion of the takeover, the typical poison pill stock becomes convertible into cash or the common stock of the raider.

This was by design, of course. In Gordon's absence, the board authorized the issuance of an additional 9 million shares of company stock. Unlike the stock repurchase first recommended by Gordon, then by Goldman, Sachs, this new plan would result in the Trust and Museum shares combined totaling less than 50 percent. The Getty board also discussed a way to shackle Gordon's control as sole trustee of the Trust (and its 40 percent).

The resourceful Winokur told the board he had contacted Gordon's brother, Paul, still living a reclusive existence in England. On behalf of the company, Winokur encouraged the heirs to file a suit seeking the appointment of a co-trustee, because he said Gordon's meddling in company affairs ran contrary to the best interests of the trust.

The company then helped Paul retain a prominent Los Angeles attorney, Seth Hufstedler, to represent Paul's son from his second marriage, the aforementioned Tara Gabriel Galaxy Gramaphone Getty, with the intention of filing just such a suit.

Winokur told the board that the company would want to support the lawsuit, or indeed file it independently if it didn't work out with Tara Getty or some other heir. The board authorized management to encourage the suit, intervene on its behalf when it was filed, or have the company itself file the action to tie a tin can to Gordon's tail.

Armed with these weapons (the authorization to issue new shares as well as file a lawsuit against Gordon), Petersen and Winokur flew to London, where Gordon and the Museum representatives were meeting. They also brought along Herbert Galant, a noted lawyer with the New York law firm of Fried, Frank, Harris, Shriver & Jacobson, who had been retained as special counsel for the company, as well as Geoffrey Boisi of Goldman, Sachs, which was still employed by Getty Oil.

Galant explained their London trip this way: "We wanted to make a last-ditch effort at a deal with Gordon. We knew that actually issuing the shares would be laying down the gauntlet. It would dilute Gordon and the Museum, but it would so advertise all the strife that a new buyer could come in, buy shares on the open market, and team up with Gordon or Gordon and the Museum and take control. So we went over to make a deal."

In London on October 3, Gordon and his lawyers met with Lipton and Williams for the Museum, and later that night with Petersen's group. No agreements came out of either of those sessions.

The next day, Lipton met with the Petersen contingent, which indicated they were prepared to take unspecified action to prevent the Museum and the Trust from acting together to take control of the company.

Under Delaware corporate law (Getty Oil, like many large corporations is chartered in Delaware), the majority of stockholders can dictate policy to the board of directors or remove the board if the members refused to comply. This was the sword hanging over the heads of Petersen and the Getty board. They could be removed at the whim of Gordon and the Museum, if they ever got together.

Lipton is a resourceful man. While unwilling to cast the Museum's lot with Gordon, he was equally disturbed by the company plan, which would cause the Museum shares to lose their pivotal status. Neither alternative served the interests of his client.

While still in London, Lipton proposed a formal cease-fire between the three parties called the Standstill Agreement. Under this proposed agreement, the company, the Trust, and the Museum would put their hostilities on hold for eighteen months; no party would take any action to alter the current balance of power. As an incentive, Gordon would be allowed to appoint someone, an ally, to the Getty board. Museum president Williams would also be added to the board. The intervening time would allow tempers to cool and give the company time to make long-range plans acceptable to all parties.

After some discussion, Gordon and Petersen tentatively agreed to the Standstill Agreement. Lipton and Patricia Vlahakis, a 27-year-old associate from his firm, drafted a two-page version of the proposed agreement and submitted handwritten copies for Gordon and the company to study and flew home.

It appeared that peace would reign in the kingdom of Getty, at least for the next eighteen months. It didn't quite work out that way.

Two weeks after the negotiations in London, a meeting was set for representatives of the three parties at the San Francisco offices of Lasky, Haas, Cohler & Munter, Gordon's lawyers. Moses Lasky was a 75-year-old attorney who had represented J. Paul Getty for many years. His firm had replaced Lansing Hays in the role of legal advisor to the Trust.

In those two weeks, Gordon and the company had each come up with changes they wanted before they would sign the Standstill Agreement.

The Trust had finally retained an investment banker, Kidder, Peabody & Co, represented by Martin Siegel. Lipton and Siegel were well acquainted, having been on the same or opposite sides of many mergers.

Siegel had represented Martin Marietta in its celebrated battle to repel a hostile takeover by Bendix, a story that gained wide circulation outside business circles because of Bendix CEO William Agee's personal association with one of the firm's vice-presidents, Mary Cunningham. Actually, it had been something of a triumph for Siegel, whose advice to his client to hang tough had seemed like a prescription for disaster until he was vindicated by a last-minute turn of events.

Siegel proposed to Lipton that Gordon be allowed to add four directors, instead of one, and that they be accomplished and independent-minded businessmen who could bring stability to the situation. Laurence Tisch, founder and chairman of the Loews Corporation, was one of the people mentioned. (Tisch would make the cover of *Time* magazine in 1986 when he took control of CBS. He played a prominent role in the unfolding drama at Getty Oil.)

For the company, Winokur and Galant had taken Lipton's two-page handwritten draft and come back with a sixteen-page single-spaced contract that Gordon flatly refused to sign. Ultimately, Lipton prevailed upon Gordon, and the original draft was signed. Gordon insisted on one change—the Standstill Agreement would only be in effect for twelve months instead of the eighteen months originally contemplated.

When his attorneys pointed out that the agreement appeared to bind Gordon before it was ratified by the company, he replied, "I don't care. Sid Petersen is a man of his word, and if he says he'll get his board to go along, then I know he will."

This statement is revealing in two ways: although Gordon sat on the Getty board, he clearly recognized that it was Petersen's board, and despite the tensions of the past year, Gordon still respected Petersen as a man of principle. Subsequent events would make it seem that Petersen took that acknowledgment of respect on Gordon's part as a sign of weakness.

Before leaving San Francisco, Petersen had reluctantly abandoned Winokur and Galant's sixteen-page draft of the Standstill Agreement. On the company's behalf, he affixed his signature to Lipton's original draft, which had been amended to satisfy Gordon's demand that it only be in force for twelve months.

A meeting of the Getty board was held three weeks later in Houston to, among other business, formally ratify the Standstill Agreement. Gordon Getty was present with the other company directors.

At one point during the meeting, Gordon was asked to leave the board room so the Standstill Agreement could be discussed. (To avoid even the appearance of a conflict of interest, directors often excuse themselves while matters that pertain to them in some other capacity are presented. As trustee, it was entirely proper that Gordon leave the room while the board discussed ratifying an agreement to which the Trust was a party.)

When Gordon left the room, attorneys Winokur and Galant and investment banker Boisi were ushered in through another door. When it was later alleged that the trio (who had not previously been present for any part of the meeting) "snuck in the back door," the indignant reply was that there was no back door, they had merely used a different door from the one Gordon Getty had exited through.

Once inside, they advised the board that the Standstill Agreement did not prohibit the board from going ahead with the co-trustee suit against Gordon. By consensus, the board agreed to intervene in the lawsuit, avoiding a formal, recorded vote. The lawsuit, filed in the name of Tara Getty, had been instigated at Petersen's instruction.

The co-trustee the suit sought to have appointed was the Bank of America, whose retired chairman was Chauncy Medberry III, a Getty board member seated at the table. There can be differences of opinion among honorable men; what one regards as a direct conflict of interest, another may consider just a slick maneuver.

Gordon Getty returned to the board room believing that, with the Standstill Agreement ratified, any hostilities had at least been put on hold for twelve months. Despite his initial reluctance to sign the document, Gordon left Houston with the feeling that something had been accomplished.

This incident was described in detail at the Pennzoil-Texaco trial in Texas two years later to illustrate the kind of "good faith" practiced by Petersen and his minions. Winokur still insisted that these actions did not violate the Standstill Agreement and were "entirely proper."

In an interview with *American Lawyer* in early 1984, Galant and Winokur listed numerous reasons why the intervention was not prohibited by the Standstill Agreement, reminiscent of the rationale employed by our government in years past to justify violating nearly every treaty negotiated with the Indians.

Another facet of the American system of justice is that no matter what you do, you can always find an attorney to get up and say you were right. Petersen's attorneys told him in advance that it was proper to sign a peace treaty with Gordon Getty and at the same time sponsor a lawsuit against him by one of his own relatives.

Granted, the back-door incident at the November 11 board meeting had no direct bearing on the later dispute between Pennzoil and Texaco, although it was an important part of the backdrop of the ensuing war. Attorneys for Texaco argued vehemently against allowing this sorry tale to be told to the jury because it might be "misinterpreted."

As we will see, when Winokur, Boisi, and Medberry took the witness stand on Texaco's behalf, the critical portions of their testimony focused on the events of January 1–6, 1984. But the knowledge of what happened at that meeting and the roles these men played in it cast a shadow over their testimony at the trial. Texaco never called Getty Oil chairman Sidney Petersen to the stand; a brief excerpt of his video deposition was offered by Pennzoil.

The problem this created for Texaco at the trial was the revealing glimpse this incident offered into the mindset of Getty Oil officials less than two months before the events of the first week of January 1984. These are the same Getty officials Texaco would insist acted honestly and in good faith in their dealings with first Pennzoil, then Texaco.

Ultimately, however, it was only one part of the complex mosaic case against Texaco.

Four days after the November 11 Getty board meeting in Houston, Geoff Boisi was at a dinner in New York with Martin Siegel, Gordon Getty's investment banker, and Martin Lipton, the Museum lawyer. He took both men aside and described what had happened when Gordon had left the room.

Lipton and Siegel had been aware that the company intervention in the lawsuit had been contemplated prior to the Standstill Agreement. Both professed surprise and dismay that the company had proceeded in the face of what they felt was a genuine cease-fire.

For the Getty Oil Company, it was the beginning of the end.

Siegel called Gordon, who was predictably "shocked, hurt, and embarrassed." Only six weeks before, he had overruled the concerns of his own lawyers, insisting Sid Petersen was a man of his word.

Lipton, who had talked a reluctant Gordon into signing the Standstill Agreement, said he felt personally betrayed. He got on the phone with Galant and Winokur, who trotted out their litany of excuses. "You snookered me," Lipton angrily told them. "No, if we snookered anyone, Marty, it was Gordon, not you," Galant replied.

Lipton gave Harold Williams a written legal opinion that stated the company's action had violated the Standstill Agreement. Williams, who had taken a seat on the Getty Oil board under that agreement, called Petersen, demanding an immediate meeting of the full board. Petersen refused, but did obtain a legal opinion from Winokur and Galant defending the company's action.

If Petersen's greatest fear was the Gordon would join forces with the Museum, this maneuver was a gross miscalculation at best; its direct result was to bring the Museum and Gordon together.

On December 5, Harold Williams and Gordon Getty acted together as majority shareholders under Delaware law and signed a document that became known as the Consent. They revised the bylaws of the Getty Oil Company and presented them to the board as a fait accompli.

The new bylaws stipulated that the company immediately withdraw its support of the lawsuit against Gordon and further required that any major action taken by the board had to be approved by fourteen of sixteen directors. This, in effect, gave veto power to Gordon Getty and the four new board members he had appointed the month before.

When the Consent was filed (it was a public document), the internal strife that had wracked Getty Oil for the past year became public. News accounts of the actions were widespread; the whispers on Wall Street quickly escalated into a full roar.

Getty Oil was wounded and the blood was in the water.

In Houston, these events were noted with more than passing interest by J. Hugh Liedtke, the chairman of Pennzoil. Liedtke was an oilman, and the company he had built traced its roots to the South Penn Oil Company. Although it was primarily known for the motor oil marketed in the distinctive yellow can, what had originally drawn Liedtke to South Penn was its oil reserves.

At the urging of a large minority stockholder of South Penn, Liedtke had bought up absolutely as much of the company's stock as he could afford, which was still a small percentage of the total. Before long, however, he was running the company with the support of that minority stockholder, J. Paul Getty. Under Liedtke, the company prospered. Three years later he bought out the old man's interest in South Penn and the modern Pennzoil Company was born.

Now, two decades later, Pennzoil was a thriving enterprise with over $2 billion in annual revenues. Like Getty Oil, Pennzoil had grown through the acquisition of other companies. Among oil companies, however, it was still "second tier," one level below the industry giants like Exxon, Shell, Mobil, and Texaco.

The blood letting at Getty Oil in 1983 had been followed closely by Hugh Liedtke. He was familiar with the rich oil reserves the company owned, and with the unique ownership situation with the majority of shares concentrated in two private trusts. If it could acquire a significant minority share of Getty Oil, Pennzoil might be able to play a pivotal role in future events. It would have to move fast, however. In the parlance of the mergers and acquisitions game, Getty Oil was "in play."

Liedtke got his financial lieutenants working on an analysis of Getty Oil under a lid of tight security. If the firm did decide to make a bid, any advance word that leaked out could cost Pennzoil millions of dollars.

Pennzoil's investment banker also got involved; James Glanville of Lazard Freres had known Liedtke for many years. Although his planning group was in place, it was clear that Liedtke would call the shots.

With Getty stock currently selling at slightly less than $80 a share, Pennzoil purchased 590,000 shares on the open market. Then, on December 28, Pennzoil made a $1.6 billion tender offer for 20 percent of the Getty stock at $100 per share.

When a company makes a tender offer, shareholders may tender or sell their shares after a brief waiting period. With Pennzoil offering a $20 premium over the current stock price, many shareholders would undoubtedly rush to take advantage. Pennzoil was only committed to buy 20 percent of the stock, although it was free to purchase more if it wanted. This could result in the tender offer being oversubscribed, with more sellers than the offer could accommodate.

This was not a bad strategic position for Pennzoil, but it spelled potential disaster for the Museum and the Getty Oil Company, who had no way of telling what would happen.

As the final days of 1983 ran out, the year of discontent for Getty Oil had ended with the situation more unsettled than when it began. The denouement would come in the first six days of the new year, a time of laughter and tears, bravery and cowardice, triumph and tragedy.

I call it the Six Day War.

PART II

THE SIX DAY WAR

January 1-6, 1984

The first casualty when war comes is truth.

HIRAM JOHNSON
SPEECH TO UNITED STATES SENATE, 1917

3

The details of a tender offer are published in a formal document filed with the Securities & Exchange Commission and made available to shareholders, potential investors, or anyone who wants to read it. Notices of tender offers are also printed in leading newspapers.

The document Pennzoil filed on December 28 stated that the tender offer had been made independent of the Getty Oil Company, the Sarah C. Getty Trust, and the J. Paul Getty Museum. The only contact between Pennzoil and the Getty entities had been a simple notification after the offer was filed. The truth of this statement is manifested in the surprised reactions of the Getty principals and their advisors, who were spread out over a significant portion of the northern half of the Western Hemisphere.

Despite his previous business dealings with the old man, Hugh Liedtke had never met Gordon Getty. His reading of the news accounts of the ongoing battles at Getty Oil intuitively told him that Gordon might welcome a strong ally. Liedtke knew that the success of the venture might hinge on convincing him that Pennzoil was that ally.

Efforts to contact Gordon soon bore fruit. A meeting was set for January 1 at the Pierre Hotel in New York City. It was about to get exciting.

January 1, 1984

Hugh Liedtke arrived in New York in the early morning hours and picked his way through the broken glass in Times Square to a company suite in the Waldorf Towers.

Martin Siegel, who had returned from a vacation in the Caribbean when he learned of the Pennzoil tender offer, met with Gordon Getty at the Pierre Hotel. They discussed a plan Lipton had formulated for Getty Oil to defeat the Pennzoil tender offer. A Getty Oil board meeting would be held the next day to adopt this strategy formally, which Gordon and his advisor Siegel had initially favored. Whatever enthusiasm he had for Lipton's latest plan had waned, however. Gordon had been bitten once too often by the machinations of Petersen and Winokur and Galant.

When Liedtke and Pennzoil president Baine Kerr showed up at the Pierre Hotel, they found a fertile ground for the seeds of their own ambitions. Liedtke, an imposing figure with a deep gravelly voice and large baleful eyes, made a good first impression on Gordon Getty. Two hours of conversation followed, with Liedtke first describing how Pennzoil was built with the initial help of J. Paul Getty, then eliciting Gordon's ideas and hopes for Getty Oil.

After only two hours, the two men struck a multibillion-dollar deal to take Getty Oil private, to buy out the Museum and all the public stockholders. After some wrangling, the price was set at $110 a share, a $10 premium over the tender offer. Gordon would become chairman of the new Getty Oil Company, freed at last from the humiliations heaped on him by Petersen; Liedtke would become president and CEO and actually run the venture.

Why would Gordon Getty cast his lot with a man he had met only two hours before? Liedtke offered him the opportunity to become majority owner and chairman of the company that bore his family name. J. Paul Getty hadn't become one of the richest men in the world without being a shrewd judge of character. The old man had correctly judged Liedtke to have something of a Midas touch—nearly every business venture he launched turned to gold.

Under the proposal they worked out, the Trust would own 57 percent of the new Getty with Pennzoil owning 43 percent. After one year the assets of the company would be divided according to that same ratio if the two parties were not getting along. This so-called "divorce clause" was also quickly agreed upon. The deal had taken shape; now they had to sell it to the Museum.

Lipton was surprised his plan to defeat the Pennzoil tender offer had come untracked, but he quickly focused on the new deal. He made several changes to the draft that became known as the Memorandum of Agreement, but altered none of its fundamental terms. At $110 a share or higher, the Museum was a seller. The deal required the approval of the Getty board.

Day One ended with handshakes in the Pierre Hotel and a deal on the horizon.

January 2, 1984

The Memorandum of Agreement was signed by Liedtke for Pennzoil and Gordon for the Trust. Museum president Harold Williams was en route from Los Angeles for the Getty board meeting scheduled for late afternoon.

Liedtke, meanwhile, was concerned about what the Getty board would do, specifically, what pressures they would bring to bear on Gordon when the Pennzoil proposal was placed on the table. He decided Gordon needed a little reserve muscle. Together with Arthur Liman, a prominent New York lawyer retained by Lazard Freres and working for Pennzoil, he contrived a letter ostensibly from Gordon to Liedtke. "Dear Hugh," the letter began. It went on promise that Gordon would support the Pennzoil plan before the board over any other proposal. It went one step farther: if the board refused to approve the plan, Gordon would urge the Museum to again act in concert with the Trust, this time to remove the board members and appoint a new board.

Gordon's lawyer penciled in three lines, and the "Dear Hugh" letter was signed. Gordon folded up a copy of the letter and put it in his pocket.

Harold Williams arrived at the Intercontinental Hotel and signed the Memorandum of Agreement as the meeting began. Copies were passed out to the board members, who had not heard about it previously, much less seen it. Their reaction was predictably negative.

Lipton had added a provision to the Agreement that stated it would expire if not approved at this meeting. This placed the directors in a quandary: if they didn't approve the deal and Pennzoil walked away, the board members could become targets of lawsuits from angry stockholders if another buyer didn't come forth and the stock price fell far below the $110 now on the table.

The directors and their advisors, some forty people in all, were crammed into a long, narrow conference room. After several hours of heated discussion, the Pennzoil proposal was rejected by a 9 to 6 vote. Williams, Gordon, and the four new board members voted in favor of the deal. All the "old" board members voted against it.

The discussion then turned to alternatives available to the board. Lipton's earlier plan to defeat the Pennzoil offer (before the Memorandum of Agreement materialized) was revived. Gordon refused to budge, however. With the requirement for fourteen of sixteen directors to approve any board action, Gordon only had to muster one additional vote to defeat any proposal. (His four new appointees were at the table attending their first Getty Oil Company board meeting.)

After a recess, the board reconvened without the advisors present and turned up the heat on Gordon. They wanted to stall for time so the entire company could be auctioned off to the highest bidder. Gordon refused even to consider it.

Finally, one of the "old" directors confronted Gordon directly: "Do you have any agreement or arrangements with Pennzoil that you haven't told us about?" he demanded.

After consulting with his lawyers, Gordon returned to the room and produced the "Dear Hugh" letter. The threatened ouster enraged the board members. "The old man must be spinning in his grave," one fumed. "Gordon was trying to sell the other shareholders down the river," said another.

After the furor abated somewhat, there was still the matter of Pennzoil to deal with. The board would support the Pennzoil proposal if the price was raised another $10 a share. This counterproposal was approved 14 to 1 (only Gordon voted against it) and transmitted to Arthur Liman who had been waiting outside the board room throughout the meeting.

At 2:30 A.M., the board recessed until the following afternoon. The meeting had specifically been recessed, not adjourned. Had the meeting simply been adjourned, the Memorandum of Agreement would have died; instead, it clung to life by the thinnest of threads.

Day Two became Day Three as the battle raged on.

January 3, 1984

Liedtke angrily rejected the counteroffer when Liman brought it to him. Hell, he had already raised the bid $10. Now they wanted another $10, but he was the only one bidding. Arthur Liman tried to come up with a compromise solution Liedtke would endorse before the Getty board reconvened that evening.

Meanwhile, Geoff Boisi was also busy. The investment banker from Goldman, Sachs had watched in dismay as repeated proposals to auction off the whole company had been rejected. Approval of such a move would have meant a huge increase in the fee Goldman, Sachs would receive.

Boisi got on the phone and started trying to find a third party to make a bid for the whole company. He contacted big oil companies like Mobil, Shell, Chevron, and Texaco. He tried General Electric and representatives of the Saudi government. Since he had not been authorized by the Getty Oil board (fourteen of sixteen directors would have to approve such a move), he said he didn't call these people soliciting a bid for Getty Oil Company. He was just making "courtesy calls." At least that's what he said when he later testified under oath.

One of those "courtesy calls" found a receptive ear at Texaco, where president Al DeCrane was monitoring the Getty contretemps. Texaco was an industry giant that was in trouble. Its oil reserves, the lifeblood of an oil company, had been declining for over a decade. Texaco's cost per barrel for finding new reserves was the highest in the industry; bad luck and poor judgment had consistently plagued its exploration program.

Texaco had also missed out on the oil and gas reserves changing hands in the new wave of oil industry mergers. It had walked away from Conoco at the eleventh hour—at Conoco's request. A long-standing Texaco policy insisted it would only engage in "friendly" mergers. When antitrust laws had been strictly enforced, this hadn't been a problem. Under those restrictions, Texaco could never have made a move on a company the size of Getty Oil. The laissez-faire attitude of the Reagan administration had sparked a wave of takeover frenzy that had yet to crest three years later. If Texaco could somehow acquire Getty, it would be the largest merger in American corporate history. Not incidentally, those same rich oil reserves that had attracted Pennzoil would solve Texaco's reserve problem for a generation—at least.

Later that day, DeCrane also received calls from investment bankers Joseph Perella and Bruce Wasserstein of First Boston. They were seeking a client to go after Getty. The duo were celebrated for their hard-ball tactics in the mergers and acquisitions field; such tactics weren't much good if you didn't have a client, however. They gave DeCrane a rundown on the status of the Pennzoil-Getty negotiations. The information was gathered from inside sources, some of whom had been present at the Getty board meeting currently in recess. DeCrane said he'd get back to them.

Trading in Getty Oil stock had been suspended on the New York Stock Exchange that morning at the company's request, fueling already widespread speculation about the unfolding drama.

By the time the board gathered that evening, Liman had sold Liedtke on a counteroffer to present to the Getty board. It was a complicated offer involving the insurance company Getty had obtained, the star performer of Petersen's diversification program, but due to be sold when Gordon and Liedtke formed the new Getty Oil. Everything over $1 billion that the insurance company sold for would be paid to the stockholders. Pennzoil would guarantee a minimum payment of $3 a share within five years.

Before the meeting started, Liman outlined the idea for Lipton. No good, said Lipton, "it's too cute." Make it $5 and we can do the deal, he told Liman.

Now Liman was in a box. The $5 guarantee Lipton wanted amounted to only $2.50 in real dollars, which shouldn't be enough to kill a deal of this magnitude. On the other hand, it had been like pulling teeth to get Liedtke to go up to a guaranteed $3, which had a present value of $1.50.

Liman told Lipton he would get the Pennzoil chairman to agree to the $5 guarantee if Lipton could get the board's blessing on a firm deal that night. Lipton made a note of the detail's of Liman's proposal and headed for the board room.

When the meeting reconvened, Petersen asked Lipton to repeat the details of the counteroffer from Pennzoil. Lipton complied. Then Harold Williams moved that the board approve the Pennzoil proposal. The measure was approved by a vote of 15 to 1. Only Chauncey Medberry voted "no." Sidney Petersen voted for a proposal that would see Gordon Getty succeed him as chairman of the Getty Oil Company.

After the vote was taken, two minor housekeeping matters remained. The board voted unanimously to indemnify Petersen and his top officers against any legal action arising out of the internal warfare of the past eighteen months. Gordon Getty voted to protect Sidney Petersen from retribution for any actions taken against Gordon himself.

The board also enhanced the employment agreements of nine top company executives, including Petersen. These so-called "golden parachutes" had become common in takeover deals, as a cushion for those in top management whose positions would not survive the merger. Again, the vote was unanimous. Lipton and Siegel were dispatched to bring the word to Liman, waiting anxiously outside the board room. "Congratulations, Arthur, you've got yourself a deal," they said. Liman got on the phone and relayed the news to Liedtke. Relieved that the deal had been approved, Liedtke swallowed the increased price Liman had given Lipton to present to the board. He also approved the "golden parachutes."

Siegel brought word of Pennzoil's acceptance and the marathon, two-day board meeting was finally adjourned.

After the board adjourned, Liman asked Siegel if he could go in the board room and shake hands with the directors. Siegel said sure, and Liman finally entered the board room.

That night back at the Pierre Hotel Gordon drank champagne with Laurence Tisch, celebrating the deal approved by the board that would make him chairman of the company that bore his family name. Liedtke, invited to the Pierre by Gordon's wife, asked for a rain check; his group was having their own celebration at the "21" club.

The elation of the new partners was understandable; they had just made the biggest deal of their respective careers. It seemed like a time to celebrate; after all, this was what was known as a "done deal."

At this point there was no way to know that this had only been Day Three of the Six Day War.

January 4, 1984

The news of the deal spread quickly as a press release went out early in the morning on the letterhead of the Getty Oil Company, but issued jointly by the company, the Trust, and the Museum.

A frantic session after the board meeting involving lawyers, investment bankers, and public relations types representing the three Getty entities and Pennzoil had produced the joint release. It detailed the terms of the agreement: 57 percent ownership for the Trust, 43 percent for Pennzoil; the "divorce clause"; and other details.

The press release described what the Getty board had approved as an "agreement in principle" that was "subject to the execution of a

The "New" GETTY Oil Company
As announced in Getty press release Jan. 4, 1984

PRIVATELY HELD
CORPORATION

(NO PUBLIC
STOCKHOLDERS)

SARAH C. GETTY TRUST
57%

PENNZOIL
43%

GORDON P. GETTY, Chairman of the Board

J. HUGH LIEDTKE, President and CEO

BAINE P. KERR, Chairman, Executive Committee

definitive merger agreement." What seemed to be a fairly straight-forward announcement of an agreement was seen by some hardball practitioners as an announcement that, in fact, no agreement existed. They interpreted it as nothing less than an open invitation to bid.

Meanwhile, Pennzoil's lawyers (from the Houston firm of Baker & Botts) were drafting the final merger document that would reflect the agreement reached the night before. Working in the offices of Paul, Weiss, Rifkind, Wharton & Garrison (Arthur Liman's law firm), the Houston attorneys were to send the completed draft to the Getty Oil lawyers waiting at the Dechert, Price & Rhoads office (Barton Wino-kur's law firm). There was a delay in the preparation of the draft document, and it was not delivered to the Dechert, Price office until early evening.

By that time, two important events had occurred. A judge in California had signed a temporary restraining order (TRO) at the request of Claire Getty, daughter of the late George Getty and a beneficiary of the Trust. The TRO prevented her Uncle Gordon from closing the deal with Pennzoil for at least twenty-four hours.

Over at Texaco, the assiduous courtship of Perella and Wasserstein had finally paid off. After first attempting to hire Morgan Stanley (which had a conflict), Texaco retained First Boston. Boisi, Lipton, and Tisch had all been pressed for details about the transaction by Wasserstein. Boisi didn't have to be pressed too hard; he had been frantically peddling the deal since the recess in the board meeting.

The notes taken by Texaco president Al DeCrane during the early phone calls and later strategy sessions with Wasserstein and Perella clearly show the evolution of the Texaco plans for the taking of Getty.

Once again, the pivotal role belonged to the Museum, their 11.8 percent of the stock and their lawyer, Martin Lipton. Texaco is a former Lipton client; DeCrane is well acquainted with the man and his reputation.

DeCrane's notes and those of Texaco deputy controller Patrick Lynch reflect an early understanding of the situation:

> "KEY PERSON—MARTY, MUSEUM," DeCrane's notes recorded Wasserstein's assessment.
>
> "1. MUSEUM, 2. TRUST, 3. PUBLIC," the notes outline the strategy: Texaco would go after the Museum shares first, then approach the Trust, and finally buy out the public shareholders.
>
> "ONLY HAVE 24 HOURS—LIPTON." Indeed, Lipton has told Wasserstein that the final documents with Pennzoil will likely be executed within 24 hours.

Texaco chairman John McKinley returned from his vacation in Alabama and called a board meeting for the next afternoon.

For Gordon Getty and the Pennzoil people, this had been a day to savor the accomplishment of their deal and make plans for the future. Gordon and Liedtke made plans to go to Los Angeles as soon as possible and walk through the headquarters of Getty Oil to reassure the employees about their plans for the future.

Rude awakenings are in store for the would-be partners, as well as those Getty employees.

January 5, 1984

John McKinley was back in his office at Texaco headquarters. With Wasserstein sitting beside his desk, he placed a call to Lipton. "I've called a board meeting for this afternoon," McKinley told him. "I hope you don't sign anything until you get a chance to hear our offer."

McKinley then called Petersen, who was eager to thwart the Pennzoil merger and the ascendancy of Gordon. Despite the 15-to-1 board vote approving the Pennzoil merger two days before, despite the Getty entities' announcement of that agreement to the world, Petersen told McKinley that Getty Oil was free to receive another offer. McKinley told Petersen that Texaco would only proceed on a friendly basis.

The Texaco board of directors met in the afternoon for McKinley to lay out the details of the move on Getty Oil. He told the board the price tag would be about $9 billion. After about three hours, the board authorized McKinley to pursue the biggest takeover in history at any price up to $125 a share. The order of authorization in the board resolution tracked DeCrane's notes exactly: Museum, Trust, public.

Back at the Paul, Weiss offices, the Pennzoil lawyers waited for their counterparts from Getty Oil. They had a merger agreement to finalize. Their draft was sent over late the day before. Finally, in late afternoon, the Getty Oil lawyers arrived and the meetings started. Progress was slow, however. The Getty lawyers kept getting phone calls, and started slipping away one by one.

Patricia Vlahakis never made it to the Paul, Weiss offices to participate in the negotiations with Pennzoil lawyers. "You can't leave. I need you here," Lipton told her. The Texaco board meeting had ended; a team had been dispatched to the Wachtell, Lipton offices.

Lipton prepared an agenda for his meeting with Texaco. After the basic issues (price) and the exotica (crown jewel options, hell and high water, favored nation) is the clincher: indemnity.

Night had fallen; Texaco's team split up when they got to Manhattan. The top four men in the company were personally handling this deal. Vice-chairman James Kinnear and general counsel William Weitzel went to visit Lipton, who presented his agenda.

All three later described a scenario that had Lipton extremely upset because they wouldn't tell him the price Texaco would offer. That price was known only to chairman McKinley, who was at the Carlyle Hotel waiting to meet with Gordon Getty. It was getting late. When the Texaco contingent finally left his office, their story went, Lipton still didn't know the price.

Harold Williams, who had returned to Los Angeles the day before, said under oath that Lipton called him in California at 4 P.M. (7 P.M. New York time) and told him the price was $125. Maybe in addition to his prodigious business and legal talents, Marty Lipton is also a psychic.

The Texaco contingent reunited with McKinley; they headed for the Pierre Hotel. All of them, that is, except Al DeCrane. The Texaco president, a key player throughout the planning and strategy sessions, was inexplicably dispatched to a midtown hotel suite at the critical moment of the move on Getty to wait for . . . what?

McKinley arrived at Gordon's suite, site of that seemingly historic meeting with Hugh Liedtke only four days before. Gordon and Martin Siegel were there, along with lawyer Thomas Woodhouse (of Moses Lasky's firm); McKinley was accompanied by Kinnear, Weitzel, and Wasserstein.

After some initial fencing that produced no results, there was a lengthy recess in the meeting. McKinley and his team retired to the hotel lobby, while Gordon conferred with his advisors. During that time, Marty Lipton and Larry Tisch also arrived at the Pierre and went to Gordon's suite. No fewer than eight of these men will testify at length, under oath, about the events of that night.

Witnesses to a significant event nearly always differ in their remembrance of particular details, small and large. About that night at the Pierre Hotel, though, everyone was in complete agreement on two points: McKinley led off by making some remarks to Gordon about their mutual interest in the arts, and, by the time McKinley left several hours later, Gordon had signed a hastily written document stating he will sell the Trust shares of the Getty Oil Company to Texaco.

This "mutual interest in the arts" that everyone remembered was finally revealed to be an exchange about Texaco's long-time sponsorship of the radio broadcasts of the Metropolitan Opera.

In *The Seven Sisters*, Anthony Sampson's definitive 1974 book on Big Oil, he traced the origins of that sponsorship to a 1940 incident that led to the disgrace of another autocratic Texaco CEO, Torkild Rieber. A twisted tale that ultimately involved Herman Goering, Texaco tankers, and a Nazi spy at Texaco's New York headquarters led to a public outcry that caused Texaco shares to slump badly on the stock market. Sampson suggests a linkage between this disgrace and Texaco's rare act of patronage. At any rate, the sponsorship of the weekly broadcasts, uninterrupted by commercials, continues today. Fortunately, Texaco said the sponsorship would not be affected by the bankruptcy filing.

This story is somehow omitted from Texaco's company history. Gordon Getty might have identified with certain elements of the tale, however. According to one recent biography, at about this same time, J. Paul Getty enlarged his collection of antique French furniture at fire sale prices by some hard bargaining—with wealthy European Jews fleeing Nazi persecution.

It was during my research for this book that I first read these stories and was struck by the juxtaposition of these facts and that "mutual interest in the arts"; obviously, they played no part whatever in the outcome of the lawsuit.

For balance, in a later chapter I will include the equally irrelevant but also intriguing story of how Hugh Liedtke got Pennzoil mixed up in a little political scandal called Watergate.

We now return for the conclusion of the Six Day War.

While the Texaco team waited in the lobby, the talk in Gordon's suite focused on the question John McKinley kept asking: Was Gordon free to consider an offer from Texaco for the Trust's 40 percent of the Getty Oil Company?

Tisch and Lipton arrived at the Pierre and joined the discussion in Gordon's suite. Lipton obviously wanted the higher price from Texaco for the Museum, which was not going to have any further involvement with Getty Oil. Gordon, on the other hand, had a deal in hand to become chairman and majority owner of the company—a deal with Pennzoil.

Or was it a deal? Gordon's lawyer pulled out the "Dear Hugh" letter Gordon had signed. The handwritten inserts his lawyer had made read "subject to my fiduciary duties" before each of the key promises the letter contained. The thrust of these was that Gordon had promised to stick with Pennzoil unless it violated his duty as a fiduciary, responsible to all present and future beneficiaries of the Trust.

The Trust instrument itself mandated that the assets could not be sold "except to prevent a substantial loss" to the Trust. If Texaco acquired the balance of the Getty Oil shares, Gordon would still be a minority shareholder, but one without hope of attaining a majority.

Lipton told Gordon that the Museum's shares were going to be sold to Texaco. Without the Museum's 12 percent, the numbers of the

Pennzoil deal didn't work. Although McKinley said he wouldn't make an offer to Gordon unless invited to do so, this news from the Museum delivers the implied threat from Texaco: Don't sell to us if you don't want to. We're buying the rest anyway, rendering you a perpetual minority.

Again, Lynch's notes of the strategy sessions had contemplated just such an implied threat: *"If there's a merger and Gordon doesn't tender, he could wind up with paper."* The reference to "paper" is the value of minority shares of stock in an interest controlled by a dominant majority owner.

The point was well taken; the "divorce clause" in the Memorandum of Agreement had given Pennzoil protection from the Trust in a similar situation. In a Texaco takeover, the Trust would be the minority shareholder; no "divorce clauses" were available.

Texaco had finally found a way to be "friendly" in a takeover situation and still exert the maximum force required to attain their goal in the face of opposition. If that meant putting the squeeze on Gordon, well . . . the Trust's 32 million shares would fetch $4 billion. What could be friendlier than that?

The Texaco team returned to the suite. McKinley presented the $125 offer to Gordon, who later testified he felt that he had no choice to but to accept it and sell the Trust shares to Texaco. But that night he was prohibited from doing anything by the temporary restraining order obtained in California by his niece, Claire Getty.

Then, a phone call came in from Charles Cohler (another lawyer for Gordon) who was in California trying to get the TRO lifted so Gordon could sign the final merger agreement with Pennzoil. Cohler reported that he had just filed an affidavit with the court from Martin Siegel that had been telecopied from New York, as well as an affidavit of his own. Both men had sworn that a merger agreement existed among Getty Oil, the Museum, the Trust, and Pennzoil. When they told Cohler the latest turn of events, he went back and filed an amended version of his affidavit with the California court. Cohler's hastily scrawled changes couldn't hide the intent of the original affidavit, however.

To get around the TRO, the ever-resourceful Lipton drafted a letter to Texaco from Gordon. "I regret that an order of the California Court prohibits me from selling" the Trust shares to Texaco, the letter began. The handwritten document, which later became known as the "I regret" letter, went on to state Gordon's intention to have the TRO lifted and sell the stock to Texaco.

It was well after midnight by the time the Texaco team departed from the Pierre Hotel.

Lipton and key members of the Texaco team regrouped at the Wachtell, Lipton offices. They worked through the predawn hours, and at

5:30 A.M., the Museum shares were signed over to Texaco. An escrow agreement Vlahakis had drafted for Pennzoil was only slightly modified and became part of the Museum's agreement with Texaco.

Another important element included at Lipton's insistence is the indemnity: Texaco must defend the Museum against any lawsuits "arising out of the Pennzoil agreement." In another clause, the Museum asserted that while it owned its stock free and clear, "no representations could be made about any claims arising out of the Pennzoil agreement."

These provisions were questioned by Texaco. If there was no binding agreement with Pennzoil, why the insistence on these specifically worded clauses?

Lipton gave a number of unpersuasive reasons before stating that he had to have these provisions to "protect the psyche" of his client. A tough businessman and former SEC chairman, Williams was not regarded as a man with a tender psyche. Indemnity provisions that identified the Pennzoil agreement by name also became imbedded in Texaco's subsequent formal agreement with the Trust.

As Day Six dawned, the tide had definitely shifted in the fortunes of the Six Day War.

January 6, 1984

Early that morning, a news release announcing Texaco's lock-up of the Museum shares moved across the Dow Jones wire. It also referred to a Getty Oil board meeting that afternoon that would complete Texaco's taking of Getty.

Hugh Liedtke had gone to sleep the night before blissfully unaware of Texaco's unraveling of his "done deal" with Getty Oil, the Trust, and the Museum. When his lieutenants read the announcement, none of them relished the idea of breaking the news to their formidable chairman. Obviously, he had to be informed. When given the news, he sat there in stunned silence.

Finally, he looked up and bellowed, "They've made off with a billion barrels of our oil."

He fired off a telex to the Getty board that demanded they live up to their agreement with Pennzoil; if not, the telex essentially threatened to "sue everybody in sight."

That afternoon, the Getty board met and quickly approved Texaco's bid after first attempting to withdraw their prior acceptance of the Pennzoil merger agreement three days before.

*   *   *

The Six Day War had ended, but it quickly proved to be a mere cur-
tain-raiser for the bitter fight of historic proportions that followed.
Ultimately, the new wave of hostilities would eclipse those events in
the first week of January 1984 that triggered them. Those keen observers who had closely followed the reports of each development of
the Six Day War with steadily increasing fascination will soon realize
they ain't seen nothing yet.

PART III

"THE WHITE KNIGHT GETS SUED BY THE DRAGON"

Pennzoil Co. v. Texaco, Inc.

4

Even as the events of the Six Day War were drawing to a close, the opening moves of the next wave of intrigue were underway.

Hugh Liedtke's telex to the Getty Oil Company had essentially formalized his threat to "sue everybody in sight." This captured the attention of the top in-house lawyers at Getty and their new masters at Texaco.

Before January 6 ended, Getty Oil (with the encouragement of Texaco) filed a motion in Delaware seeking a declaratory judgment that no binding agreement existed between the Getty entities and Pennzoil. The Delaware court refused to issue such a judgment.

Four days later after the Getty filing, Pennzoil filed suit in Delaware against Getty Oil, the Museum, the Trust, and Texaco. The lawsuit asked the court to compel the Getty entities to live up to the agreement approved only a week ago by the Getty board.

At a hearing in the Delaware Chancery Court on January 25, John Jeffers, a trial lawyer from the Houston law firm of Baker & Botts, served notice that the typically low-key approach utilized in corporate civil cases would not prevail in these proceedings.

Jeffers didn't wave his arms or shout. In a gentlemanly fashion, he presented a detailed and devastating account of duplicity and deceit: how the Getty entities and their lawyers and investment bankers had essentially sold the company twice. The takeover lawyers and investment bankers especially felt the sting. Previously lionized as the new princes of Wall Street, their whole way of doing business was now being called to account.

During the discovery process, critical notes and documents produced from the files of Texaco and Getty had provided Jeffers with plenty of ammunition to prove Pennzoil's case. Among the documents was Getty general counsel R. David Copley's minutes of the Getty board meeting, which contained the 15-to-1 vote to approve the Pennzoil proposal as well as the subsequent transmittal to the board of Pennzoil's acceptance.

This compelling evidence was combined with the Memorandum of Agreement, the January 4 Getty press release, and the "Dear Hugh" letter to establish the contract claimed by Pennzoil. Other helpful additions were the California affidavits of Martin Siegel, Gordon's investment banker, and Charles Cohler, Gordon's lawyer. Only hours before Texaco put the squeeze on Gordon, both Siegel and Cohler had submitted sworn affidavits that "there was a merger transaction agreed to" among the Trust, the Museum, and Pennzoil.

Testimony about Gordon's celebratory champagne toast with Larry Tisch the night of the board's approval bolstered Pennzoil's contention that it had concluded a merger agreement with the company, the Trust, and the Museum.

Additional testimony was offered about the planned Los Angeles trip for Gordon and Liedtke to tour Getty Oil's headquarters and reassure anxious employees. Why would Gordon, much less an experienced businessman like Hugh Liedtke, contemplate such a trip if it were not a "done deal"?

Lawyers for Getty Oil responded with the same arguments that would be made in the Texas trial. One major theme went something like this: "That 15-to-1 board vote? Oh, that wasn't to approve the deal, only the price to begin negotiations. Gosh, I don't know why that little fact wasn't recorded in Copley's detailed minutes."

Texaco's lawyer echoed some of the arguments espoused by the Getty attorney, but placed great stock on "the name and reputation of Martin Lipton" and the fact that "Laurence Tisch is one of the most respected businessmen in the United States."

Michael Schwartz of Wachtell, Lipton spoke for the Museum in no uncertain terms: "Your Honor, the whole situation was, I think—if you look at it as a contract—laughable." It would soon be demonstrated to Lipton and Schwartz that this was no laughing matter.

After hearing extensive arguments, Judge Grover Brown retired to take the matter under advisement.

On February 6, Judge Brown handed down a forty-nine-page opinion that denied Pennzoil's request for an injunction barring the Texaco takeover of Getty. In an unusual move, he devoted the bulk of his opinion to an evaluation of Pennzoil's claim that it had a contract. "I can only conclude that on the present record Pennzoil has made a sufficient preliminary showing that in all probability a contract did come into being between the four parties," the judge wrote.

Further, he ordered, the case would proceed to give Pennzoil a chance to prove that contract in a full trial.

In Delaware, the case would be tried before a judge. G. Irvin Terrell, like Jeffers, a trial lawyer from Baker & Botts, had drawn the assignment of researching the jurisdiction problem. When he discovered that Texaco had failed to file a formal response to the Pennzoil suit, Irv Terrell suddenly realized he had found an opening. With Liedtke's approval, he engineered what many would later regard as a pivotal point in the whole legal battle.

Texaco's failure to respond formally to the lawsuit in Delaware allowed Pennzoil to have the suit dismissed "without prejudice," that is, without making any acknowledgments or giving up any rights in the dispute with Texaco.

Fifteen minutes later, the case of *Pennzoil Co.* v. *Texaco Inc.* was filed in the district court of Harris County, Texas. According to the filing, the lawsuit sought money "in such amount as is proper, but in no event less than $7,000,000,000 in actual damages and $7,000,000,000 in punitive damages.

Besides the staggering damages sought (at least $14 billion), there was another surprise in the filing. Listed as lead counsel for Pennzoil was Joseph D. Jamail, a flamboyant plaintiff's lawyer whose success in personal injury cases had made him rich and famous. His reknown had extended to the *Guiness Book of World Records*, where he was celebrated for the record settlement he won in a case involving a man who had been paralyzed by a Remington rifle that had misfired due to a design flaw.

Pennzoil had managed to get the case moved to Texas where it could be tried before a jury. It had also bolstered the vast resources of Baker & Botts by hiring the meanest legal street fighter on the block to lead the Pennzoil team.

Texaco continued to believe the conventional wisdom among most investment bankers and takeover lawyers, which basically said, "Pennzoil didn't have a contract; therefore they don't have a case."

Conventional wisdom can be a dangerous thing, however, if one doesn't understand that events have a way of overtaking the conventional wisdom of today. The revisionist conventional wisdom of tomorrow will then reign until it, too, is rendered inoperative by subsequent events.

5

In 1985, Joe Jamail was easily the most famous lawyer in Texas who had never represented a dubious but wealthy defendant in a spectacular murder trial. Governor Mark White was technically a lawyer by trade, but he counted himself among Jamail's friends. And Joe Jamail had a lot of friends. Widespread tales of his intimate association with such legendary Texas folk heros as Willie Nelson and Darrell K. Royal (the Bear Bryant of Texas) are well documented and undoubtedly true. Even if the stories were merely apocryphal, the linkage of Joe Jamail on an equal basis with these cultural touchstones served as a powerful intoxicant indeed.

Combined with his burning courtroom presentations, this mixture frequently proved lethal for the hard-line insurance company lawyers and corporate apologists he routinely encountered.

Jamail's considerable fame did not come from representing the rich and powerful, although he had surely done that as well. He was reknowned for his efforts on behalf of people who had been hurt or maimed or had had their loved ones killed by accidents caused by the failings of men and machines. Jamail made these cases and defendants come alive for juries, who invariably granted large damage awards. These clients and their families were special to Jamail. More than just a lawyer, they regarded him as a powerful friend in their time of need.

Jamail had another friend, J. Hugh Liedtke. The two men had met many years earlier when their children had attended the same private school. Liedtke, the son of a judge who became chairman of a major corporation, found a curious soulmate in Jamail, the son of Lebanese immigrant grocers who became a multimillionaire "sore-back" lawyer. The unlikely pair would vacation together with their families on the beaches of the Gulf of Mexico. The two men savored the conversation and companionship of that time-honored custom of drinking euphemistically labeled "fishing."

In early January, not long after the bitter climax of the Six Day War, Jamail commiserated with his old friend. As Liedtke described in detail the events of those six fateful days, Jamail began to realize "this case would play beautifully to a jury." His total lack of familiarity with the arcane specialties of mergers and acquisitions law wouldn't be a problem for Jamail. He could command the vast resources of Baker & Botts to handle that aspect of the case.

What Jamail instinctively grasped was the presence of the simple concept he had relied upon throughout his career: a client had suffered a grievous wrong, and it was his moral duty to right that wrong. His success in enlisting jurors to his cause had made Jamail the most formidable figure of the plaintiff's bar.

Jamail's initial reluctance to get involved, his feeling that this was not his kind of case, began to give way in the face of his evolving analysis of the dispute and Liedtke's grim determination to see it through. "I tried to talk him out of it," Jamail would later say; Liedtke would not be dissuaded.

Finally, in late January, Joe looked at his old buddy and said, "Hell, if you're going to stand up to those sons of bitches, I'm with you."

Texaco had cried foul when Pennzoil had managed to get its Delaware litigation dismissed and refiled in Texas. After much finger-pointing and a round of fervent pleas to the Delaware court, the realization finally set in: the lawsuit would be tried in Texas.

Although Texaco's corporate headquarters are in White Plains, New York, its domestic operations (Texaco U.S.A.) are based in Houston, about six blocks east of Pennzoil Place. In-house lawyers for Texaco began looking for some outside legal muscle to help offset what they were already calling Pennzoil's hometown advantage.

Several of the big downtown law firms in Houston had done extensive work for Texaco over the years, including Baker & Botts. As Pennzoil's lawyers in the Getty deal, Baker & Botts attorneys had been severely criticized for their delay in preparing the final merger agreement. One principal Texaco argument at the Delaware hearing had been that the delay by those Baker & Botts lawyers had created the opening that Texaco had taken advantage of in strictly fair competition.

Needless to say, the stately old firm of Baker & Botts bristled over these accusations. If anything, it strengthened its resolve to see this lawsuit to a successful conclusion, for vindication if nothing else.

One of the costs of this vindication would be the legal work it had previously done for Texaco. In 1983, Texaco paid over $900,000 to Baker & Botts for personal injury and property damage defense work.

Texaco found a number of firms willing to take on the legal team assembled by Pennzoil, but the name of one firm kept coming up. Miller, Keeton, Bristow & Brown was a fairly new firm, with only thirteen lawyers. But the rock-solid reputations of the senior partners had been made at the largest firms in town.

Richard Keeton, formerly a partner at Vinson & Elkins, was a cerebral litigator with something of a Texas pedigree: his father had been dean of the Law School at the University of Texas and his sister had been mayor of Austin, the state capital.

Partner W. Robert Brown had been a principal in another Houston firm, Liddell, Sapp, Zively, Brown & LaBoon, and counted among his long-time clients the *Houston Chronicle*.

But it was Richard B. Miller who had cemented the star credentials of the fledgling firm. Dick Miller was regarded by many as the most feared litigator in town, an image he had cultivated during his twenty-five years at Baker & Botts.

Defending corporate clients against the kind of lawsuits filed by plaintiff's attorneys, Miller had built a reputation as a "scorched earth" trial lawyer. He didn't come to the courthouse to hand out settlements. By coincidence, he had never tried a case against plaintiff's attorney Joe Jamail. In fact, just two years before he had actually represented Jamail in a libel suit.

When Miller left Baker & Botts in 1982, he had been the head trial lawyer in the firm. Among the attorneys there he counted as his proteges were John Jeffers and Irv Terrell, now co-counsel with Jamail on the Pennzoil side. Miller was initially reluctant to get involved in a case opposing his friends and former co-workers. Baker & Botts, for that matter, was dismayed by the prospect of taking on their former top litigator.

The exact circumstances of Miller's voluntary departure from Baker & Botts are mired in a complex tangle of contradictory events. After beating a life-threatening health crisis, Miller began to reassess his future and ultimately opted to leave the firm. The fierce will and intellect that made him such a formidable trial lawyer were not necessarily conducive to office politics.

None of this was remotely relevant to the issues of this case, but there is little doubt that the deeply personal nature of the relationships among the attorneys heightened the tremendous animosity already generated by their respective clients.

John Jeffers called Miller and expressed his belief that the work of one of Miller's partners (Darrell Bristow, yet another Baker & Botts alum) disqualified the firm from taking on Texaco as a client. Bristow had briefly worked on a matter involving Pennzoil before leaving

Baker & Botts. Miller checked out this latest claim, rejected it, and Miller, Keeton, Bristow & Brown took the case. Dick Miller became lead counsel for Texaco.

The fact was, that for all the differences in their backgrounds and styles, Jamail and Miller both approached a trial in a single-minded fashion: it was nothing short of war, no prisoners would be taken. Their talents and reputations were put to the test every time they walked into a courtroom. They would have it no other way.

The heat was immediately turned up when Miller, Keeton entered the case. Texaco made good on its threat to pull all its legal business from Baker & Botts. Dick Miller and Richard Keeton showed up at the Baker & Botts offices and gave instructions for all pending Texaco business to be transferred to other firms.

Then they dropped another bombshell: Texaco intended to disqualify Baker & Botts from representing Pennzoil in the lawsuit over Getty Oil by invoking the lawyer-witness rule. This rule is generally used to prevent an attorney from representing a client he may be called to testify for or against.

Because of the issues raised about Baker & Botts lawyers' delays in drafting the Getty documents, Miller asserted that Pennzoil would present testimony from these lawyers.

Pennzoil had not raised the lawyer-witness issue, and Baker & Botts responded that its lawyers "were not important to the case, would not be called, and were not adverse to the Pennzoil position in any event." It rejected Texaco's attempt to have it disqualify itself. This issue didn't die, however. Miller made sure of that.

Case Number 84–05905, *Pennzoil Co. v. Texaco, Inc.*, was filed in Houston, Texas, at the Harris County courthouse. It was assigned to Judge Anthony J. P. Farris of the 151st State District Court for pretrial motions.

A conservative Republican in a traditionally Democratic state, Farris had been appointed U.S. attorney for the southern district of Texas by President Nixon. He had served as the prosecutor in the Sharpstown Bank scandal, the watershed Texas political scandal of the early 1970's with far-reaching implications that ended the careers of an incumbent governor, lieutenant governor, speaker of the house, and numerous lesser lights. Rather than gaining an image as a reformer, however, Farris was roundly criticized for approving a lenient plea bargain for the central figure in the case, financier Frank Sharp. These mixed reviews and Nixon's distraction by a watershed national political scandal killed Farris's chances of a lifetime appointment as a federal judge.

Then, in 1978, opportunity knocked again. For the first time since Reconstruction, Texas elected a Republican governor. Although all judges in Texas are elected, the governor appoints judges to serve in vacant or newly created seats. One such appointment went to Farris. When he ran for a full term, he managed to hang on and win election to the bench in his own right.

Nineteen eighty-four was another election year. The Democrats had reclaimed the governor's office two years earlier, sweeping out a number of Republican judges in the process. Although the presence of Ronald Reagan on the ballot would be a source of comfort to Texas Republicans, memories of the Democratic sweep of 1982 were still fresh. The length of Reagan's coattails depended on who was measuring.

The Re-elect Judge Tony Farris Steering Committee was raising funds for the campaign effort. Although he faced no opposition in the primary, Farris would undoubtedly draw a Democratic opponent in the fall. Judges in a large city like Houston must compete for recognition on a long, complex ballot. This creates a situation where numerous congressional representatives, county commissioners, and state senators are competing for campaign funds. The dozens of judicial candidates without significant personal wealth have to raise campaign funds from people they know, most often the lawyers who practice in front of them. Sitting on the Re-elect Judge Tony Farris Steering Committee was Joe Jamail. He had made an early contribution of $100.

The big downtown law firms (like Baker & Botts) had political action committees set up that methodically raised funds from the hundreds of lawyers in their employ and spread it around according to the inclinations of the powers-that-be at the particular firm. Such contributions also served as a signal to other potential contributors, in the same way the blessings of the power-brokers always did.

The plaintiff's lawyers like Jamail invariably encountered the big downtown firms when they sued insurance companies and large corporations on behalf of their clients. For years, they had worked to offset the political influence wielded by the larger firms. Trial lawyers outside the big firms formed their own political action committees. An attorney like Jamail could individually contribute as much to a single candidate as the political action committees of the big downtown firms—and frequently had. This was how, over the years, some balance had been injected into the antiplaintiff tendencies of a frequently elitist judiciary.

This preamble does not dilute the impact of the events that followed.

On March 7, 1984, Joe Jamail contributed $10,000 to the political campaign of Judge Anthony J. P. Farris. This brought his total contributions to the reelection campaign of the judge hearing the pretrial motions in *Pennzoil* v. *Texaco* to $10,100. Case closed, right? Pennzoil's lawyer bribed the judge; that explains everything. Hell, you can just throw this book away right now. The cat's out of the bag.

Except that it's not that simple, of course. Jamail's campaign contribution was not only perfectly legal, it was a matter of public record. Bribes are delivered in cash in brown paper bags or white envelopes (or briefcases, if you're a notoriously successful speculator).

Jamail's $10,000 contribution was reported to the county clerk as required by the election code. When the contribution was made, it was duly reported and became a matter of public record as he knew it would.

When Joe Jamail had agreed to serve on the Re-elect Judge Tony Farris Steering Committee, he was not involved in the *Pennzoil* v. *Texaco* litigation because it didn't exist. Although a Democrat like Jamail was not your typical Farris supporter, the relationship between the two men went back three decades to a time when they both had offices in the old Sterling Building in downtown Houston.

Jamail's agreement to serve on the committee was not a guarantee of any specific contribution, but there was an implied understanding that one would be forthcoming.

The size and timing of the contribution were at best unfortunate. The fallout from this contribution was spectacular in a courthouse sort of way. The events were duly recorded in the Houston newspapers.

Somehow, I missed the whole thing. In the weeks after the trial more than a year later, however, there was no shortage of reporters willing to recount the details of the $10,000 contribution. Having been recently exposed to the gap between reality and what George Orwell aptly dubbed "Newspeak," I pursued the story on my own with grim determination.

In most cases, I found the story these reporters were spouting was Texaco's version of the facts that got the number right ($10,000) and everything else wrong. This was a bit of overkill that wasn't really necessary. Like so many elements of this case, the truth was much stranger than fiction.

6

By the time Miller, Keeton took on the Texaco defense of the lawsuit in Texas, they were at something of a disadvantage. The first wave of litigation in Delaware Chancery Court had already generated reams of documents and testimony that would also be utilized in this proceeding.

For Pennzoil, continuity was assured since Baker & Botts had been in on this deal from the beginning and had handled the subsequent Delaware court action. Although Joe Jamail wasn't named as lead counsel until after Chancellor Brown's opinion was handed down on February 6, he retained the backing of the legal team in place at Baker & Botts.

Texaco filed its formal response to the Texas lawsuit in early March, signaling the commencement of the pretrial phase of the case. Under the rules of the courts, this involves "discovery" by the litigants of the evidence possessed by the other side. Relevant documents are produced, along with sworn statements about the time, place, and circumstance of their origin. In addition, the testimony of persons who had knowledge of or participated in the events can be compelled by the court. This testimony of these individuals is taken in a proceeding known as a deposition.

Having been called to give a deposition in a case where I had witnessed an automobile accident, I knew it could be an unsettling experience. Although usually conducted in the offices of one of the law firms, a deposition is similar to testifying in court. The witness is under oath and is subjected to questioning by both sides. A court reporter records the questions and answers. Objections are raised to questions considered improper by one of the lawyers. No judge is present, but upon review of the deposition, the witness may be compelled to answer the question.

Extensive depositions were taken from almost all the participants in the Six Day War pursuant to the Delaware action. Now that the playing field had shifted to Texas and Pennzoil's litigation had focused on Texaco, a whole new round of depositions was in order.

When Dick Miller found out Pennzoil had scheduled depositions with twenty-nine Texaco witnesses for later that same month, he went into court and demanded that Judge Farris impose a more favorable schedule for the discovery phase of the case.

The way Miller saw it, Baker & Botts had a three-month head start on his firm. He asked the court essentially to delay the discovery process for three months—to give him a chance to catch up. The flip side of this argument, of course, was that Texaco had been represented by counsel throughout the Delaware court actions. The knowledge and evidence gathered by those attorneys would surely be provided to Texaco's new lawyers. Judge Farris refused Miller's request and the discovery went forward.

The ruling did not mean the process was going to be a smooth one, however. If anything, it guaranteed just the opposite. The litigants were back before Judge Farris twice in the ensuing months. Having been denied the delay he sought from the court, Miller seemed to do what he could to slow the discovery process. On two separate occasions Pennzoil asked the court to require Texaco witnesses to produce certain documents they had refused to release.

This wrangling was mild compared to what occurred in some of the depositions, as the rancor already present among the lawyers and their clients at times rose to a fever pitch. Pennzoil videotaped most of their depositions; all were duly recorded by court reporters. Insults flew freely between the opposing lawyers, often to the amusement of the witnesses. Occasionally these insults degenerated into thinly veiled threats. During one deposition, Irv Terrell of Baker & Botts objected to the way Bob Brown was phrasing his questions. This led to an exchange that culminated with Terrell telling Brown if he wasn't "polite, I'm going to throw you out of here." Brown challenged Terrell to do just that. And so it went. Jamail and Terrell also took turns baiting Miller during other depositions. Miller replied in kind. Thus, in true barroom fashion rarely seen in civil cases, the opposing parties brawled toward the impending trial.

In late August, Miller discovered the $10,000 campaign contribution Jamail had made to the Re-elect Judge Tony Farris Steering Committee over five months before and immediately thought he smelled a rat. Miller quickly concluded that Judge Farris's initial refusal to delay the discovery and all his subsequent rulings were directly linked to the contribution.

Under the court procedures in effect at the time, after the pretrial phase was concluded, the case would be assigned to a judge for trial. With twenty or so civil court judges in Harris County, that meant there was about a 1 in 20 chance that the case would ultimately be tried before Judge Farris. With the approval of his client (Texaco), Miller filed a motion to have Judge Farris disqualified from presiding over the pretrial proceedings.

Given the climate of the case, it is understandable that Miller's motion was severe. In it, he alleged that Pennzoil had been behind the contribution: "The sheer size of the gift, far beyond the resources of one lawyer to give to one judge when so many judicial candidates are seeking election and reelection, suggests that Jamail's client Pennzoil is involved." The motion went on to suggest that the case had been moved to Texas so Pennzoil could "employ a lawyer who was known to have contributed heavily to judicial races."

Miller's motion to remove Farris didn't spare the Judge either: "His acceptance of such a massive contribution from a lawyer who is known by him to be the lawyer for a litigant in an important case just filed in his court—and at the very commencement of the case—is highly questionable, but his failure to disclose the gift to the other litigant and its counsel is a clear breach of his legal, ethical, and moral responsibilities as a judicial officer."

Although the contribution had been duly reported to the county clerk and was a matter of public record, Miller's motion incredibly seemed to accuse Judge Farris of accepting a bribe and Pennzoil of engineering the alleged bribe, which would have been a direct violation of the election laws prohibiting corporate contributions. The risk to public officials violating these election laws invariably outweighs any gain from the violation.

Miller also took this into account in his motion: "If one assumes Judge Farris was ingenuous enough not to recognize the true purpose of the gift, nevertheless, the Jamail-Pennzoil contribution appears to have succeeded, appears to have affected the outcome of the hearings, appears to show bias, and appears to the bar and the public as an affront."

While Dick Miller's entrance in the case had turned the heat up, this motion ensured it would soon boil over.

Pennzoil filed a response to the motion to disqualify Judge Farris that directly attacked Miller's motives. "Why would he do this? Perhaps it is as a precautionary alibi for the ultimate outcome; the evidence of Texaco's liability in this case will be difficult even for him to explain to a jury and the potential damages are of an order of magnitude he has not previously faced and cannot face now."

Despite this claim, the number of attorneys who had previously faced potential damages like the $14 billion Pennzoil sought could have held a meeting in a phone booth with room to spare.

The stage was set for another bloodletting. A hearing to rule on the motion to disqualify Judge Farris was set for October 25, 1984.

Retired State District Judge E. E. Jordan of Amarillo was assigned to preside over the hearing. The opening minutes set the tone for day.

When Judge Jordan called the proceeding to order, he customarily inquired if each side was ready. He then ordered that the witnesses be sworn. Miller replied that he didn't have any live testimony to offer. Jamail said he would like to have Miller placed under oath. Judge Jordan waived the oath, common practice for attorneys who are considered officers of the court.

Miller offered documents and depositions that outlined the $10,000 contribution and could be considered circumstantial evidence to support the allegations in his motion. These included the Pennzoil reply to his motion just quoted. He then rested his case.

John Jeffers rose to speak for Pennzoil, and the following exchange took place:

JUDGE JORDAN: "Do you have anything you wish to offer at this time?"

JEFFERS: "Your Honor, is it your desire to have evidence received before arguments by either side?"

JUDGE JORDAN: "I would think so. I wouldn't know what arguments you were making."

Jeffers then offered a certified copy of similar motions that had been filed in an unsuccessful attempt to disqualify three justices of the Texas Supreme Court in another case.

JEFFERS: "That's all we have to offer."

Joe Jamail had let Jeffers take the lead for Pennzoil, but immediately spoke up when he concluded.

JAMAIL: "I have a few questions to ask Mr. Miller if we can get him to take the witness stand."

MILLER: "I understand Pennzoil rests. Who does Mr. Jamail represent?"

JAMAIL: "We didn't rest. I'm putting you on."

MILLER: "I heard . . ."

JUDGE JORDAN: "Counsel, he has not rested."

JAMAIL: "Perhaps he's fearful of taking the stand."

MILLER: "We will see."

JAMAIL: "And I would like to have him sworn, Your Honor."

Jamail finally prevailed and the oath was administered to Miller. It is considered an insult to ask that a lawyer involved in a proceeding be sworn, but it was understandable in light of the serious allegations Miller had made against Jamail and Pennzoil in his Motion to Recuse.

JAMAIL: "Tell us your name, please."

MILLER: "Richard B. Miller."

JAMAIL: "And you claim to be a lawyer?"

MILLER: "Well, yes, I do claim that."

JAMAIL: "And you're licensed to practice?"

MILLER: "I claim that as well."

Jamail then asked Miller if he was familiar with the Canons of Ethics provision that imposes sanctions on one who files a frivolous and knowingly false motion.

MILLER: "Yes, I am."

JAMAIL: "And how do you understand that? What do you understand that to mean?"

MILLER: "I understand it to mean what it says."

JAMAIL: "Will you tell us what you think it means?"

MILLER: "Yes. Would you hand it to me so I can read it verbatim?"

JUDGE JORDAN: "There is no point in that."

JAMAIL: "Let me ask you this: You have filed a motion to recuse Judge Farris and have alleged I, Joe Jamail, have performed an act of bribery by taking money from Pennzoil?"

MILLER: "That isn't what I have alleged at all."

JAMAIL: "Well, it's in your pleading, sir."

MILLER: "Well, no, sir. If you will just read it to me."

JAMAIL: "I'll read it to you, of course I will."

MILLER:	"I don't agree with you. I don't agree with you at all."
JAMAIL:	"Perhaps you can recall that you referred to it as the Jamail-Pennzoil gift. Do you recall using those terms?"
MILLER:	"I referred to it in that manner after having recited, prior to that, that it was my belief based upon information available to me that there was an understanding, whether tacit or stated, between yourself and Pennzoil, that you would in some manner be reimbursed for this gift in whole or in part."
JAMAIL:	"Do you have any evidence of that, sir?"
MILLER:	"I have no direct evidence. I have a great deal of circumstantial evidence."
JAMAIL:	"Where is it, sir?"
MILLER:	"Well, it's based on what I know about Mr. Liedtke and his past."
JAMAIL:	"It's your opinion then, is it?"
MILLER:	"Certainly it is my opinion. I think I'm right."
JAMAIL:	"I asked you to swear to this, and you have refused to do that. Do you swear today that Pennzoil has in some way agreed to reimburse me or was in any way apparent in my contribution to Judge Farris?"
MILLER:	"If you have read the supplemental motion . . ."
JAMAIL:	"Sir, the question . . . Your Honor, I would like an answer to the question."
MILLER:	"I'm going to answer your question. The supplemental motion very clearly states that we have no direct evidence of that. Those kinds of deals are not made by direct evidence, and you know that just like I do."

Jamail had elicited from Miller an admission that his motion was not supported by direct evidence. Jamail then hammered at that point, referring again to Miller's pleading.

JAMAIL:	". . . you say, under 'Signed, Counsel for Texaco,' meaning you, Miller, believe and therefore allege that the Jamail ten thousand dollar contribution was made in cooperation with Pennzoil and the gift was known to Pennzoil chief executive officers. Do you have any evidence of that, sir?"

MILLER:	"I certainly do."
JAMAIL:	"Where is it, sir?"
MILLER:	"It's right there in your pleadings. When you filed your sworn motion, you did not deny that he [Liedtke] knew about it and that's very compelling to me."
JAMAIL:	"Sir, that is so cowardly."
JUDGE JORDAN:	"Wait a minute. I'm not going to tolerate that."
JAMAIL:	"I apologize for that, but this gentleman . . ."
JUDGE JORDAN:	"You can curse one another all you want to somewhere else, but not in this court, and this is not going to be permitted to degenerate into an argument between counsel."
JAMAIL:	"I'm asking for an answer now under oath."
JUDGE JORDAN:	"He says he will not."
JAMAIL:	"He will not swear to that?"
MILLER:	"I can't. I don't have personal knowledge, and I'd have to have personal knowledge to swear to it."
JUDGE JORDAN:	"Let's don't beat around the bush."
MILLER:	"I stated that I believe it."
JUDGE JORDAN:	"Yes. That's not what he asked you."
JAMAIL:	"I asked him if he had evidence."
JUDGE JORDAN:	"He said he had nothing but circumstantial evidence that leads him to so conclude."
JAMAIL:	"I'd like to know if we may see the circumstantial evidence, if you will tell me . . . where can I find it, other than your opinion?"
MILLER:	"Well, my opinion is based on a great deal of information that I know about Mr. Liedtke. I also know about some of the activities that you have engaged in in the past. If you want me to state that, I will."

Jamail started to object, but was cut off.

JUDGE JORDAN:	"Counsel, I'm going to at this point say I don't see the materiality of the point in this court. Now, you may wish to pursue this in another forum, but at this point it has no bearing on the fairness or unfairness of this judge."

JAMAIL: "I was making that remark, Your Honor, and trying to elicit this testimony to show that this could not have influenced Judge Farris."

JUDGE JORDAN: "I don't believe it anyway."

JAMAIL: "Thank you. That settles that. I have no further questions, Judge."

Judge Jordan's observation that he didn't believe the contribution could have influenced Judge Farris was a clear indication of which way the wind was blowing in this hearing.

Miller's inability to support his motion with anything other than "what I know about Mr. Liedtke and his past" made this whole action seem strategically risky, at best. It is easier to understand when you consider the fact that Miller was at Baker & Botts many years before when the firm became the principal counsel for Liedtke and ultimately Pennzoil.

Although he hadn't really been involved in representing him, Miller had viewed his current foe from an inside vantage point for over two decades. In the process of building Pennzoil, Hugh Liedtke had been involved in a number of controversial deals.

Pennzoil's 1965 acquisition of United Gas, a company ten times its size, was accomplished through an unsolicited, or "hostile," tender offer. The practice is common today, but was virtually unheard of in 1965. When Liedtke later spun off some of United Gas' operations into independent companies, the deals favored Pennzoil, at the expense of the surviving operations. The details of the United Gas affair are intricate and not particularly relevant to this story, but it resulted in Pennzoil returning a $100 million dividend of preferred stock to United Gas. After signing a consent decree with the Securities & Exchange Commission in 1974, Liedtke and several other insiders had to pay back a total of $100,000 in profits from trading in Pennzoil shares prior to the spinoff announcement.

Another thorny problem for Liedtke arose when it was revealed that a Pennzoil jet had been used to fly $700,000 in Nixon campaign funds to Washington two days before a new disclosure law took effect. Liedtke approved the use of the plane. "It seemed like the neighborly thing to do," he later recalled. This was not illegal, but some of the money was later traced to the bank account of Bernard Barker, one of the Watergate burglars. This aroused the interest of a congressional committee and a federal grand jury.

No criminal violations were found in either case (United Gas or Watergate), but his awareness of these events had led Miller to make what was turning out to have been an ill-advised attack on Liedtke, Jamail, and Judge Farris. What he knew about Liedtke's past was not evidence that directly supported this motion.

With the testimony completed, each side was allowed to make a final argument.

JUDGE JORDAN: "I'm going to limit you to thirty minutes to a side. I never turn a lawyer loose without a bridle on him because I've been where you are and I know that when you get through saying all you can say, you think 'there must be something else I can say.' And just in fairness to me, I'm going to ask you to limit your argument to thirty minutes to a side."

MILLER: "I'd like to try to make my presentation to Your Honor in fifteen minutes."

JUDGE JORDAN: "You may. I'd appreciate it. I think you can do it in fifteen minutes. I'll give you thirty if you need it."

MILLER: "I appreciate that very much, sir. It's an important matter to Judge Farris and to all of us, and I'd like to have an opportunity to state our position fully and completely and at the same time not waste the court's time."

JUDGE JORDAN: "Please do."

MILLER: "Now this is a Motion for Recusal. The basis for the motion, the law under which the motion is brought . . . states quite plainly it's brought under the Code of Judicial Conduct of the State of Texas.

"There is a particular section of that code on which we are relying that's set out fully on page 4 of our brief, Judge Jordan. Canon 3C(1) of the code provides as follows: A judge should disqualify himself in a proceeding in which his impartiality might reasonably be questioned . . .

"A judge must not serve, shall not serve, must disqualify himself in any proceeding in which his impartiality might reasonably be questioned. That's an important rule because—I say this as an aside—if I learned anything in three years as an enlisted man in the Marine Corps, if you treat everybody the same, you're going to get along okay; but the first time you show favoritism in any manner or if there is the appearance of favoritism, you are in trouble. That's the cornerstone of our case."

Miller staked his legal claim on the provisions of the Code of Judicial Conduct he cited. He then turned to the Texas Constitution, which Pennzoil had cited in its reply to his original motion.

MILLER: "If the judge has previously represented one of the parties to the litigation, he's disqualified under the Constitution. If a judge has a financial interest in the judgment that might be entered in the case, he's disqualified. And if he's related by blood or marriage to one of the litigants, he's disqualified. Those are the only three disqualification grounds set out in the Constitution.

"Now, we are told in this case—and I'm still not where I want to be in terms of describing what I think the law is— but we are told by counsel for Pennzoil and Judge Farris that those three grounds for disqualification that are set out in the Constitution are the exclusive grounds. 'Inclusive and exclusive' is the language used. . . . As I understand their argument, it's no ground to disqualify a judge if he is biased or prejudiced or if he appears to be biased or prejudiced. They say—and they use the word, well—that is 'mere bias or prejudice,' as if to denigrate it, and that it's okay to go to trial before a judge who is biased and prejudiced against you but not fair to go to trial in a case in which a juror is biased or prejudiced. The anomalous, and I say archaic, argument: You can challenge a juror for bias or prejudice, but you can't challenge a judge. Now, that offends my sense of fair play. That's the law on which we rely.

"Now, what facts do we rely upon, which we ask the court to find suggest that the impartiality of Judge Farris might reasonably be questioned? This lawsuit was filed in Harris County in mid-February of 1984, this year. I don't have the precise date in my mind. We filed an answer on behalf of Texaco on March 5, 1984. That same day we received a bale of paper, I mean literally a bale of paper just about this high, from counsel for Pennzoil of discovery matters. And two days later Mr. Jamail personally delivered to Judge Farris a ten thousand dollar campaign contribution.

"One of the most important hearings in this case was held on March 14, a week after that contribution. It was a hearing in which we requested relief from the very onerous discovery schedule which Pennzoil had thrust upon us. . . . We asked for reasonable time to become familiar with the case. Judge Farris wouldn't give us the time of day. He wouldn't delay discovery one day."

Miller then recited details about Jamail's role in the Farris campaign, including fundraising letters from Jamail soliciting contributions from other attorneys. He also produced a letter from Judge Farris to Joe Reynolds, an attorney who was chairman of the Re-elect Judge Tony Farris Steering Committee. In that letter, Farris had referred to the fact that all the steering committee members had made contributions and "one Plaintiff's attorney in particular has contributed a princely sum."

His argument went on to address the question of the state constitution versus the Code of Judicial Conduct at some length, elaborating on relevant case law.

He concluded his argument with a frankly emotional appeal:

MILLER: "I'd like to say in closing—I have fooled around and taken my thirty minutes—I want to particulary make the point to the court that . . . if you do not feel, if a litigant does not feel, if the public does not feel that there is going to be equality under the law, and if there is not at least one branch of this government that is above suspicion, which is above graft, which is above handouts, which is above influence-peddling, then we are in bad shape. Thank you, sir."

After the initial clash when he put Miller on the stand, Jamail had sat and listened to his argument in silence. When he rose to begin the argument, his client was Pennzoil but the attack had been on him personally.

JAMAIL: "Your Honor, I'm Joe Jamail and I want to respond to some of what has been said to you. Mr. Miller claims they're facts. They are falsehoods. When Mr. Miller tells this court that I have never supported a Republican, it's a falsehood. I'm on the Steering Committee of District Republican Judges, have been, have contributed money to Republican judges. It's a falsehood, as everything else he has stated to you is a falsehood.

"It is a disgrace for an officer of the court to come in and with suspicion and surmise smear the judicial system. I have contributed money to almost every judge, county, criminal (court judges). I don't practice criminal law (but) they come to me and they ask me. If someone is going to tell me it was wrong to make contributions, *reported* contributions—nothing was hidden. There was nothing done surreptitiously. Mr. Miller's facts come from his mind, jumping to conclusions from suspicion he himself forms. To say that I have never given a sitting district judge ten thousand dollars before is a lie. I have. And he knows it. And they have been published as they are duly required to be published. But to stand here and deliberately perpetrate falsehoods on this court is unbecoming, undistinguished, and not respectable, and it's an attack on the judicial system.

"I suppose he wants to say if you give a hundred dollars it's all right, but if you give more it's not. Well, those of us who have been successful at the Bar, unlike Mr. Miller, are called upon to do more."

For Jamail, always a forceful advocate, this wasn't just another argument. After repeatedly describing Miller a liar (well, a perpetrator of falsehoods) and a stream of other unflattering terms, Jamail focused on the allegations in Miller's motion.

JAMAIL: "Now to say I took a check to Judge Farris is a lie. I never handed Judge Farris a check. Mr. Joe Reynolds, who is Steering Committee Chairman, called me—and I want him to speak with part of my allotted time—and he asked me to get a check over to his office for ten thousand dollars. And he will tell you why it was done. And I didn't do it immediately and I did wait, and he can explain that to you. But never did we try to conceal anything.

"Now, as far as sending a letter out, yes, I've sent letters out for many judges, many judges, because when we have responsible judges, I think that the lawyers owe the public and the Bar a duty to the bench to try to support them and help them. And we have never supported a judge based on whether he had a prefix of a political party in front of his name or not. And that is a lie to say that I have never supported a Republican judge, and he knows it. And it's also a lie to say that I have never given ten thousand dollars to a sitting district judge, and he knows that, too.

"It's all, judge, this: It's a long whine and a pout by Mr. Miller who has not been able to run over Judge Farris. Most of the hearings we have attended there have resulted in an agreement. The last hearing that was there we attended, Mr. Miller was given a time schedule. He became surely and churlish because he couldn't have his way, and he left. The judge had no alternative but to adopt the time schedule we had given him. Their partner calls, 'Can we revise it?' Certainly we can. Mr. Miller's problem is that he can't run over Judge Farris.

"Now, for Judge Farris to disqualify in the face of this scurrilous motion—I hate to even dignify it with the term motion—a conglomeration of lies and half-truths that Mr. Miller filed and signed and refused to swear to, he thinks should disqualify the judge. Now, if the judge says, 'Okay, if he's done this, that changes me, then I have to get out,' what do we do with him judicially? He can pick and choose who he wants at that point if Judge Farris had disqualified.

"And, Your Honor, Judge Farris won't try this case. This will go on the central docket and it will go back to be reassigned. Mr. Miller, if he can achieve his absolutely despicable tactic, as he has attempted to do in this case, then no judge can sit with impunity because judges rely on lawyers to make campaign contributions. How else are they to advertise? The media doesn't give them that for free. If that's wrong, they need to tell us it's wrong. It would save me a lot of money.

"But to allow Mr. Miller to hide behind a frivolous motion and say, 'Okay, I don't like this judge because he hasn't ruled with me,' it would never occur to the peevish Mr. Miller that he had nothing the judge could approve of. He's brought no time schedule that was within reason. His whole purpose in this case, Your Honor, from day one has been to delay it. Delay, every time we've ever had a motion. He does not want to go to trial. We don't have enough cowboys in Texas to whip him with to make him come try the case. So this is another tactic. Mr. Miller wants to delay the case."

Jamail then attempted to pass the floor to the aforementioned Joe Reynolds, who didn't represent Pennzoil, Texaco, or Judge Farris. Miller objected that this was more testimony, clearly improper during arguments. Reynolds responded that he had an argument to make.

In fact, Reynolds had filed a motion for sanctions against Miller for taking frivolous depositions from three of his clients who had also made contributions to Judge Farris. Reynolds wanted Judge Jordan to order Miller to reimburse his clients for their expenses.

Judge Jordan declined to hear the motion. John Jeffers rose to complete the argument for Pennzoil.

JEFFERS: "Your Honor, I'm of two minds as to how to present this
            argument. On the one hand, I'm of the mind to say we don't
            have a serious matter here. It's not often, at least in my ex-
            perience with lawyers of this caliber, that a contested mat-
            ter is brought before a judge when the Supreme Court of

Texas has just spoken to precisely the same point. I mean, the law is established on the matter of political contributions. So I'm inclined to begin by saying, 'What is it we are talking about here?'"

Jeffers reviewed the recent cases ruled on by the Supreme Court in some detail.

JEFFERS: "Even though the law is clearly established, I must say that it's still a most serious matter because you have here a charge in effect that a bribe of a judge has been offered and accepted in order to gain influence.

"As our sworn response says, in the first place, Mr. Jamail was on the Judge's Steering Committee and obligated himself before the case was filed, long before the case was filed and long before anyone knew it was going to fall in Judge Farris' court.

"But, in any event, when a charge like this is made, it is not to be taken lightly. Mr. Miller says, 'Well, this is simple enough. There are lots of judges here. The judge ought just to recuse himself since I've accused him of this impropriety.'

"What he's saying is the judge ought to admit that he was wrong, ought to permit a public inference that would then become a horrible blot on his otherwise distinguished career, Pennzoil ought to admit that it was wrong all because of these groundless charges which he won't swear to, which haven't been supported by any evidence, and which aren't going to be sworn to or supported by any evidence.

"Now, if he were successful in this tactic, not only would the judge's career be tainted, Pennzoil's reputation in the community would be tainted for no reason, on no grounds. And moreover, we would face, to the extent that Mr. Miller succeeds and Texaco succeeds in tarnishing Pennzoil's reputation in the community, they make it difficult for Pennzoil to get a fair trial by jury or to find jurors who have not heard about these allegations. Bear in mind, these allegations went immediately to the press. The press somehow got ahold of these motions and published stories the very next morning after these motions were filed. So I think that's a good indication of what is really going on here. And it's for that reason a very serious matter to Pennzoil, even though the law is very clear.

"The other thing that makes this a serious matter is Mr.

Miller, if he gets his way, is putting judges in an untenable and impossible situation. The trial lawyers of Texas to my knowledge have historically and overwhelmingly supported the elective system. I have never known Mr. Miller, for example, to raise his voice in favor of any other system or in favor of the abolition of the elective system.

"By his motion, he is saying that the judges are going to have to run for office, they are going to have to stand before the people, they are going to have to put their faith with the people to do that. Oh, yes, they are going to have to raise money, twenty-five dollars here, a hundred dollars here; but, by the way, anytime that contribution gets above a certain size, and he's not willing to say what that level is, but anytime it gets above a certain size, it's going to be prima facie evidence that there is a bribe involved, that there is influence peddling.

"I don't know where the cutoff is. I read his motion and he said in his original motion that no other contribution besides the ten thousand from Mr. Jamail was given to Judge Farris larger than a thousand dollars. And I looked at his own exhibit which had the list of contributions, and there were four of them on one page for twenty-five hundred dollars. And I said, 'How on earth did he miss something like that?' Four people gave together what Mr. Jamail gave. How on earth could he have missed that? A few days later, I found out what the method to the madness was. He had assumed anyone that would give twenty-five hundred dollars must have been infuence-peddling in some way."

"So we took off all day Tuesday and noticed these four people for their depositions and he found some plaintiff who had a case in Judge Farris' court at about the same time and who had won the case in Judge Farris' court in the same, more or less, time frame that these four contributions were made and he [Miller] said, 'Well, there has got to be some hookup there. These four people must somehow be tied into this plaintiff; otherwise, they wouldn't have given twenty-five hundred dollars.' Mr. Miller didn't offer those depositions in evidence this morning. They are all right here, and they have been transcribed. And so Your Honor can infer what happened at those depositions, that the four contributors never heard of that plaintiff and that plaintiff never heard of them. It turned out two of those contributors had known Judge Farris thirty years and gone to school with him. They were all brought down here to be subjected to the innuendo that they were influence-peddling. They had nothing to do with Pennzoil or Texaco."

Jeffers concluded his argument by taking Miller to task for accusing Judge Farris of unfair treatment, then failing to produce transcripts of the hearings where the alleged mistreatment took place. His cerebral tone was an effective complement to the bombastic rhetoric of Jamail. Jeffers seemed to have effectively snuffed what was left of Miller's motion.

Julius Glickman was an attorney representing Judge Farris. As he would later describe his role in the hearing, "I just sat and watched while they went at each other. I finally did get to talk for a minute there at the end."

Glickman spoke about the case law on the disqualification of judges and pointed out that Jamail's contribution amounted to less than 20 percent of the funds Farris raised. He then concluded with this:

GLICKMAN: "Your Honor, I think there are certain realities in this case. Judges must run for office, there is no broad-base support for the judiciary, there is little name identification in a judicial race, and we have no public financing of campaigns. In November of 1982 in Dallas and Houston, 51% of the judicial races were contested. Campaigns are becoming more expensive; and there is simply a necessity for judges, like other political candidates, to raise money. I think these attacks not only diminish Judge Farris but diminish the Bar and the judiciary. And in spite of these attacks—and these are innuendos only—I hope this court will say very strongly that the price of judicial integrity is not ten thousand dollars and I hope the court will make short work of this motion."

Miller would have the last word, since the burden of proof was on Texaco in this proceeding.

MILLER: "I'd like to take two or three minutes to summarize."

JUDGE JORDAN: "I'll give you ten minutes."

MILLER: "Thank you. One thing that made me finally realize that I ought to never let the other side represent my views is they always want to recast the question. The suggestion is made that this Motion for Recusal is because we have alleged that a bribe was taken. And the word 'bribe' always is used to put the position in the most extreme way. That word appears nowhere in our motion. What appears in the motions is a recitation of the known facts and of the argument—and it's an honest, legitimate argument—that it's like Sherlock Holmes says to Mr. Watson, 'Watson, I don't believe in coincidences.'

"This ten thousand dollar gift was given because the Pennzoil case fell in Judge Farris' court. That's why it was given."

Miller then addressed the charge that his motion would throw the entire court system in disarray.

MILLER: "I have no interest in answering these charges. I know about myself. I know what I did, and this is the first time I ever filed such a motion. And this idea there is going to be some kind of a landslide batch of motions to disqualify judges is baloney. The Code of Judicial Ethics that we are trying to enforce in this case is enforced in every federal court in this land, the identical Code. And the claim that judges are going to be put at the mercy of the lawyers is so much eyewash.

"I say that in the circumstances of this case, that any reasonable person—I'll just put it like this: If it had been *my* case, if I was a party to this case and I found out that the lawyer on the other side had given ten thousand dollars to the judge as a political contribution, I would be in grave danger. I don't think anybody in this courtroom—if they want to say 'Yes, I'd be willing to go trial in front of a judge who has taken ten thousand dollars from my opponent or his lawyer, I'd like that. I'd particularly like it if I didn't know about it.' And if you could somehow keep it a secret, file a paper over in the clerk's office.

"We found out about it by accident. If offends my sense of fair play. I'm disappointed to see that it does not offend the sense of fair play of these other counsel.

"To me, if a citizen of this state has to go to trial before a judge who has accepted a 'princely sum' from his opponent and has to take his chances in that court to rule on motions, to pass on evidence, then there is something wrong somewhere. Something is rotten in Denmark. Now, what it is, I don't know. But I do know this, that whether or not Judge Farris knew the purpose of the gift, whether or not he knew it, Mr. Jamail knew it. But he certainly has by his failure to have disclosed it, when the gift was given right after the case was filed in his court, right after. I don't know, maybe Watson is right and there is a coincidence, but nobody is going to believe it, no member of the public is going to believe it. I think it's a shame.

"I ask Your Honor to rule that the judge is recused and to appoint a substitute judge."

Immediately upon the conclusion of Miller's argument, Judge Jordon began to deliver his ruling.

JUDGE JORDAN: "I have been a student and practitioner of the law for a long time and one time I remember an appellate court telling me that I made the right decision for the wrong reason and it didn't make any difference what reason I gave, whether it was right or wrong, as long as the decision was right. And perhaps I ought not to say a word here except to make a ruling on this motion, yea or nay, but I think that this is so important a case that I ought to have an opportunity to express my feelings about this matter.

"It's most unfortunate that the system has risen where the judiciary is dependent on the generosity of the public, and particularly the Bar, to pay the expenses of campaigns. I don't know how it can be amended, how it can be changed, how it can be improved. I regret it. I have never been unfortunate enough to have to go to counsel and ask them for their help. I did have one contested election and I told them 'I don't even know who contributes or how much. Give it to my campaign manager.' I've got the records at home. To this day I haven't looked at them because I didn't want anybody to be able to do to me what has been done to this judge today.

"From what I've heard here today, the matters that have been brought before this court would be more proper before the Grand Jury. If there is any evidence of political contribution by a corporation to which might be in violation of the law, the Grand Jury ought to investigate it rather than this court. And counsel declines to use the word 'bribery.' That's what it amounts to, in my opinion."

Judge Jordan then addressed the clash between Miller and Jamail.

JUDGE JORDAN: "And accusations back and forth between the lawyers associated in this case have become so acrimonious that, well, it's just disgraceful in my opinion what they have said about each other. And if they are true, the Grievance Committee of the Harris County Bar ought to be interested in seeing which one of them is right and which one is wrong. But I'm just saying what I think ought to be done.

"You are going to tear up the whole judicial system in this state if any lawyer at any time feels like that in his judgment there is an impropriety on the part of the court to even file the Motion to Recuse that judge. It's going to play havoc with the judicial system, that's just all there is to it. Because then the judge is placed in the unenviable position of having to say, 'Yes, I'm a crook,' or fight it, which the judge in this case is doing. I think had counsel gone to him in a gentlemanly way and privately said, 'Judge, do you feel like you can be fair in this case?', given him an opportunity to recuse, it's possible he might have done so. But this judge has been placed in a position where he's got to fight it. He can't recuse at this point. I wouldn't. No other judge would.

"I learned a long time ago, and I'm persuaded this is still the law, the Constitution of the State of Texas or of the United States means the same thing today that it meant the day it was written. It doesn't change just because some judge or some people might think it ought to be changed."

Judge Jordan particularly refuted the argument that the Code of Ethics was a de facto amendment to the state constitution and then made his ruling.

JUDGE JORDAN: "It serves no good purpose for me to contend on this. I don't think this motion should be granted for two reasons: One, I think the Constitution still controls, the Code of Ethics does not. Number two, there is not only insufficient evidence, there is *no* evidence in my mind to justify recusing this judge. It's so ordered.

"The court is in recess."

JAMAIL: "I would like to order a transcript of this, certified, so that I may forward it to the Grievance Committee for the Grievance's consideration of Mr. Miller's conduct."

MILLER: "May I have a copy of it, please?"

The hearing ended as it began, with a sharp exchange between Jamail and Miller. Judge Farris was back on the case, which had been in a sort of legal limbo after the Motion to Recuse was filed.

In an article in *American Lawyer*, it was reported that a disciplinary complaint was filed against Miller by one of the Pennzoil attorneys, but not Jamail personally. In turn, Miller filed a counterclaim alleging that his accuser had filed a frivolous complaint. Both complaints were summarily dismissed without a hearing. In such instances, the filings do not become a matter of public record unless sanctions are imposed.

The district attorney did not present the matter of the $10,000 contribution to the Grand Jury. After reading the transcript of this hearing, it is understandable why the D.A. didn't want to get involved.

Although the Motion to Recuse had been covered by the Houston newspapers, it was not a headline grabber. When the motion was denied, it is inevitably received even less notice. Sensational accusations are considered more newsworthy when they are made than when they are dismissed.

These allegations seemed to have a life of their own, however. When I first learned of the controversy in the weeks after the trial, I obsessively gathered as much information about it as I could. I obtained a copy of the transcript of the hearing, which was heavily excerpted in this chapter.

The $10,000 campaign contribution became a central element of Texaco's public relations campaign after the verdict. Texaco's new Chief Executive Officer James Kinnear continued to refer to it after Texaco filed for bankruptcy in April 1987.

In reflecting on Judge Farris and his reelection campaign of 1984, it is difficult to follow the logic of the motion Dick Miller filed on behalf of Texaco. Jamail's contribution ultimately amounted to less than 20 percent of a campaign budget in excess of $50,000. The total amount is not exorbitant in light of Republican judicial losses in the 1982 elections. Judges who mounted only token campaigns may have had reason to worry. But Jamail's $10,000, while undoubtedly important, was not a make-or-break proposition for the Farris campaign. Ronald Reagan would be on the ballot again in 1984, and while his coattails had not proved as long as Texas Republicans had hoped, a replay of the 1982 debacle seemed farfetched.

As Jamail repeatedly pointed out during the hearing, the contributions were duly reported as required by law. As has been noted, the reports were a matter of public record. Jamail's sizable campaign contributions to other judicial candidates long before Pennzoil became his client were also well documented.

The notion that the Jamail contribution was the pivotal factor in the case is patently absurd. It is a notion that is still given wide currency by Texaco, but it crumbles under the weight of the evidence of what took place long before Joe Jamail and Judge Farris came into the picture.

In early 1985, the rules of the district courts in Harris County were changed. The same judge who heard pretrial motions would now preside over the trial itself. This meant that Judge Farris would preside over the trial of *Pennzoil Co.* v. *Texaco, Inc.*

Texaco renewed its motion to disqualify Baker & Botts from the case. On July 5, 1985 Judge Farris denied the motion. Texaco appealed, and the Texas Supreme Court upheld Judge Farris's ruling on July 9, clearing the way for the trial to begin that day.

Despite Jamail's blast at Dick Miller nine months earlier ("We don't have enough cowboys in Texas to whip him with to make him come to trial"), the trial was about to begin. This is where I come in.

PART IV

A SUMMONS FROM THE SHERIFF

My Decade of Jury Duty

In suits at common law, where the value in controversy shall exceed twenty dollars, the right to a trial by jury shall be preserved, and no fact tried by a jury shall be otherwise re-examined in any court of the United States, than according to the rules of common law.

AMENDMENT VIII
UNITED STATES CONSTITUTION

7

The envelope is actually a multipart, computer-addressed self-mailer like you might get from your insurance company. The return address, however, reads "Harris County Sheriff."

When I received my juror summons in late June 1985, it was not unexpected; I had been called for jury duty at least a dozen times since I first registered to vote in 1971.

The trip downtown to the jury assembly room in the courthouse complex was a familiar ritual. One constant in the process was Deputy Clerk Marion Cleboski. Twice a day on Mondays, Tuesdays, and Wednesdays, he took responsibility for greeting the arriving throngs of summoned citizenry. Several hundred people a day would crowd into the room as he called the roll, assigning potential jurors to seats sequentially numbered according to the number on their summons. He also provided a little humor and good cheer to people who had their lives disrupted for at least a day to discharge this civic responsibility.

The potential jurors are selected at random from the approximately 1.5 million registered voters in Harris County. This process is now conducted every three years, utilizing a computer program developed with the help of NASA and the University of Houston. "I'm not a computer-type person," Cleboski says, but he admits the present system works better than the one employed prior to 1981, when the random selection was done every year.

The change was welcome. In one 20-month period during 1978–1980, I was summoned for jury service three times and actually served as a juror for the first time in two separate felony criminal trials.

When I had first answered a jury summons in 1974, I was a newly wed 21-year-old working at a graphics firm. At the beginning of each session, the judge who had charge of the juror pool addressed the assembled candidates. Automatic exemptions are provided for persons over age 65 and those who care for children under a certain age. In addition, a potential juror may be excused for medical reasons and economic hardship. The judge explained this and heard applications for exemptions. I joined the line at the bench to present my case: when I had informed my boss I had jury duty, he instructed me to get out of it, reminding me that I would not be paid for any time I missed. Losing three days pay at that point in my life could have meant a failure in making the rent.

Judge Andrew Jefferson heard my story and kindly but firmly denied me an exemption. "The man on trial in one of these courts today may not be able to afford to miss a day of work either, but under our laws, he's entitled to a trial by a jury of his peers." A rather obvious answer, and one he no doubt gave to many others that same day. But it made a lasting impression on me, and I faithfully showed up with each new summons.

This was not a particularly sterling act of courage, however. At that time, young men with hair down to their shirt collars were not often picked to serve on juries; many prosecutors would routinely strike me along with any young blacks on the panel. Sometimes I made it home by lunch.

Apparently, over the years I drifted closer to the mainstream. Hairstyles changed; my hairline receded on its own. I went to work for the City of Houston in early 1975 as a communications specialist, working on a wide range of projects in print and electronic media for the Model Cities Department.

The federal program that established Model Cities was one of Lyndon B. Johnson's Great Society projects in social engineering that was frankly on its last legs when I came on board. While the program was roundly criticized (with some justification), a lot of good things were done. Through my participation in it, I felt a sort of spiritual linkage with Franklin Delano Roosevelt and the New Deal. I realize that putting this declaration in a book about the battle of the corporate giants may weaken my credibility with some ideologues, but there it is.

My unwitting move toward respectability had another effect, however. When I was first selected to serve on a jury, my reactions were mixed. Although financial hardship was no longer a factor (government employees are paid for time missed for jury duty), I had landed on a felony case of Indecency with a Child.

It was a wrenching trial for me, with two days of testimony by four little girls and their mothers. The defendant was a fellow named Harvey who seemed to be regarded as the neighborhood weirdo, a generally harmless sort who lived alone with his 5-year-old daughter.

The judge was Neil Caldwell, a former member of the Texas legislature and then a visiting judge from down in Alvin, Texas (hometown of Nolan Ryan). An affable, downhome type, Caldwell would lay back during the breaks with his feet propped up, cowboy boots sticking out from under his robe, and try to relieve the considerable tension generated by the testimony with some front porch–style conversation with the jury.

Harvey and his daughter lived in the same apartment complex as the other girls and their mothers. In the evenings when he would come home from work, he would take his daughter down to the playground. I got the feeling that most of the daddies in the complex didn't spend much time down at the playground. The other little girls loved to play games with Harvey. He would accommodate them as best he could. When he would take his daughter for a piggyback ride, the other girls all wanted to ride, too.

Some of these girls were a few years older than his daughter; the oldest was around 9. The allegation of Indecency with a Child involved the position of Harvey's hands when they were locked together to support the rider; were his fingers pointing up or down?

The girls testified, one by one. They each took the stand and told their story. When it came to the critical question of the prosecution's case (During these piggyback rides, "where did Harvey touch you?"), each little witness answered in the same words: "Where I go to the bathroom."

Harvey's lawyer objected to the testimony of each of the girls, saying they were not competent witnesses. Their mothers also testified, but not at any great length. They were there to offer an adult, if mostly secondhand, version of the same story.

For his defense, the lawyer put Harvey on the stand. He seemed somewhat bewildered, but denied every charge. His lawyer did manage to get the information across that since the charges were made, Harvey had sat in the county jail. His child's mother had returned to get the girl and then left the state. Other than Harvey himself, there was not a single witness put on the stand on the defendant's behalf. The obviously court-appointed lawyer did not produce one person to say "Yes, I know Harvey. He's a good man who loves his daughter."

The prosecutor did his job with considerably more enthusiasm. His closing argument was a passionate plea to rid society of this "menace." This was at a time when the rise in the reported incidence of child abuse was still a couple of years away from the cover of *Time* magazine.

The judge delivered his charge to the jury, the instructions that tell the jury what the law is, and what is necessary to deliver a "guilty" or "not guilty" verdict. The burden of proof in a criminal trial is always on the state. The prosecutor has to establish the guilt of the defendant "beyond a reasonable doubt," which is distinctly different from "beyond any possible doubt." The charge spoke to this point directly.

With that, the jury retired to consider the verdict. After electing a foreman, we took our first vote by secret ballot: 8 to 4 for conviction. At that point, deliberations began in earnest. In many ways, the jury room is as close to the fulcrum of democracy as a citizen can get. Each member has a voice in the outcome. In criminal trials, the consent of all twelve is required; the verdict must be unanimous or there is no verdict and a mistrial is declared.

Trying to reach a consensus on the guilt of Harvey the alleged child molester would consume the better part of that afternoon and the next day. With some discussion, the tide soon shifted against the defendant—after an hour of deliberation the vote had moved to 10 to 2.

Almost invariably, there is a person or persons on every jury that plays the role of devil's advocate. While they might personally be inclined to favor conviction, they will try to leave the defendant cloaked in the presumption of innocence for a while and challenge the other jurors to remove it. In this way, they can satisfy their own view (for conviction) and still tell themselves they gave the defendant a fair hearing.

These jurors will search for any other possible scenario that explains the facts of the case; the idea of convicting an innocent man is still repugnant to most Americans. This is where "reasonable doubt" comes into play. Are the alternative explanations possible? Well, there are circumstances under which almost anything is possible. But to find that one of these explanations constitutes "reasonable doubt," it must be more than possible. There must be a reasonable likelihood that it could have occurred.

With the case of Harvey and the piggyback rides, I had serious problems with the testimony of the little girls. The fact that they had all used exactly the same wording in describing the critical point of the state's case was troubling indeed. Had they been coached? Were they, in fact, saying what they knew the mother or some other adult wanted to hear?

These questions were not kindly received by some of my fellow jurors. "Absurd," said one juror. "Do you have any kids?" asked another. When I responded that I did not, she snorted, "I didn't think so." "Kids don't lie," contributed the foreman.

The jury in this case was composed of a typically mixed bag of people covering a wide range of age, race, and social backgrounds. My sole ally at 10 to 2 was an older gentleman who was a cab driver. As the pressure increased in the cramped quarters of the jury room, his discomfort became visibly apparent. With a half-apologetic shrug to me, he changed his vote. We had reached a critical juncture: 11 to 1 for conviction. I silently cursed Harvey's lawyer for not giving us any evidence to mitigate the prosecution's case, other than the hapless (but I believed truthful) testimony of the defendant.

Although I was only 25 years old, I was not the youngest juror. A hot-headed young man about a year my junior was one of the most outspoken, if inarticulate, advocates of conviction. All he wanted to do was, as he put it, "get this thing over with and get the hell out of here." He would blow up, then allow himself to be calmed down by a middle-aged juror (and a mother!) who could not resist the urge to mother him. She lectured me as if I were a particularly recalcitrant student at vacation bible school. This was great.

We failed to reach a verdict on that first afternoon of deliberations, but it was still far too soon to call it a hung jury. We would have to return the next morning. As you can imagine, there were at least a few hard feelings toward your humble narrator. In my view, however, there was not only reasonable doubt, there was serious doubt.

When the court reconvened the next morning, the judge ordered the courtroom cleared for the jury's use. The cramped quarters of the jury room had possibly exacerbated the tensions of the deliberations thus far, and the air conditioning wasn't working too well either.

When we resumed our discussion, attention was quickly focused on the reasons for my holdout. Harvey had never tried to get any of the children alone, I reasoned. He had never placed his hands on anything other than their clothing, as his lawyer had reminded us. Was this sufficient evidence to brand a man as a convicted sex offender and banish him to a lengthy term in the penitentiary?

Among the panel, these arguments could find no purchase. As the day wore on, the reactions to my position were divided among three camps: total indifference, rampant disgust, and sympathy toward what I had managed to communicate were deeply felt reservations, not mere obstructionism.

"I want to go home as badly as anyone at this table," I wearily declared, "but I'm not going to send a man to prison based on what we heard in that courtroom." Instantly, there was a reaction from the foreman. "Nobody said we had to send him to prison," he said.

In Texas and many other states, the accused may elect to have his sentence set by the jury or the judge. The determination is made before the trial begins. If the defendant is convicted, another phase of the trial is held to determine punishment. The rules of evidence are different in this second phase; evidence of past crimes may be introduced by the prosecution, as well as testimony sympathetic to the defendant that was not admissible during the guilt phase. We had been informed at the outset that the jury would set the punishment, if any, in this case.

Although initially taken aback by the foreman's remark, I begin to evaluate the implications of a hung jury. The consensus for conviction had formed so quickly I had little hope that another trial would produce a more sympathetic jury. It had taken all the will I could muster to resist the societal inclination to "go along to get along"; if my doubts were shared by a member of Harvey's next jury, would they be strong-willed enough to hold out?

With grave misgivings, I proceeded to cut a deal in the jury room. In exchange for the minimum sentence of probation, I would change my vote to "guilty." The judge's charge had contained nothing that prohibited such an understanding, although it was clearly contrary to the question of guilt or innocence.

As I write these words almost ten years later, I still harbor mixed feelings about what transpired in that case. While I can derive some comfort from sparing Harvey an almost certain prison sentence, my vote to convict him of a felony helped to forever brand him as a convicted sex offender.

I have told myself that if I had to do it all over again, I would not compromise my conscience. My duty was clear: if not convinced of the defendant's guilt beyond a reasonable doubt, my vote must be for "Not Guilty." To this day, I do not believe that Harvey was guilty of Indecency with a Child. However, I am equally convinced that another trial would have resulted in conviction, unless his attorney conducted a much more thorough and spirited defense.

Perhaps my duty was to hang the jury, force a mistrial, and then contact the attorney and offer to do whatever I could to bolster the defense. The problem with that plan is that in our case Harvey was only tried for the offense relative to one of the girls. A sufficiently motivated prosecutor could try him for each girl as a separate offense, greatly increasing the odds of conviction.

When my wife picked me up at the courthouse, I was reminded that we had tickets to a long-awaited concert that night. Although the seats were good and the music was excellent, the musicians made no impression as I replayed the events of the trial over and over in my mind.

My first experience as a juror had been most instructive indeed. I learned that even though "ours is a government of laws, not of men," it is men and women who make it work, with all the imperfections that make human beings such a damned interesting species. I also found out that serving on a jury can be a hellish experience, one I was determined not to repeat.

Less than eighteen months later, I ended up on another jury. This case involved a woman who was bookkeeper for a doctor, accused of stealing money from him. The evidence was a paper trail of checks for cash that had been altered. After playing devil's advocate for a while, the guilt of the defendant was apparent. After the verdict, we learned that she had been convicted of the same offense (also involving a doctor) in another state. While out on bond, she had gone to work for yet another doctor, who had also filed charges on her for theft.

These experiences were still fresh in my mind when I received that jury summons in June 1985. Determined not to be part of another jury, I brought along a copy of a thick nonfiction book, *At Mother's Request: A Story of Murder, Money & Madness*. I figured that either the defense or prosecution would strike me from the panel as a crime buff who might prove dangerous on a jury.

This tactic might have worked if the panel of jurors to which I was assigned hadn't been sent to a civil court. What transpired there would ultimately eclipse the story of Harvey in the spiritual and emotional toll it exacted, both during and after the trial.

The case, you see, was *Pennzoil Co.* v. *Texaco, Inc.*

PART V

"THE MOST IMPORTANT CASE IN THE HISTORY OF AMERICA"

A Legal Marathon in a Texas Courtroom

8

In the jury assembly room that July afternoon, it was announced that the next panel sent to a courtroom would be made up of one hundred prospective jurors, instead of the usual thirty or forty. I turned to the guy sitting next to me and said, "Either we're going to a capital murder (death penalty) case or it's that Pennzoil-Texaco thing."

Actually, my knowledge of the case was quite limited. I had watched Tom Brokaw on NBC in the first week of January 1984, as he announced a Getty Oil merger with Pennzoil. Two days later, he seemed somewhat exasperated as he announced that the Getty Oil merger would go through, but it would be with Texaco. Shortly thereafter, the Houston newspapers reported Pennzoil's lawsuit against Texaco. Not being in a regular habit of reading the business section, that was the last I'd heard of it.

On the way to the jury assembly room that morning, I had heard a news story over my headphone radio that reported the Texas Supreme Court had just denied Texaco's motion to disqualify Baker & Botts from the case, clearing the way for jury selection to begin. There are several dozen courts in Harris County, so I never even considered the possibility that I would be sent to that court—much less end up on the jury.

At that point, I had never heard of J. Hugh Liedtke or Gordon Getty and knew nothing of the $10,000 campaign contribution, the turmoil at Getty Oil in 1983, or any of the other personalities and controversies I would come to know so well in the weeks and months ahead.

Curiously, I had recently heard the name of J. Paul Getty come up in another courtroom context entirely. In the weeks preceding my jury service, I had watched portions of Cable News Network's live coverage of Claus Von Bulow's second trial for the attempted murder of his wealthy wife. One interesting point of contention had been Von Bulow's tenure of employment as an executive assistant to J. Paul Getty.

The debate concerned whether he had been a mere flunky or a key executive. The prosecution pointed to the relatively modest salary Getty paid him as evidence of his lowly stature. On the other hand, photographs showed Von Bulow functioning as a Getty emissary to King Faud. Ironically, I would come to realize that both views were correct. Von Bulow's responsibilities could have far exceeded his pay as a personal aide to J. Paul Getty, a miserly bargain hunter to the end of his life.

This information didn't provide me with any insight into the Pennzoil-Texaco battle over the Getty Oil Company, however. My knowledge of corporate takeovers was almost as slim, even though my wife had been employed at Conoco when that company was bought out by DuPont. Subsequent layoffs had cost her and many others their jobs, but the oil industry had yet to bottom out. That summer, she was employed on a lengthy contract job for Sohio at a salary level slightly higher than she had been making at Conoco. Her current employment was duly recorded on my juror information card; I had no way of knowing this simple fact would itself become the subject of legal action and a front-page story in *The Wall Street Journal* before the end of the year.

The courthouse complex at Harris County is composed of many buildings with architectural styles ranging from the turn of the century to the present day. Oldest by far is the old county courthouse, now the exclusive domain of the civil courts. Completed in 1900, the old courthouse is a massive, brown stone structure capped with a dome. It could be the forlorn capitol of some none-too-prosperous state. A decade or so after it opened, it was reportedly the site of Houston's last public hanging. The mostly forgotten controversies of the past are buried in the dusty old county records; their significance would be further diminished when compared to what transpired in that old building in the last half of 1985.

The sheriff's deputies assisted the bailiff from the 151st State District Court in the logistical nightmare of taking one hundred people lined up in numerical order on the three-block march to the old courthouse. Luckily, we didn't lose anybody to fast-moving traffic along the way. District Attorney John Holmes had been hit by a pickup truck when he stopped to light his pipe while crossing the street a few years before. Fortunately, the only serious injury was to his dignity. Holmes had prosecuted judges and numerous other elected officials over the years, but had declined to get involved in the wake of Judge Jordan's denial of the Motion to Recuse seven months before. The outcome of that little drama had the effect, as I've said, of relegating the events to virtual obscurity, far beyond the reaches of public interest.

The panel of one hundred was taken to the largest courtroom in the building on the third floor and seated according to our assigned numbers. A center aisle divided the seating; numbers 1 through 50 were seated on the right-hand side, with numbers 51 through 100 seated on the left. My assigned number was 52, so I had a front row seat.

We were instructed to sit quietly and wait. I returned to my book. After a prolonged wait, a line of attorneys filed into the courtroom. I turned to see what legends of the local Bar would try this case, half expecting to see Leon Jaworski himself come forward. I then remembered that the distinguished Jaworski, whose fame stretched from Nuremberg to Watergate, had passed away the year before.

Although not a trial buff or courthouse observer, I followed the news reports of our town's colorful attorneys. I didn't recognize any of the anonymous blue suits that walked in that afternoon until there, at the end of the line, grinning from ear to ear, was Joe Jamail.

Like many Houstonians, I had seen Jamail on television news announcing the verdict or settlement of a wrongful death or personal injury lawsuit. He had a way of summing up the case for the reporter that needed no translation. What in the world was he doing here in his sport coat with all these blue and gray suits? Slowly the reason behind Jamail's involvement dawned on me: if you're going to sue somebody, go get the best damned sue-er you can find. I don't want to exaggerate the extent of the reputation he brought to court that day, but the impression I had formed over the years of watching him on TV was that, like the Mounties, Jamail always got his man.

Well, if we couldn't have "Racehorse" Haynes or Percy Foreman, maybe Joe Jamail could provide some entertainment. Actually, I had always been an avid follower of the genre of nonfiction crime reporting that Truman Capote had greatly advanced with *In Cold Blood* in the 1960s. Thomas Thompson's *Blood and Money* was the definitive account of a spectacular murder among the Houston rich in the early 1970s. It was a huge success and had spawned a host of imitators. Any criminal case that gained some notoriety was now apt to end up being retold in the pages of a book.

At Mother's Request, the book I was reading in the courtroom that day, was from this genre. In fact, Thompson himself had been researching the same case (the story of Frances Schreuder, wealthy Manhattan socialite who had persuaded her son to go to Utah and kill her father) when he died after a brief illness. His research would become the basis of *Nutcracker*, Shana Alexander's book on the same case.

But there was no murder here, just a civil suit between a couple of oil companies. How exciting could this be?

*   *   *

The judge ascended the bench. Anthony J. P. Farris was a self-described "sixty-four-and-a-half year old ex-Marine" who looked every inch a judge. With his thinning but still leonine gray hair, luxuriant moustache, and booming voice, he cut an impressive figure up on the bench, elevated above us mere mortals.

Of all the participants in the trial, Judge Farris was the one with whom I had the closest previous contact. Twice before in judicial elections, I had pulled the lever of his Democratic opponent. When you get to the long list of judicial races at the end of a Texas ballot, you make your choices based on whatever scant information you have assimilated about the dozens of candidates, including the stated party affiliation. Farris's connection with Richard Nixon (who had appointed him U.S. attorney) shot his chances with me.

I'll admit I have voted for at least one Republican in my lifetime, which disqualifies me as a "yellow dog Democrat," a term that is either an epithet or a badge of pride in this traditionally Democratic state. While I wouldn't necessarily vote for a yellow dog just because he was running on the Democratic ticket, I would feel compelled to at least seriously evaluate his qualifications against any Republican.

But in the courtroom, Judge Farris seemed like anything but a lingering limb of the Nixon White House. After telling us that this was indeed the case of *Pennzoil Co.* v. *Texaco, Inc.*, he gave us another detail.

JUDGE FARRIS: "I will confirm to you now that this is a civil case. But this isn't your average civil case. This is the largest case ever filed in anyone's knowledge in Harris County.

        "And you prospective jurors are here to be examined by the attorneys and myself to see whether you will be the ones who will hear the largest case ever filed in Harris County.

        "At this time, I am going to ask the attorneys . . ."

Dick Miller's partner, Richard Keeton, broke in.

KEETON: "Your Honor, may we approach the bench?"

The trial had been underway for about one minute and had already come off the track.

KEETON: "Your Honor, on behalf of Texaco, we object to the continued reference of largest case . . ."

JUDGE FARRIS: "I have not mentioned the amount."

KEETON: "I understand, but you said it's the largest case, Your Honor, that's ever been filed in Harris County. That tends, in our view, to give credence to the claim . . ."

JAMAIL: "I don't see that you are prohibited from saying what you want to say . . ."

KEETON: "We view, Your Honor, that as a comment by the Court we think is not proper."

JUDGE FARRIS: "I think that's a fair request. I will abide by that."

Miller had let Keeton press Texaco's objection, but quickly jumped into the fray.

MILLER: "We would like also for the Court to advise the jury that when he says this is the largest case, what he means is that this is a case in which the largest damages are being claimed, and it may or may not be the largest case when [it's] over."

JAMAIL: "Your Honor, I object to that."

JUDGE FARRIS: "Mr. Miller, I think that if I do that, it will just throw the spotlight on it. That will bring us right back to the largest case."

After a couple more verbal exchanges, the conversation at the bench abruptly concluded with this:

MILLER: "That's fine. We want our objection, but . . ."

KEETON: "Well, your Honor, I want to make a formal motion for mistrial because of the comment that has been made up to now and let's get another panel."

JUDGE FARRIS: "Your motion, your oral motion for mistrial, is denied."

Keeton seemed determined to achieve the delay the Texas Supreme Court had refused to grant to Texaco by having a mistrial declared.

After continuing his toned-down introduction, Judge Farris asked the attorneys to introduce themselves, beginning with Pennzoil. The plaintiff always has the first and last word, since he is charged with proving the case.

Jamail introduced himself and the four senior members of the Pennzoil legal team. Then, Judge Farris called on Miller.

MILLER: "Ladies and gentlemen, my name is Dick Miller. I'm one of the lawyers for Texaco and I will be taking the lead, although you might not be able to tell it after I'm through."

Judge Farris then read the instructions to jurors in civil cases to members of the panel: don't talk to the lawyers or witnesses, offer or accept any favors from them (this includes rides, food, or refreshments); don't discuss or even mention this case to anyone or permit it to be mentioned in your hearing (this applied to spouses, parents, etc.).
The judge had a further instruction on the last point.

JUDGE FARRIS: "If anyone attempts to discuss it with you, tell the bailiff of this court at once. That's the bailiff right there with the curly hair. I'm the one with very little hair."

He then moved on to the hardship issue, reading a statement that the trial of this case was expected to take from six to ten weeks. Prospective jurors could not be excused unless there was some specific and unusual hardship, which did not include inconvenience, minor personal sacrifice or loss of pay, unless jury service would cause severe economic hardship.
Judge Farris requested that any members of the panel who felt they had such a hardship come up to the bench one at a time. The bailiff then asked who wanted to come up.
The first two people on the panel claimed job-related hardships that were not particularly specific or unusual.

JUDGE FARRIS: "Bring the next one up."

    BAILIFF: "Your Honor, a lot of people are confused as to exactly the reason they should be coming up. If you could re-state it . . ."

    JEFFERS: "I think the microphone wasn't on when you read part of it."

JUDGE FARRIS: "I will repeat again that portion about those who may come up if they wish. I am having a hard time with the microphone. It's a modern one, and the one in my court is an old one. I am not used to these modern contraptions."

The parade to the bench continued. When they made it to the second half of the panel, the guy next to me went up. Number 51 was a CPA with the big accounting firm Arthur, Andersen and had confidently told me he was as good as gone.

NUMBER 51:   "I'm with Arthur, Andersen & Company. Pennzoil is a client of ours, and it would be against my ethical standards to sit on this jury."

JEFFERS: "Well, I think both companies are [clients], aren't they?"

MILLER: "Sounds like a tie to me. You know in fact that Arthur, Andersen also represents my client, Texaco."

NUMBER 51:   "I wasn't aware of that."

MILLER: "Well, I guess that you should have been but it is a fact. Given that circumstance, do you have any reason to believe that you would treat them any differently?"

NUMBER 51:   "None at all."

MILLER: "Okay. Thanks a lot."

JAMAIL: "I don't understand. You say you have an ethical problem?"

NUMBER 51:   "Well, I'm not sure it would be in compliance with the AMA-CPA standards to sit on a jury of clients."

After he returned to his seat, Number 51 turned to me and said, "They're both our clients."

Other panel members who didn't relish the possibility of coming back the next day joined the procession to the bench. Several had compelling reasons for claiming hardship, including some of those who operated their own small businesses. Still others had significant medical problems.

Over an hour passed while they made their case to Judge Farris and the attorneys. I passed the time in the pages of my book until a break was called so these hardship claims could be individually reviewed.

When they came to Number 51, the judge had a comment.

JUDGE FARRIS: "That's the one with the ethical problem. Mr. Jamail [said he] didn't know CPAs had that. I'm going to tell every CPA in town you said that."

The review continued, with the attorneys arguing about what constituted a real hardship. Judge Farris refereed and finally agreed to excuse twenty panel members the first afternoon. Judge Farris then repeated the instructions he had read at the beginning of the session and asked if there was anything else from counsel.

Miller requested the panel be specifically instructed not to read newspaper accounts or listen to television reports on the case.

Judge Farris turned to Jamail.

JAMAIL: "We have nothing and are ready to go home, Judge."

JUDGE FARRIS: "Thank you, Mr. Jamail."

MILLER: "You can sit down."

JAMAIL: "I take my orders from him, not you, Miller."

Miller's attempt at jocularity elicited a sharp response from Jamail. Despite his initial smiles, Jamail's demeanor throughout the trial would remain intense, his constant reminder that to his client this case was no laughing matter. For the prospective jurors, this exchange capped what had been a dull afternoon with a promise of fireworks to come.

After a stern reminder from the judge to be on time, the bailiff gave some tips on parking. The $6 a day the county paid jurors wouldn't even pay for a parking space anywhere near the crowded courthouse complex.

The judge began to dismiss the panel, but there was a question from the second row.

UNIDENTIFIED PANEL MEMBER: "Your Honor, sir, for those of us that are working, we just call our jobs and tell them we have to report back to court?"

JUDGE FARRIS: "That is correct. Furthermore, if any of your employers fire you because you are on jury duty, guess who will take care of them? Me. I already fined and jailed one [employer] for firing a juror."

UNIDENTIFIED PANEL MEMBER: "What if they just don't pay you but don't fire you?"

JUDGE FARRIS: "I will tend to them in my own ugly way."

Bailiff Carl Shaw turned his back to the judge and stage whispered, "And it's very ugly indeed."

Day One had ended, but the quirky nature of the participants left little clue of what was to come. We still hadn't heard one word about the case itself. That oversight would be remedied with a vengeance the next day—and for twenty weeks after that.

9

The jury selection phase of a trial is known as voir dire, which translates as "speak the truth." This is when attorneys from both sides put questions to prospective jurors, who are required to "speak the truth." The questions are designed to uncover bias or prejudice toward their respective clients.

Sometimes, bias is uncovered with little or no effort on the part of the attorneys. One example that springs to mind occurred during jury selection in the trial of John Wayne Gacy, the notorious serial killer in Chicago. When one middle-aged gentleman was called upon, the first question was "What is your name, sir?" "He's guilty," the prospective juror replied.

Under normal circumstances, some probing is required to show prejudice. Few citizens will take the stand and declare "I'm prejudiced." If the bias is demonstrated to the satisfaction of the judge, the potential juror is disqualified "for cause."

Each side is also entitled to a certain number of peremptory challenges, which they may use to disqualify or "strike" prospective jurors for any reason or for no reason at all. The limited number of peremptory challenges available to each side intensifies the effort to elicit responses that will cause the judge to dismiss those prospective jurors they don't want, thus enabling them to save their strikes to use on those who are otherwise qualified that they want keep off of the jury.

In criminal cases, some prosecutors have routinely used peremptory challenges to exclude blacks or those opposed to the death penalty, if it is a death penalty case. Defense attorneys often strike police officers, prison guards, or their relatives, who might automatically accept the testimony of law enforcement witnesses.

In civil cases, the criteria are frequently not as clearly defined. Attorneys are generally give wide latitude in voir dire questioning; the prin-

ciple that the case be tried before twelve citizens with no ax to grind is
the cornerstone upon which our jury systems rests.

The voir dire in the case of *Pennzoil Co.* v. *Texaco, Inc.* was unusual.
As *American Lawyer* later reported, "Observers call the three-day
event one of the most extraordinary jury selections they have seen.
Said one, 'I was amazed at the latitude the judge gave [the lawyers] to
talk about evidence and documents that weren't in evidence yet. It was
the broadest voir dire I had ever seen in Texas.'" Actually, the voir dire
lasted for six days, not three. But it was extraordinary indeed.

On the morning of the second day, Joe Jamail rose to begin Pennzoil's portion of the voir dire.

JAMAIL: "Ladies and gentlemen of the jury, yesterday Judge Farris
            told you that my name was Joe Jamail and that I represent
            Pennzoil, and I do. This lawsuit that we are here about arose
            out of evidence that you are going to hear of Texaco's execu-
            tives and just a few people who wrongfully interfered with
            Pennzoil's contract for Getty Oil Company.

            "There are going to be a lot of issues that you are going to
            hear, but after you sift through all the issues, only one thing
            is going to be clear to you. And that is, this is a case of prom-
            ises, and what those promises meant to Pennzoil, what they
            ultimately meant to Texaco.

            "Pennzoil had a promise with Getty Oil and the Getty in-
            terests to purchase 3/7ths of the assets of the Getty Oil Com-
            pany. That promise was made by people, not companies.

            "Hugh Liedtke, who I represent and who is my friend, who
            is seated here today, was the chief executive officer and
            chairman of the board of Pennzoil. Wasn't a building that
            made those promises, it was people."

Jamail's reference to a building may seem cryptic until you consider
the impact Philip Johnson's landmark architectural design for Pennzoil Place made on the Houston skyline in the early 1970s. Most people
couldn't tell you anything about Pennzoil beyond the heavily adver-
tised motor oil they marketed, but that building was an early symbol
of the Houston boom. Although it has been subsequently eclipsed by
larger, more spectacular buildings (many also designed by Johnson),
Pennzoil Place retains much of its symbolic value. Significant credit
must go to superdeveloper Gerald Hines, whose company actually
built and owns the structure, but there is little doubt its magnificent
headquarters enhanced Pennzoil's stature in the community.

JAMAIL: "The question ultimately that you are going to have to decide, those of you that get chosen to serve on this jury, is what a promise is worth, what your word is worth, what a handshake is worth, what a contract is worth. Because that's what a contract is, a promise.

"The twelve of you that get to sit and listen to this evidence are going to be very fortunate. You are going to learn more about, I suppose, whether or not people ought to keep their word when they give it and excuses for not keeping your word.

"Whether or not you can rely on exchange of promises, evidenced by handshakes and there is something else, evidenced by writing. A memorandum of agreement. All of this happened.

"After all of this happened and my client, Pennzoil, really my client, Mr. Liedtke, relied on these promises, Texaco entered the picture.

"[Texaco] never had any interest in Getty until Pennzoil made its agreement with Getty. And Pennzoil and Getty are not large companies in relation to what Texaco is. Texaco is the third largest, I suppose, in the world.

"Now, I will come back to this agreement in detail because we are going to be together for a while. I know these seats are hard, but you can appreciate my burden and it's—I think—the most important case ever brought in the history of America. There isn't any question about that."

While this was surely an important case, it is doubtful that *Pennzoil Co.* v. *Texaco, Inc.* would cause legal scholars to forget *Marbury* v. *Madison, Dred Scott* v. *Sandford, Plessy* v. *Ferguson,* or *Brown* v. *Board of Education.* Joe Jamail was undeterred, however.

JAMAIL: "Whether or not there is morality in the marketplace, whether or not people can at will break their contracts at the instigation of some other third party who insinuates themselves into the contract and offers more money after the deal is done. That's what it's about.

"These promises I have been talking to you about, which I believe are enough the way I grew up and the way I am sure most of you did, were more than just promises. There was a memorandum of agreement that was reached between Mr. Liedtke and Gordon Getty, two people."

Having sketched the origins of the deal, Jamail stated that it had come undone at the hands of "what I say was a conspiracy between

Texaco and a group of New York investment bankers and New York lawyers."

JAMAIL: "Mr. Liedtke made the offer and it's simple and I'm going to
            try to keep it simple because that's the only way I can under-
            stand it. They made an offer. The Museum and the Trustee
            accepted that offer, shook hands. Those are the promises and
            the evidence of it [is] in writing, which you'll see."

Jamail's remarks about keeping it simple so he could understand it were not as disingenuous as they might at first seem. Frequently throughout the trial he would stumble over some arcane point of the tangled transaction. This confusion was shared by some of the lawyers, the judge, and all the jurors. He never missed a beat, however, when he was hammering away at the central issue of the case, Texaco's inter-ference with a "done deal."

After detailing how the deal had been approved by a 15-to-1 vote of the Getty board on January 3, 1984, Jamail turned to another docu-ment: the Getty press release.

JAMAIL: "On the morning of the 4th, Getty, mind you on Getty letter-
            head and stationery, a press release was issued announcing
            to the world that the Getty Oil Company, the Getty Museum,
            and the Getty Trust had entered into an agreement in princi-
            ple with Pennzoil. Now, I ask you, each one of you at this
            point, is there any one of you who would not accept that as
            evidence in considering whether or not the Getty interests
            thought they had a binding agreement. If there's anyone who
            could not accept that as evidence in this case, *powerful* evi-
            dence, just show me by raising your hands and we can dis-
            cuss it."

Not a single hand was raised.

JAMAIL: "I take it each of you would do that."

He had managed to tell his story in a way that seemed to require the jury panel to speak out if they did not agree with it. This was the first account I had heard of the events described in Part II of this book, "The Six Day War."

Jamail now set his sights squarely on Texaco and its undoing of the Pennzoil deal.

JAMAIL: "We believe the evidence will convince you that the Texaco
            advisors, both legal and financial and I think top level, Mr.

McKinley [and] Mr. DeCrane, decided that 'What we need to do is isolate Gordon Getty to make a minority stockholder out of him.'

"'The way to do this is for us to go get the Museum's shares, that 11.8%, then get the board's shares, then Gordon Getty's got to tumble.' I believe the proof will show this. Pretty slick.

"'Okay. How do we do that? Well, this is the way we do it. The Museum, Getty, the Trust, Gordon, we're going to pay you $125 a share for your stock over the $112.50 Pennzoil's agreed to.' And does the Museum, does the trustee, Gordon Getty, say that's all right? Uh-uh. No.

"Does Mr. Williams or Mr. Lipton for the Museum think that's all right? No.

"Does the Getty board say that's all right? No.

"Why? 'Well, the price is all right, but what about Pennzoil? What if they sue us?'

"Texaco says okay. After these people insist to Texaco—and this will be in evidence—and I'm going to ask you to listen closely because this is the inducement, this is the fraudulent inducement that Texaco put on Getty to make Getty crumble.

"'If Pennzoil sues you, then we promise you and we'll write it in a contract . . . we'll indemnify you, Gordon Getty and the Trust. We will indemnify you, the Museum and Mr. Williams. And not only that, but all of your representatives are going to come under that guarantee and you too, board of directors. We're going to indemnify you if Pennzoil sues you.'

"And they spell it [out] specifically. If they didn't have Pennzoil on their minds, I don't know who they did. They didn't name anybody else. 'We will indemnify you.' Guarantee, if you please, is what indemnity really means.

"'That if Pennzoil sues you and you have to pay any damages, we, Texaco, will pay them.' And that's why Texaco is in here all by themselves, because they own Getty now lock, stock, and barrel, and they've agreed to pay any damages that Getty or any of their representatives might incur as a result of inducing Getty to knowingly breach a contract that they knew they had made with Pennzoil on January the 3rd."

Since the trial, some legal experts have contended that Jamail grossly misstated the effect of the indemnities Texaco had given to the Getty entities. These experts point to the lawsuits Pennzoil had filed in

Delaware against Getty Oil Company, the Trust, and the Museum, actions that had yet to be adjudicated. If the indemnities applied, the argument went, it would only be in actions brought against the Getty entities themselves. Over the most strenuous Texaco objections, Judge Farris had refused to permit any mention of the pending Delaware lawsuits, citing established legal precedents about collateral litigation.

The indemnities were an important part of Pennzoil's case. Not only were they strong circumstantial evidence of Texaco's knowledge of the Getty-Pennzoil deal, the indemnities also provided a logical reason why Getty Oil, the Trust, and the Museum would back out of a deal they themselves had "announced to the world" only three days before.

Unbridled by the rulings of the Court, Jamail continued to pound away at the indemnities.

JAMAIL: "I want to ask each one of you, and this is so vital, is there *any* member of this panel who could not and would not accept [those] indemnities that Texaco gave at the insistence of the Getty people as evidence of the fact that Texaco had knowledge of the Getty and Pennzoil agreement and binding contract?

"Is there anyone who would not accept that as that kind of evidence? If there is, I've got to know that now, obviously, because that is our proof. Raise your hand if there's anybody.

"I see no hands. I didn't expect any."

These questions are what made Jamail's presentation qualify as a legitimate voir dire. Addressed to the panel as a whole, his questions focused on the critical elements of Pennzoil's case. The next point was Getty's refusal to represent to Texaco that Pennzoil had no claim on any shares of Getty stock.

JAMAIL: "That still wasn't enough for Gordon Getty and the Trust. Still wasn't enough. That's fine so far. 'But Texaco, we've got to do something else.' You see, they had to warranty, which is guarantee, that they owned the stock they were selling to Texaco.

"They said 'When we transfer this stock, we've got to guarantee we own it, and I'm not sure we can do that.' So what did they do? They put a reservation in it: 'We guarantee that we own this stock except for whatever claim Pennzoil might have against it.'

"Now, why would they do that? And why would Texaco

accept it under those circumstances and with those restrictions?

"My question to you here is: Is there anyone here who would not accept that as evidence of Getty's belief that they had a binding contract? Is there anyone who would not accept that as binding proof that Getty thought they had a contract? If there is, I need to know. We all do.

"Secondly, is anyone here who has any reason to believe you could not accept as evidence that which I've explained to you as an intentional act on Texaco's part to forgo the usual warranty and to give indemnity as an intentional, knowing act on the part of Mr. McKinley and Texaco to take Getty away from Pennzoil?

"Is there anyone who would not consider that as evidence and use it when coming to conclusions at the end of this case?

"I didn't think I would see any hands, and I don't."

Jamail's repeated questions also had the effect of challenging prospective jurors to accept not only the evidence itself, but the specific interpretation of that evidence that supported Pennzoil's case. Before Dick Miller had said one word on behalf of Texaco, Jamail had secured a quasi-commitment that the jury would "accept" as "powerful evidence" the documents that bolstered his version of the facts.

He then addressed a relatively technical issue, the "hell-or-high-water" provision Martin Lipton had insisted on before the Museum would deal with Texaco.

JAMAIL: "So they had their meeting and they still haven't satisfied Mr. Lipton. He knows how to play. Mr. Lipton gets with Mr. McKinley and the Texaco group and says, 'Listen, all that's just sweet and wonderful but I've got to have something else.'

"Remember the $112.50 I talked to you about earlier? That was the price Pennzoil was going to pay to Getty. Mr. Lipton then insists and gets from Texaco, 'If the Texaco deal for any reason falls, I want you to guarantee right now you will pay me $112.50 for my Museum stock.'

"We are going to prove that to you. Why $112.50, you might ask yourself? Well, that's the very price that they were guaranteed [under the Pennzoil deal]. That's the bird they had in hand.

"Now my question to you, is there any person on this panel who could not, would not accept that as evidence that

the Museum knew they had a binding contract with Pennzoil? If there is, please let me know. Just raise your hand at
any time."

Jamail was on a roll, and what happened next was a clear indication
that he had no intention of stopping.

JAMAIL: "Well now, we have to go back and I need to talk to you for
           a minute, ladies and gentlemen. It's a rare thing when the
           opposition side described their case just like we described
           their case. But that has happened here. I want to tell you
           what has happened [so] you can search your conscience, to
           see whether or not you agree with this statement. Obviously
           I don't agree with it and Pennzoil doesn't agree with the
           statement that they made. But I need to know whether you
           did so I will be accurate and I will have to read it to you, if
           I can find it."

Walking over to the counsel table, Jamail picks up a document and
reads from it.

JAMAIL: "The question was asked of Texaco's lawyer, 'What's a hand-
           shake worth these days out in the oil patch? And is it a bind-
           ing agreement?'
           "'The Pennzoil Company thinks it's worth $14 billion and
           is suing Texaco to get that money.' And this is Channel 11
           [local CBS affiliate news broadcast]. Mr. Miller's reply 'Yet if
           they want to say that there is some old tradition in the oil-
           field, huh, Jesus Christ, they were in New York.'
           "My question to you . . ."

That was too much for Miller, who had thus far suffered in silence
during Jamail's lengthy diatribe. He leaped to his feet and cut Jamail
off in a loud, clear voice.

MILLER: "Your Honor, excuse me. I was guilty of gross profanity
           when I said that. I was speaking privately . . ."

Sensing danger, Judge Farris abruptly cut Miller off.

JUDGE FARRIS: "Approach the bench. Come up to the bench."

A heated conference at the bench followed, just out of the panel's
hearing.

MILLER: "We are going to ask the Court for a mistrial. That's the grossest kind of error for the counsel for Pennzoil to read to the jury part of the statement which I made to a reporter who was interviewing me, and I had no idea it would ever be abused in this courtroom.

"And I object particularly to the recitation of a curseword I used which I should not have used. Obviously it is designed to try to create prejudice against me and against my client, and it's of total irrelevance to any issue in this case.

"We respectfully move for a mistrial. This case has just commenced and that statement has done us grievous harm."

James Kronzer was the senior member of the Pennzoil team, an attorney with over forty years' experience. Although he never spoke aloud in the hearing of the jury, Kronzer's encyclopedic knowledge of the law was frequently utilized in making arguments to Judge Farris.

KRONZER: "Your Honor, it's clearly an extra-judicial admission by a person under the scope of his authority and Rule of Evidence 801-A. We mean to offer it in evidence at the appropriate time, too."

MILLER: "I don't object to what I said concerning this oil patch handshake. What I object to is reading the curseword that I used."

JAMAIL: "There was no curseword."

MILLER: "Well, I disagree with you completely."

JAMAIL: "You said 'Jesus Christ.'"

MILLER: "If that's not a curseword in the context in which it was said, I don't know what it was. I'm glad to hear your view that it is not a curseword."

JUDGE FARRIS: "Consider it a phrase that I would not have used, but I do not consider it a curseword. Motion for mistrial is denied."

When Jamail resumed his presentation in open court, he was emboldened by Miller's vehement objection and Judge Farris's nonchalant reaction to it.

JAMAIL: "The statement that I just read to you, [that] Mr. Miller as spokesman for Texaco made, is the position Texaco takes that somehow the promises and morality of the marketplace are different than they are here.

"My question to you is there anybody on this panel who has an opinion at this time, or a feeling at this time, that is in agreement with the statement that I just read to you that Mr. Miller says is Texaco's position in this, a handshake in New York is meaningless? If there is anyone who has such an opinion? I need to know that."

Again, not a single hand went up.

JAMAIL: "I take it that you do not. Now, Texaco has raised some excuses as to their actions. I need to ask you about those.

"One of them is, 'Well, we didn't start it. Getty's investment banker came to us first.' The evidence will show that Mr. McKinley and others at Texaco had knowledge of this agreement in principle that I have been talking to you about, this contract, that even if Getty came to them, not Getty but an investment banker for Getty, came to them and said 'Listen, would you like to get in on the act' or whatever it is they said.

"Is there anyone that would accept that as an excuse or a defense on Texaco's part? Is there anyone here who would accept that just carte blanche as an excuse for what Texaco did, after having knowledge? If there is anyone who would accept that, I need—everybody needs—to know about it.

"We say it's meaningless because Texaco's actions are independent of that. They acted on their own after that. And the motive will become clear as we go along."

Jamail made a connection between the investment banker for Getty Oil out trying to recruit bidders to buy the whole company and the handwritten notes of Texaco president Alfred DeCrane and deputy controller Patrick Lynch that detailed Texaco's knowledge of the Pennzoil-Getty dealings and Texaco's subsequent strategy to thwart that deal.

JAMAIL: "Is there anybody who would not accept those written notes we will offer in evidence and which you will have from some of Texaco's people as evidence that Texaco knowingly interfered with Pennzoil's agreement and contract with Getty? Is there anybody who would not accept that as evidence? Of course, it's imperative that you let me know that.

"Is there anyone here who would not accept that as evidence of Texaco's bad faith and motive we will prove in this case? We have sued for punitive damages and actual damages. Is there anyone who would not accept it as relating to Texaco's motive?

"I take it you all would."

There were several other points Jamail covered in this initial outburst. The dealings between Texaco and Getty were done in secret. Would the jury accept that as evidence that Texaco knowingly interfered with Pennzoil's contract? He again beat on the fact that Texaco accepted a contract where Getty refused to provide a warranty that Pennzoil had no claim on the stock.

Jamail then asked some questions that would be a standard part of every voir dire examination. Did anybody know any of the principals in the case? Had any member of the panel ever been represented by any of the attorneys in the case? Did anybody own oil company stock? If so, which company? The answers to these routine questions predictably produced no bombshells.

He then addressed the standard of proof required in civil cases.

JAMAIL: "I see that there are many of you who have not served on a civil jury before. I need now to explain to you, in order to ask the question, what's a 'preponderance of the evidence,' what it means, what it is.

"It means the greater weight and degree of credible evidence. It does not mean the number of witnesses. It means the believable testimony.

"I think I can give you an example of this best by asking you to visualize the scales of justice, or just scales. And you will see that preponderance of the evidence relative to those scales means whatever it is tilts just the slightest on one side."

Preponderance of the evidence is the standard in civil cases and is distinctly different from "beyond a reasonable doubt," the standard in criminal cases. He later elaborated on that point with another assertion.

JAMAIL: "It is not incumbent upon Pennzoil or me to make Texaco confess. This is a civil case. Is there anyone here who believes that it is incumbent on us to wrench a confession out of Texaco before you can come to your conclusions at the end of this case?

"If there is, I sure need to know that."

If anybody thought Jamail had to get Texaco to confess, they weren't talking.

JAMAIL: "I take it that none of you have such beliefs or feelings, that you will follow the preponderance of the evidence."

After addresing the question of expert witnesses, Jamail asked who on the panel had serious business or home reasons that would prevent them from concentrating on the evidence in this case. This was the question that triggered the gradual disintegration of the panel of one hundred.

The jury panel assembled for this case represented a broad cross section of the residents of Harris County, a large, diverse area that covers almost 2,000 square miles. One hundred prospective jurors had been sent to the courtroom, selected according to the sequential number on their juror summons form. Their occupations ranged from bank president to minister to police officer. Five women listed "housewife" as their occupation; seven panel members admitted they were unemployed.

Exxon had two and Shell Oil had three employees among the panel of one hundred; Amoco, Arco, and a number of smaller oil companies and oil industry–related firms were also represented. None of the panel members was employed by Pennzoil. Number 58 was a 24-year-old woman who was a clerical worker at Texaco. Dick Miller argued that as a matter of law she was not necessarily disqualified from the jury, but she was ultimately excused by Judge Farris.

Since the composition of the jury has become something of an issue in the postverdict fallout, I took the juror information cards for the panel of one hundred and subjected them to some demographic analysis.

The average age of the prospective jurors was 40. They had been residents of Harris County for an average of nineteen years, although more than half had been born outside the state of Texas. I was born in Mississippi, but by age 32 I had lived in Houston for twenty-five years and had long considered myself a Texan at heart.

Sixty-seven of the panel of one hundred indicated they were married; a number of others indicated they were divorced. The average number of children was one.

Race is not indicated on the forms, but the racial makeup of the panel was roughly proportional to that of the county as a whole, with Hispanics slightly underrepresented.

Religion was another space indicated on the juror information card.

The largest number of panel members (twenty-one) left this space blank or wrote "None." This doesn't mean that these people were atheists, necessarily. Perhaps many just felt that the question of their religious affiliation is none of the government's damned business. And, of course, it isn't. Nevertheless, only one of those who declined to state a religious affiliation made it to the final sixteen—and he was as the last of the four alternates chosen.

The rest of the panel registered this way: seventeen listed the generic Protestant, another seventeen listed Baptist. Catholics were next, with sixteen. Other specific Protestant denominations represented were thirteen Methodists, five Presbyterians, and three Lutherans. Two panel members were Jewish, and two others were simply "Christian." Seventh Day Adventists, Assembly of God members, Episcopalians, and Hindus each had a lone representative among the panel.

The other principal information on the card included details of previous courtroom experience. Had the panel member ever been "an accused, complainant or witness in a criminal case?" Six panel members responded affirmatively. Had the panel member ever served on a criminal jury? Fifteen answered yes. Had the panel member ever served on a civil jury? Twenty-four answered yes to this question, but ten of these had also indicated they had served on a criminal jury. This meant one-fourth of the panel members had some jury experience.

This was the extent of the information available to the attorneys when they started the jury selection process. With the advent of "scientific selection" of jurors, one wonders what additional investigations were undertaken. In a case of this magnitude, it was not unthinkable that either side would run the names of the panel of one hundred through whatever computers were available. The resources of the respective parties would have made it a simple (if not entirely legal) matter to obtain job histories, credit records, military service records, high school and college transcripts—not to mention police records accurate down to the last parking ticket. After the trial, attorneys for Pennzoil and Texaco both specifically denied they had done this, but suggested that their counterparts on the opposite side probably had.

Dick Miller recounted how an in-house attorney for Texaco had visited the addresses listed to see how the prospective jurors lived—and if their cars sported any revealing bumperstickers. Neither side employed psychologists of any stripe to analyze the body language of the prospective jurors.

Dick Miller and Joe Jamail had both picked a lot of Texas juries during their long careers, and had definite ideas about the type of jurors they had to have—or to avoid at all cost. Of course, it was far from an exact science. With only eight peremptory challenges (or strikes) available to each side, the trick for the lawyers was to ferret out the unwanted jurors and elicit responses that would cause them to be dismissed by the Judge Farris. At the same time, they had to prevent their opponents from doing the same thing to the jurors they wanted.

These discussions were largely held at the bench, out of the hearing of the rest of the panel. Since the whole first afternoon had been taken up with the parade of hardship claims, Jamail hadn't begun his voir dire presentation until the second day.

After the first two hours of presenting Pennzoil's view of the case, he reluctantly said he hoped he wasn't "opening a can of worms" by asking if there was anyone with a business or home problem that would prevent them from concentrating on the evidence of the case.

As the lines formed to approach the bench, I noted a number of panel members who had gone up the day before to claim hardship and had not been excused; now they had another "bite at the apple." Jamail's question had effectively opened the hardship issue up again for yet another rehearing for those who frankly didn't want to serve. Yesterday's experience had shown that it would be a slow process. Miller half-jokingly told Judge Farris that Jamail "should be disbarred for asking that question."

Fortunately, I had a book to pass the time while the individual panel members conferred at the bench. I had finished *At Mother's Request* that morning, and had picked up Larry McMurtry's new novel, *Lonesome Dove,* an 834-page epic western by one of our finest writers and my favorite American novelist.

McMurtry's passionate prose would surely help the hours pass by in fine style. When the discussion was at the bench (out of the hearing of the panel), there was literally nothing else to do.

Meanwhile, up at the bench, there was little harmony. Number 26 was an Assemblies of God pastor who said he had congregation members on both sides of the case, which put him in a "difficult position." This was not adequate grounds for disqualification, however.

Judge Farris asked the attorneys for some input.

> JAMAIL: "Both of us have talked about this. We both think preachers are nuts to begin with and putting one on the jury is flirting with insanity."
>
> MILLER: "Well, he is not excused for cause. That's clear. There is no business excuse either."
>
> JAMAIL: "We agree to strike him, to let him go."
>
> MILLER: "No."

JUDGE FARRIS: "The only thing worse than somebody like this is a school teacher who's picky and holds the jury up for three days on some intersection case. There is nothing worse than that.

"I am not going to excuse that one."

MILLER: "The preacher claims he doesn't know if he can be fair. Preachers all can be fair."

JUDGE FARRIS: "If he doesn't know, who does?"

MILLER: "That's right."

Number 28 was not excused. The next juror to come up was a repeat from the day before. Number 29 was a 29-year-old woman with her own design firm whose client's included Exxon. Her request for hardship had been denied the day before, so she took a different tack this time.

NUMBER 29: "I am not sure this is appropriate, but my fiancé works for Mayor, Day & Caldwell and he has a case pending against Mr. Jamail and he's worked with Mr. Miller's firm before and I have never met anybody on either side. . . . It's over the three years we've been together [that] I've heard lots of stories."

JUDGE FARRIS: "Good or bad?"

NUMBER 29: "Both."

MILLER: "Well, sounds like a tie to me."

JAMAIL: "Who is your fiancé?"

NUMBER 29: "Mark Simon."

MILLER: "That doesn't scare me any. I am willing for her to stay."

JAMAIL: "You say he is suing me?"

NUMBER 29: "No he is . . ."

JAMAIL: "On the other side of the case? Is it an airplane crash case?"

NUMBER 29: "That was one I had heard about before. But he had told me a couple of weeks before he was in another one."

JAMAIL: "I thought I was in retirement. I didn't know I had any other cases. . . . Let me ask you this. Apparently you

	are very close to him. If he said something bad about me, are you going to let that slop over on Pennzoil?"
NUMBER 29:	"I am not really sure. I don't mean to be rude, but . . ."
JAMAIL:	"No, you need to give us the truth. I need to know . . . you be honest. You are the only one that knows it."
NUMBER 29:	"Oh, I know."
JAMAIL:	"You have some reservations now that you would not be able to do that?"
NUMBER 29:	"I know that he respects you. But I don't know if I could just negate all the stuff I've heard over the years."
JAMAIL:	"That you heard about me?"
NUMBER 29:	"Yes."
JAMAIL:	"It would make some difference to you about this case?"
NUMBER 29:	"My first loyalty, I guess, is to him and that's where my prejudice lies, I guess."
JAMAIL:	"My question is, would it make some difference to you in this case? And only you know the answer to that."
NUMBER 29:	"I wouldn't want it to, but I don't know if I could totally forget about it."
JAMAIL:	"Are you saying that it would make some difference as you stand here now?"
NUMBER 29:	"I think so. I hate that, but I think so."
JAMAIL:	"I know, but you had to tell us because this is so important."
NUMBER 29:	"I know."
JAMAIL:	"And are you saying to the judge that you believe it would influence the way you viewed the evidence in this case?"
NUMBER 29:	"I think so."
JAMAIL:	"In other words, we're not starting out even at this point, are we?"
NUMBER 29:	"No."
JAMAIL:	"I am behind?"
NUMBER 29:	"Yes."

JAMAIL: "And I would have to put on more proof or evidence to overcome whatever prejudice you now have in your mind?"

NUMBER 29: "Yes."

MILLER: "Could I ask her a question or two?"

JAMAIL: "I move that—"

JUDGE FARRIS: "Wait. He wants to ask."

MILLER: "Let me be sure I understand what you are telling us. Now, this is not any kind of airplane case."

NUMBER 29: "I know."

MILLER: "And your fiancé is not in the case."

NUMBER 29: "Right."

MILLER: "And you don't know me."

NUMBER 29: "Right."

MILLER: "And you don't know Mr. Jamail. Now, people talk about lawyers just like they talk about doctors and everybody else, and most of that gossip you hear is not true. You know that?"

NUMBER 29: "Why?"

MILLER: "We are talking about here a case in which claims are being made concerning a contract. Mr. Jamail is not a witness and I am not going to be a witness.

"We are just going to ask questions of people who are. Don't you feel under those circumstances that you can decide the case according to what you hear from the witnesses? That's the test."

NUMBER 29: "I am just not totally positive. I am not totally positive that I wouldn't be crowded at some point."

MILLER: "What point would that possibly be?"

NUMBER 29: "Well, personalities are involved in it, whether you want it to be or not. Your personality versus Mr. Jamail's personality."

MILLER: "Well, we can't do anything about our personalities."

NUMBER 29: "And I can't do anything about how I feel about either one of those things either."

MILLER: "You know we hate to lose good people just because they happen to have heard something about one or the other of the lawyers. Now, if it was one of the parties, it would concern me. But we are not talking about any feeling you have got against either of the parties, are we?''

NUMBER 29: "No."

MILLER: "You could give each of these parties a fair trial, couldn't you?"

NUMBER 29: "I hope so."

JAMAIL: "She said she hopes so."

MILLER: "You think you could listen to the evidence and decide the case according to the evidence? That's what I would like for you to do, if you can do that. If you took an oath to decide the case according to the evidence, you would decide it according to the evidence, wouldn't you?"

NUMBER 29: "Yes. I would try."

MILLER: "That's what I thought. I don't think the lady is excused."

JAMAIL: "She said she would try. Thank you very much."

Number 29 returned to her seat, but remained very much the topic of discussion at the bench.

MILLER: "I think that's just a motion for rehearing on this hardship case she filed yesterday."

JAMAIL: "Judge, she has brought up something totally different, as she has stated, and he tried to rehabilitate her. I move to strike her for cause."

JEFFERS: "It wouldn't be fair for [Pennzoil] to use a strike on her."

JAMAIL: "She said we would have to bring some extra evidence to overcome an opinion she has fixed in her mind. That relates directly to Pennzoil. That was the way the question was phrased."

MILLER: "I don't think she is a person who can be challenged for cause and that's the test. She says she could—"

JUDGE FARRIS: "She stated in answers to the questions as to whether she could be fair, it was either 'I hope so' and 'I will try.' She never said 'Yes,' she could."

MILLER: "She did at the last. That's what she said. Why don't we defer this until we get the transcript and we can see what she said."

JUDGE FARRIS: "I will wait until I get the transcript."

JAMAIL: "It's my position that it makes no difference. Once she said she was prejudiced, and I couldn't get him to overcome it."

JUDGE FARRIS: "Let's wait till I read it. You read the transcript and he will read the transcript."

The preceding exchange reveals something about how the voir dire process works, with Jamail taking the initial statement by Number 29 and asking questions to show that she should be excused by the judge. Miller then steps up and tries to undo the answers Jamail elicited with some questions and answers of his own. Miller doesn't necessarily want somebody with an anti-Jamail bias on the jury, but he is trying to make Pennzoil use one of their eight strikes to eliminate her.

The day before she had asked to be excused because she was self-employed: "I have an answering machine and myself and that's it." She was not excused from the panel.

Today, she still wants off the jury but she is now equipped with a new rationale. A number of other members of the panel of one hundred had also made hardship applications that were denied the first day. As the voir dire continued, many of these failed hardship applicants would miraculously start to develop some random bias or prejudice. Since the discussions at the bench were out of the hearing of the panel, one was never sure what the magic words were to have oneself disqualified.

Each time a question was posed to the panel that allowed an individual response, many of the same men and women would again join the march to the front. At one point, Jamail observed a woman whose hardship application had been denied pointedly refusing to listen to his presentation. Miller said he had been watching this particular panel member, and he noted for the record that she paid close attention whenever Jamail presented some new material. Of course, Miller added, she wouldn't pay as close attention to something he repeated over and over and over. (Miller always added at least three "overs" when he used this particular phrase.)

These repeated applications for hardship or contrived bias greatly prolonged the voir dire process, but they also point to a danger with the current attitude too many have toward jury service.

Granted, the personal sacrifices from serving on a jury for six to ten weeks (as was projected at the beginning of the trial) are considerable for anyone. Those chosen to serve as jurors and alternates in this case have no claim on any special virtue; indeed, an even greater number survived the week of voir dire and were not selected. But the voir dire proceedings were protracted greatly by those individuals who were unwilling to endure *any* personal sacrifice.

10

Joe Jamail's presentation had started fitfully on the second day of voir dire, but he had at least outlined a basic map of his case. The next day started with a new controversy among the lawyers. Miller's motion for a mistrial following Jamail's recitation of the "Jesus Christ" comment had been reported in that morning's *Houston Post*. Miller renewed his motion for a mistrial based on the story being leaked to the press, where it might contaminate the jury panel.

At Texaco's request, the voir dire transcript had been sealed until the conclusion of the trial. The implication from Miller was that someone from Pennzoil had leaked the story. Jamail vehemently denied that.

Then Miller requested Judge Farris to admonish Pennzoil public relations staffers who he said were lingering around the panel during breaks and reporting conversations back to the Pennzoil lawyers. One hapless Pennzoil flack was hauled before the bench and questioned under oath. He denied having any contact with the jury at all, which didn't stop Judge Farris from grilling him about the *Post* story and giving him a recap of his power to hold him or anyone else in contempt and put them in jail for six months with no bond and fine them $500.

This was one of the judge's favorite topics, it seemed, for he mentioned it regularly. The day before, he had given the same speech to a tardy member of the panel.

When the proceedings with the panel finally got started that morning, several members lined up to resume conferences at the bench. After the first two, the judge got a message.

JUDGE FARRIS: "The bailiff has told me a reporter-looking person has sat down among the jurors."

JAMAIL: "Who?"

MILLER: "That's the *Post* reporter."

JUDGE FARRIS: "How do you know?"

JAMAIL: "I never saw him, but I disclaim him."

As luck would have it, there was indeed a reporter from the *Houston Post* named Jim Simmon. Fortunately for him, the offending story had been written by another reporter on the paper.

JUDGE FARRIS: "Are you a member of the media?"

SIMMON: "I'm with the *Houston Post.* I'm filling in for some-body. I didn't talk to anyone."

JUDGE FARRIS: "Well, the problem is if you sit among the jurors and then report what they say, that's a no-no."

SIMMON: "I did not."

JUDGE FARRIS: "Obviously, the courtroom is open to the public."

SIMMON: "Yes, Judge."

JUDGE FARRIS: "And the public, obviously, includes the press but not sitting with the jurors."

Like a constable patrolling his beat, Judge Farris was keeping the peace in his own peculiar way.

Jamail resumed his presentation to the panel by inquiring if there was anyone who had any business dealings with Texaco? "I'm not talk-ing about buying gas from them, because their service station people were not involved in this," he added.

A surprising number of the panel had significant business dealings with Texaco, although most of those in the oilfield service industry re-ported they also did business with Pennzoil.

This questioning was done row by row in open court, but some ju-rors with more involved working relationships presented those in de-tail at the bench.

When Judge Farris stopped for a midmorning break, the intrigue that had started the day seemed to be positively contagious. This time, the court clerk was brought in and placed under oath and questioned about two members of the jury panel he had overheard discussing the validity of Pennzoil's alleged contract. This was a clear violation of the judge's instructions, and court personnel had been instructed to keep their eyes and ears open after Miller's allegations about the Pennzoil PR people.

Before the jury panel was let back into the courtroom, the bailiff was sent to bring the first juror identified by the clerk up to the bench.

The judge ordered Number 65 to be sworn. After questioning him about his alleged conversation to no great effect, Judge Farris ordered the clerk to repeat what he heard. Number 65 stammered out a bland explanation.

The bailiff was then dispatched to bring up the man on the receiving end of the remarks. Number 18 was also sworn and questioned, also to no great effect. When Judge Farris had the court reporter read back Number 65's recollection of their conversation, Number 18 said he didn't remember it exactly that way.

Nevertheless, they were going to be discharged by Judge Farris, who couldn't let them leave without a few parting shots.

JUDGE FARRIS: "Jurors Number 65 and 18, please rise.

                  "I'm going to discharge you from jury duty. You have not followed my instructions to keep your mouths shut and not discuss anything even remotely connected with the case. I'm terribly disappointed that you could not understand me. I speak in English and I speak it well enough that most people understand me and yet you persist in communicating with each other."

Chastised but free, 65 and 18 left the courtroom. Apparently, there was still further intrigue afoot. Miller asked that Judge Farris read into the record some comments made by another judge on that floor of the courthouse.

JUDGE FARRIS: "When Judge Phillips was looking at his bulletin board, he heard seven or eight jurors in this case talking about this case, and, as he put it, they looked like business types, whatever that means. I guess he meant they were wearing ties."

     MILLER: "Or shoes."

After the rest of the jury panel returned to their seats, Judge Farris launched into a blistering diatribe clearly intended to put the fear back into the panel. He recited the instructions, described the violations he had uncovered, and launched into a by-now familiar refrain.

JUDGE FARRIS: "I have the power, which I do not relish, to put people in jail if they violate my instructions, to fine them $500 and put them into jail for up to six months without bond. I do not like that because obviously, it's enough that you are here for $6 a day without having to be in contempt, put in jail for six months after you shoot your mouth after being told not to do so ... the next time it happens, some members of this panel will be found in contempt and the lawyers will tell you that I'm an old Marine and a mean old man and if you do not follow the law you will suffer, so please knock it off.

"Mr. Jamail, are you ready?"

JAMAIL: "I know it's tough to get up here and follow that admonition, but it applies to all of us, the lawyers as well. Most of you have never been on a jury and it's incumbent that you do get instructed because that's the only way we can have a fair trial. And I know you can understand that, and he is tough and he is an old—"

JUDGE FARRIS: "You can tell them old."

JAMAIL: "I need to get you concentrating back on this lawsuit, okay, because I've been here far too long and I'm sorry about it, but there's nothing I can do about it because we get people that come up and it has to be so carefully addressed because it's an important case.

"I don't know of anyone's case that's ever been this important and I know the seats get hard and I thought I would be through by now. I thought I would be through yesterday, but I wasn't and I'm still here and I may become the object of bad feelings on your part, but, gosh, I hope not, 'cause I didn't ...

"There's questions that have to be asked and I know you all understand that, but lawyers are all half crazy and we get nervous and paranoid about what, you know, if a juror scratches his head one way, what did he mean by that, you know. We are, but I've got such a heavy burden because the burden (of proof) is on Pennzoil ..."

Jamail went on to remind the panel that what the lawyers say is not evidence, and that he would prove what he had been saying on behalf of Pennzoil. In turn, he asked the panel to "make Texaco prove their excuses and defenses."

Then, the questioning returned to business contacts with Texaco. Again, some of the same panel members who had been up at the bench returned for another try. Miller was getting restless as the third day of the trial was sliding away and the beginning of his presentation was still nowhere in sight.

MILLER: "Am I ever going to get to say anything to this panel?"

JAMAIL: "Sure."

MILLER: "Am I going to be on social security when I do it?"

JAMAIL: "You are already ready for that."

Other questions were directed to individual panel members, with Jamail seeking details about the nature of their listed occupations. Number 17 was a clerk at Infinite Records, which I knew to be a specialty record shop in the Montrose area. When Jamail asked him exactly what he did, Number 17 provided a detailed explanation, saying that he was a writer working on his second book and had taken the job to help a friend who owned the record store. He also volunteered that he had been a newspaper reporter for an unspecified daily newspaper on the East Coast, covering the police and courthouse beats. Ultimately, he did not make the jury, but it didn't seem to have anything to do with his journalistic background. When the issue of punitive damages was addressed later in the voir dire, Number 17 said he couldn't envision any evidence that would convince him to award any punitive damages.

Miller tried to rehabilitate him by offering the conjecture that Pennzoil could have some evidence that he didn't know about that would justify the awarding of punitive damages, which prompted this exchange.

NUMBER 17:   "If the president of Pennzoil had a heart attack right here—"

JUDGE FARRIS: "Don't mention a heart attack here!"

NUMBER 17:   "Sorry, sir. You know, as a result of this [heart attack], maybe I could, but then, not to the corporation, but to him."

MILLER: "There is no reason to believe he will have a heart attack."

NUMBER 17:   "No, I don't think so."

JAMAIL: "Judge, if we are going to have a chatting session, we need to do it some other time with Mr. Miller."

*   *   *

Number 17 was sent back to his seat. After a brief recap of the voir dire presentation he had made thus far, Jamail turned his attention to the matter of damages.

JAMAIL: "As I told you earlier, this is a case for large money damages. Before we get to that, let me ask the general question. Is there anyone on this panel who has any reason to believe that you might have some sort of feeling against a party such as Pennzoil bringing this kind of lawsuit for money damages against Texaco, considering the facts as I gave them to you yesterday?

             "Is there anybody who has any such belief about people just suing people for their wrongs in general?"

This question prompted responses from two panel members who had been up to the bench before. Number 33 thought "too many people take too many other people to court" and said she couldn't award that amount of money.

NUMBER 33:   "I don't know how much money it is, to be truthful."

MILLER: "Nobody knows how much money it is that's in this courtroom. That's not the point. They have a right to sue for the moon with a fence around it."

NUMBER 33:   "But there's a possibility they can get the money and it's going to be the consumer that pays for it in the end."

She went on to say that she wouldn't award the damages even if they were proved.

The next panel member approached the bench.

NUMBER 62:   "As a Christian, I'm not allowed to sue anyone for any reason and my personal opinion is the name of this whole game is greed, and I believe every party is guilty and I'm really having a problem with the whole thing . . . as a Christian, I take my problems to God and to Jesus Christ and try—"

JUDGE FARRIS: "And you never served on a jury before, have you?"

NUMBER 62:   "No, I haven't. I've come down here five or six times but I've never been on a jury. This is the closest I've gotten."

JAMAIL: "Would it be hard for you to believe testimony you heard from the witness stand from witnesses for Pennzoil?"

NUMBER 62:	"Oh, definitely. I would believe that 99% of it was a lie."
JAMAIL:	"We could do nothing to make it believable to you?"
NUMBER 62:	"Not that I know of, because of human nature. I don't have any faith in human nature."
JUDGE FARRIS:	"Ma'am, do you think that everybody is greedy?"
NUMBER 62:	"Most people are."
MILLER:	"Could you except Your Honor that?"
NUMBER 62:	"No, I wouldn't say everybody."
JUDGE FARRIS:	"If I were greedy, I would be practicing law rather than sitting here."
NUMBER 62:	"Well, the real winners are the lawyers."
JUDGE FARRIS:	"Do you have any questions?"
MILLER:	"I'm kind of afraid to take you on."
NUMBER 62:	"I'm sorry."
MILLER:	"You feel the same way about Texaco witnesses?"
NUMBER 62:	"Any witnesses. People."
MILLER:	"I'm not saying you're right or wrong on either side, but you know, we have to have responsible people trying these cases. We can't have a bunch of rag-pickers decide this case. If you had this case, you wouldn't want some unemployed people deciding it."
NUMBER 62:	"I wouldn't have a case here. . . . See, I'm having a lot of trouble with a lot of the law of the land on especially like abortion issues."
MILLER:	"There's no abortion involved in this case. As far as I know, that's the only thing they're not claiming."

After Number 62 returned to her seat, there was a brief exchange at the bench.

JAMAIL:	"Your Honor, do I have to make a motion on this one?"
MILLER:	"Here, do you want to use my pen?"

Another potential juror was on the way home; rag-pickers had joined the ranks of preachers, teachers, lawyers, and CPAs, all worthy professions that had been slandered in the first three days of voir dire.

The central element of Pennzoil's damage claim was the loss of three-sevenths of the proven reserves of Getty Oil, something over 1 billion barrels of oil. Pennzoil valued those reserves at $7.53 billion, the amount of actual damages sought in this trial.

Jamail explained that this is based on the fact that there is only a certain amount of oil on the planet, which makes replacing reserves a difficult—and expensive—proposition.

The damage figure is based on Pennzoil's historical finding and development costs, which were calculated at $10.87 per barrel.

Under their deal with Getty, Pennzoil would have acquired reserves for $3.40 a barrel. This cost differential is $7.47. When multiplied by the 1 billion and approximately 50 million barrels of reserves Pennzoil lost when it lost the Getty deal, the figure is $7.53 billion.

JAMAIL: "Now, Pennzoil lost this invaluable property right and that's what it is. We say Pennzoil is entitled to receive from Texaco, who we say wrongfully interfered with their contract, the full value of what they lost. Not a compromise value, but full value.

            "To prove these damages, Pennzoil will present expert testimony by people knowledgeable in the field that will present evidence to you in this regard."

He then went row by row, asking if anybody had any belief that because this is such a large amount of money that regardless of the loss and the proof, they would not be able to assess these kinds of damages if Pennzoil proves it in this case?

JAMAIL: "If there's anybody who has any such feeling, obviously I cannot have asked you a more important question while this case is proceeding. If you do, you are the only ones who know it.

            "It's not wrong or an embarrassment to have such a feeling. The only thing that would be wrong is if you had such an inhibition or a feeling would be not to tell us about it, because we would be at a totally unfair disadvantage. So you need to tell us. Is there anybody on this first row who has any such feeling about this case or the money damages in it? Anyone on the second row?"

Maybe the panel had been desensitized by hearing the big numbers during the first three days of voir dire; another possibility was that Jamail had convinced them that maybe, just maybe, Pennzoil had a case. The fact that Miller had yet to make any sort of presentation for

Texaco made me adopt a wait-and-see attitude. I was sure he would have a lot to say about this claim. Like everyone present in the courtroom except probably the executives from the two companies, $7.53 billion was more of a concept than a reality.

Three of the jurors employed in some part of the oil business came to the bench to discuss the validity of the method of calculating damages or some other point. Two of these were repeat visitors. Number 38 managed to arouse the ire of the judge.

NUMBER 38:   "It's not clear to me why reserves that you purchase through another company would be included with the cost of drilling and finding of reserves. I think it's common knowledge that the finding cost for replacing reserves is higher for most companies than the purchase of reserves."

JUDGE FARRIS: "By common knowledge, you mean in the petroleum industry?"

NUMBER 38:   "Yes."

JUDGE FARRIS: "Are you saying that one hundred of these jurors are intimately involved in the petroleum industry?"

NUMBER 38:   "No."

JUDGE FARRIS: "Well, it's not common knowledge then."

NUMBER 38:   "Then I stand corrected."

JUDGE FARRIS: "As a matter of fact, I didn't know it."

MILLER: "A lot of people in the petroleum industry may not know it either."

Number 38 soon returned to his seat, the last juror to respond to Jamail's row-by-row questioning. Number 38 had other obstacles in his path to the jury box. A former Getty Oil employee, he was married to another former Getty employee who had recently become a former Texaco employee.

Late in the afternoon, Jamail handed the baton to his co-counsel, Irv Terrell.

TERRELL: "Ladies and gentlemen, my name is Irv Terrell. I am co-counsel in this case with Joe Jamail, as is my partner John Jeffers. We are both at the Baker & Botts firm here. We have both participated fully with Mr. Jamail in the preparation of this case and will participate fully in the trial of this case. We are proud to represent Pennzoil. Our firm has

represented Pennzoil for many, many years and we are de-
lighted to be here in Pennzoil's most important case.

"Now, I hope I'm not scaring any of you by getting up
here, thinking that you are going to be here for some great
number of hours further with me. You won't be. However,
I'm going to have to ask for your attention even though late
in the day, because I've got to cover a few specific matters
with you. I'd appreciate it very much if you could give me
the same attention you've given Mr. Jamail.

"I'd like to start with what I believe to be very hard evidence of Texaco's motive, and I mean statistical evidence
that will come to you from the witness stand.

"I will mention to you that Texaco moved within 72
hours of Pennzoil's announced agreement in principle to
take all of Getty, 72 hours from the time Texaco read the
January 4 press release from the Getty parties.

"Now, why would they do that? Texaco will tell you that
it did that because it had a good-faith belief that Pennzoil's
agreement was not worth anything. They will try to say
that in fifty different ways. They will try investment bankers up here. They will try lawyers up here. They will bring
people from all over this country, some of whom will be
their Getty indemnities, most of whom will be. Then, they'll
bring some people from inside Texas, but what I'm about
to tell you right now is in my heart, what I believe was
really in the minds of these men right over here—these
three men that run Texaco. They're the chairman. They're
the vice-chairman, and they're the president."

Terrell is a tall man, and a skillful, exuberant trial lawyer. Following
Jamail's prolonged tour de force would be a challenge for any attorney,
but within minutes, Terrell had extended a long arm to point a finger
at McKinley, Kinnear, and DeCrane, and three top Texaco executives
he had named. They had sat in a row behind their attorneys without
visible emotion throughout the first three days of voir dire, largely ignored by Jamail.

Terrell acted like a prosecutor in criminal court, singling out the
accused. The message he had to deliver was more complex and largely
technical, so a little theatrical flourish didn't hurt.

TERRELL: "Texaco's worldwide finding and development cost in 1981
            started to go out of sight, $20 [a barrel] and up. That's
            worldwide, even with their interest in Aramco in Saudi
            Arabia, $25.71 in 1983 per barrel costs, finding and devel-
            opment; $37.58 U.S., the most secure reserves.

"Compare that to what they ultimately paid for Getty, less than $5 a barrel.

"Now, why was Texaco so interested in Getty? Why did they move so fast? Why in secret? Why did they give indemnities?

"Well, if you look at this worldwide cost of $25.71 per barrel and subtract from that what they paid within 72 hours at $5, you come to $20.71, and if you multiply that times 2.351 billion proved barrels . . . you are looking at a savings to Texaco of forty-seven billions dollars."

This is the motive that Terrell was talking about.

TERRELL: "Whatever happens in this trial, they can't deny these numbers. They come from their annual reports.

"Now, Texaco may tell you, these executives may get up and testify to you, that they're not concerned. They weren't thinking about it on January 4 through 6, when they were talking to Marty Lipton for the Museum and to Geoff Boisi, the investment banker for Getty Oil, and ultimately they went to see Gordon Getty in a suite at the Pierre on the night of the 5th to steal it all.

"They may tell you they weren't thinking about this. Well, I'll leave that to you. All I ask you to do is you tell me, you ask yourselves, are smart men like that really not thinking about these numbers?

"They're not thinking about what they have to pay to find oil? I think that plays a critical role when you get down to thinking about Texaco [and] these indemnities."

Again and again, those indemnities that Texaco had to give to the Getty entities are put at issue in this lawsuit.

TERRELL: "I have heard in deposition testimony that I've taken in this case from the president of Texaco, Mr. DeCrane sitting over here, that he says when he negotiated these indemnities with these Getty parties, that he thought it was just going to pay legal fees. Just legal fees, that's all.

"Well, I'd like for you to consider this evidence and I need to find out if you will. . . . If you take Texaco's $5 per barrel daily cost and if you then take what Pennzoil's asking for in this case, seven point fifty-three billion dollars— that's actual damages under our finding and development model that we're talking about.

"We're looking at it the same way they did, no different, and they know this. For every one billion dollars that you award Pennzoil in damages, we believe the evidence will show under these indemnities that only increases Texaco's per barrel cost by forty-three cents a barrel.

"I will also tell you we are asking for punitive damages. Because we think what they did to us was really, frankly just raw. Just raw. And we're asking for seven point fifty three billion dollars in punitive damages."

Following Terrell's calculation that each billion dollars in damages assessed added 43 cents to the cost of each barrel of Getty Oil, the $7.53 billion in punitive damages added another $3.24 to the cost of each barrel.

The logic of this damage argument took the approximately $5 a barrel Texaco paid for Getty, added $3.24 per barrel for Pennzoil's actual damage claim, then added another $3.24 for the punitive damage claim.

TERRELL: "You add that up and where are you? My math is not completely off. You are somewhere around $11.50 [a barrel]. That's less than half that number [Texaco's per barrel finding and development cost].

"Now, I will tell you that for this reason, and I want you to consider whether it was in the minds of these men when they gave these indemnities. But they didn't give these indemnities because they were just going to have to fight off a bunch of lawyers.

"They are going to have to do that, but that's not the only reason they did it. They know that if you as a jury come in and you find what Mr. Miller calls 'the world with a rope around it'—"

MILLER: "Fence."

TERRELL: "Moon with a fence around it, the whole universe, fifteen billion dollars. [Even if Texaco is forced to pay damages of] fifteen billion dollars they've got one heck of a deal when they took Getty away from us. They know that.

"We've got to prove it. These numbers I am telling you are real numbers. Will you consider this evidence when you consider the testimony of those Texaco men from their good-faith belief—'Pennzoil didn't have an agreement'?

"Will you consider this evidence when you consider the testimony of their investment bankers, their lawyers, their Getty indemnities, people who are tied to them?

"Will you consider that in judging their credibility? That's so fundamental to our case.

"If there is anybody here who has got a doubt, and I am not going to be upset with you if you have got one, if you can't consider this kind of evidence. It's not wrong for you to say that you won't consider it.

"What would be wrong is if you don't come forward now and tell us. I mean, if you are shocked by the numbers in this case, there is nothing we can do. We can't prove our case, because our case is based on huge numbers. The numbers we are asking for are huge. The numbers they say are overwhelming, and will be, no matter what happens in this case.

"Thank you. I take it you will listen to this evidence."

Terrell's presentation had run nearly to the end of the afternoon, and his assertion that "this was still a good deal for Texaco even if they pay fifteen billion dollars of damages to Pennzoil" didn't arouse much dissent from the weary citizens. Which is not to say that a couple of panel members didn't make a twilight trek to the bench.

Most of the panel was already resigned to coming back for the fourth day of jury selection. Maybe Pennzoil would finish its presentation tomorrow and we would hear something from Texaco.

11

Before the voir dire resumed on Friday morning, Judge Farris met in chambers with the attorneys. Several points of contention were raised during this session, which ultimately became part of the trial record.

JUDGE FARRIS: "First, Mr. Jamail, do you know how much time your side has left?"

JAMAIL: "Mr. Terrell can better address the Court with that, Your Honor, because he knows what he needs to cover."

JUDGE FARRIS: "I know we started July the 10th at 10:53."

MILLER: "Is that 1985?"

JUDGE FARRIS: "1985, yes . . . and so far it's only been one side."

JAMAIL: "Your Honor will recall that most of that time has been taken up with silly questions by Mr. Miller at the bench."

MILLER: "In view of that remark, we'll have a transcript of the voir dire and we'll count the number of pages Mr. Jamail talked and I talked, and we'll see how that comes out."

JAMAIL: "I make sense when I talk."

MILLER: "Whenever Pennzoil finishes their voir dire examination, I would like the Court, because of the upcoming weekend, to allow me ten minutes or so to speak to the jury in the sense of our case. Not to interrupt the voir dire, but to mention two or three things that are important to us so we don't start the weekend with the jury having nothing before it but Pennzoil's case.

"I think Pennzoil has deliberately strung out their voir dire to make that come about. They're legally entitled to do that. I'm entitled to ask Your Honor's indulgence in making a brief statement to the jury on our behalf in connection with that.

"I can't imagine what Mr. Terrell could say for two hours this morning that Mr. Jamail hasn't covered in one way or another. I don't want to have to get up and object to repetition.

"I don't propose to object to anything Mr. Terrell says unless it's completely outrageous or unless he seeks to repeat in one way or another what's already been covered.

"And then I'd like simply to ask him not to do that so we don't have to interrupt him and don't go over these things over and over and over.

"Mr. Jamail has been guilty of that. I haven't been of the view that it helped him as much as it's hurt him. How about that?"

JAMAIL: "Your Honor, I'd like to respond. First off, I've never had this happen. Mr. Miller wants to interrupt our voir dire to tell them whatever he wants to tell them. I object. There's no way we will agree to that.

"To say that we have drawn this out needlessly is not so. I don't think Mr. Terrell will repeat anything I've done. Mr. Terrell is a fine lawyer, but to allow Mr. Miller, just because he wants to say some prayer to these people, I've never heard of this and, of course, I object to it and there's no provision to allow him to step in in the middle of our voir dire."

MILLER: "In the middle?"

JAMAIL: "Hush. He can do all that on opening statement, Judge. He knows that's a ploy. There's no provision in our rules for that and we object to that. Why should we let them go home listening to him? There will always be days when they will go home listening to both sides. Perhaps Monday night and perhaps Tuesday night. Are we going to be allowed every Monday evening and Tuesday evening, when he's through, to get up and tell them some other things?

"We demand or insist on that, if he gets this."

JUDGE FARRIS: "Give me a quick summary of what you want to say to them for ten minutes."

MILLER:	"Yes, sir. I would like to ask simply to tell the jury that we want to be certain that no juror makes up his mind about any aspect of the case until the case is concluded and that they not form even any tentative conclusions until they have heard from both sides . . . remarks of that general nature. I don't propose to speak to the merits of the case."
JAMAIL:	"Well, Your Honor, I still object to that. What precedent is there for this and what need is there for it?

"If he gets this, on Monday when he finishes, I want the same opportunity. I just think it's wrong and none of it should be allowed. We have an orderly procedure set out by our rules." |
JUDGE FARRIS:	"I think it's a reasonable request and I will grant it. And if next Friday, you want to make a similar—"
JAMAIL:	"Monday I do."
JUDGE FARRIS:	"All right. If Monday you feel you want to make a similar—"
JAMAIL:	"I would like to make a speech every time he gets up, Judge."
JUDGE FARRIS:	"For ten minutes, I will allow it."
JAMAIL:	"Now, is this going to be strictly confined to what he just said? Because if not, I want to rebut it today."
JUDGE FARRIS:	"It will be confined."
JAMAIL:	"I still object to it."
JUDGE FARRIS:	"Your objection has been noted for the record."

The latest clash in the ongoing war of nerves between Jamail and Miller had seemingly been resolved. Advantage this round went to Miller.

JUDGE FARRIS:	"What else do we need to take up before we go down and continue this marathon?"
JAMAIL:	"Your Honor, I hope I'm not being blamed for the length of time that is being taken, because so many jurors have been to the bench. I get the uneasy feeling that the Court is angry with me about that."
JUDGE FARRIS:	"Mr. Jamail, believe me, when I get angry, everyone knows. All right?"

JAMAIL: "It appears that you are on the verge of that and I don't want to do that. I really don't."

JUDGE FARRIS: "The only verge that I am on is I am getting bored with the voir dire because I want to get on with the trial."

Indeed, the jury selection process was still on the opening side on the fourth day. This was unheard of outside of highly publicized murder cases. Although this case had received substantial pretrial publicity due to the large amounts of money involved in the mergers and the lawsuit, the very nature of a complex civil suit made it a difficult case for television news to sensationalize.

The detailed presentations here were not designed to eliminate jurors tainted by what they had heard about the case, but rather to identify jurors who could not consider the type of evidence that would be presented.

While Jamail had outlined the basis for Pennzoil's damage claim during the first two days, he had left it to Irv Terrell to provide the nuts and bolts on damages during the voir dire. Terrell resumed his presentation on Friday morning. He identified the damage model Pennzoil was presenting as a "replacement cost analysis" that he said represented the minimum amount of money Pennzoil would have to spend to achieve the 1.008 billion barrels of proved reserves.

To prove these damages, Pennzoil would present three witnesses. Dr. Ron Lewis and Clifton Fridge were both senior employees of Pennzoil. They would be called to testify about the three-sevenths of Getty's proven reserves Pennzoil claimed to have lost and about Pennzoil's cost to replace those "lost" reserves.

Terrell asked the panel members if any of them were acquainted with either Fridge or Lewis. None were.

He then focused on the critical third witness.

TERRELL: "Now, the man who will testify about this generally and who is Pennzoil's primary expert is a man who does not work for Pennzoil.

"He is a man who Pennzoil believes is a man with great experience, great skill in the oil business. He's an oil man. He's not from Wall Street. His name is Thomas Barrow.

"Mr. Barrow, until June of this year, was vice-chairman of Standard Oil Company of Ohio, also known as Sohio. Prior to becoming vice-chairman of Sohio, Mr. Barrow, for years, was the chairman of Kennecott Copper Company, the largest privately owned copper company in the world.

"Prior to that time, for about 25 years, Mr. Barrow was with Exxon . . . and had spent, I believe 5 or 6 years on the board of directors of Exxon, and I'm talking about the parent company, Exxon Corporation."

Pennzoil's lawyers would continue to tout Barrow's credentials; his resume would actually be introduced as an exhibit when he testified at the trial.

TERRELL: "Mr. Barrow has studied, made a great study of the Getty reserves. He had some personal knowledge of them before he began this study. He has been made available to Texaco. They have taken his deposition on two separate occasions.
            "Mr. Barrow will testify that this measure of Pennzoil's loss of bargain, this replacement cost model is an appropriate way under the circumstances to measure Pennzoil's loss. He will testify to this seven point fifty-three billion dollar number.
            "There's one other fact I need to tell you about Mr. Barrow. Mr. Barrow is not doing this for money. Mr. Barrow is not paid a cent. Mr. Barrow is not friends with Mr. Liedtke. Mr. Barrow is not friends with Mr. Kerr. He knows who they are. He has no personal relationship with either one of them, and you may judge for yourself why Mr. Barrow is testifying in this case.
            "All I want to ask you now is do any of you know Tom Barrow?"

In addition to asking these legitimate voir dire questions, Terrell is serving notice that Pennzoil's damage claim will be supported by independent evidence.

TERRELL: "Let me mention another thing. Pennzoil has another expert whose name may come up, and I need to ask you if any of you knew him. His name was Pendleton Thomas.
            "Mr. Thomas was Pennzoil's original expert witness. He unfortunately died in February of this year. His deposition has been taken and portions may be read into evidence in this case. Mr. Thomas also believed in this model. Mr. Thomas at one time was president of Sinclair Oil Company. At one time he was vice-chairman of Arco, Atlantic Richfield. And his last job was chairman of B. F. Goodrich.
            "Did any of you know Mr. Thomas?"

The odds of prospective jurors being acquainted with these high-level corporate officials was admittedly slim. The question had to be asked, however, to qualify this exposition as a legitimate part of the voir dire.

TERRELL: "Okay. While Mr. Barrow replaced Mr. Thomas after his death, we are proud to have Mr. Barrow.
            "Mr. Barrow is a man of great skill and high integrity. We're proud to have him.
            "There will be quite a bit of disagreement, I suspect, from our friends at Texaco about this damage model. They have their own expert witnesses. We've taken their depositions; they don't like this model."

Among the experts on Texaco's witness list was William Haynes. Terrell informed the panel that in addition to being personal friends with Texaco vice-chairman Jim Kinnear, Haynes was the former chairman of Chevron, the worldwide partner in over sixty companies with Texaco.

To complete his part of the voir dire, Terrell moved on to deal with the defenses and excuses of Texaco.

TERRELL: "I'd like to turn to some of the things I think Texaco may say in response to Pennzoil's [case]. Now, I don't know everything that's in Mr. Miller's mind, but I heard quite a bit in the pretrial depositions. I've heard various theories they have. I'm not going to try to exhaust the subject, but I think there some things you should know. . . . These defenses are the after-the-fact excuses that these people are raising to justify what they did."

One curious Texaco defense, Terrell said, was the assertion that it was somehow wrong for Pennzoil to make an unsolicited tender offer for 20 percent of the outstanding shares of Getty Oil Company stock. (Pennzoil offered a 25 percent premium over what the stock was selling for at the time, Terrell pointed out.) Among other things, Texaco had claimed that this tender offer put the Getty board and its advisors under duress.

TERRELL: "I just want to ask right now, is there anybody here who believes that thirty grown intelligent individuals representing a company bigger than Pennzoil could have been forced against their will to do anything?"

It's ironic that Texaco would make the argument that an unannounced tender offer is somehow wrong. Terrell pointed out that one of Texaco's expert witnesses, Robert Greenhill of the investment banking firm of Morgan Stanley, was widely considered to be one of the fathers of the unannounced tender offer.

Next, Terrell turned to the notes taken by senior Texaco personnel during strategy meetings with First Boston's Bruce Wasserstein and Joseph Perella, where the battle plans for the taking of Getty were formulated. The handwritten notes of Texaco's Alfred DeCrane and Patrick Lynch (president and deputy controller, respectively) had been produced during the discovery phase of litigation. The DeCrane notes would later be characterized as the "smoking gun" on the knowledge and interference issue.

TERRELL: "We're going to put these notes in evidence. We're going to ask you to carry them back into the jury room, and when you decide this case, we want you to read these notes. You don't have to, but I'm going to ask you now: If we get these notes in evidence, will you consider them when you decide this case?

                    "When you decide what Texaco knew, will you read these notes?"

His assigned portion of the voir dire wound down shortly after these remarks, and Terrell passed the baton back to the Pennzoil team captain, Joe Jamail.

JAMAIL: "I noticed a little barb by co-counsel about the time I have taken with you people and I thought that was kind of shabby of him because I have been here as long as you folks have and longer, and I have felt like what I was doing was important.

                    "We had so many things we had to take up at the bench that I hope there is nobody in there that's going to hold it against my friend, Hugh Liedtke, and Pennzoil because you had to sit there while we went up there.

                    "Wasn't just me talking up there, you know. And that's the last apology I am going to give you, because it's so important and you understand that."

The next message delivered by Jamail has been spoken in some form by some attorney in every trial I'm acquainted with.

JAMAIL: "I started out by saying to you that I am an officer of this court. I am duty-bound to be honorable, and I feel that way anyway. But what I say to you is not evidence, and I have said it many times. What the lawyers say is not evidence. What the Pennzoil, what the Texaco lawyers, whoever they are, when they stand up and talk to you, what they say is not evidence.

"I have told you what I know we can prove by a preponderance of the evidence. I have given you my word. Now, I am going to keep my word, and I want you to hold me to it, each one of you.

"All, I guess, I am saying to you is this: Is there anyone who would not hold the Texaco lawyers to the same standard that I am asking you to hold Pennzoil lawyers to? And that's the standard of the law.

"I think that everybody would give me that kind of break, make them talk about the case."

To conclude Pennzoil's portion of the voir dire, Jamail touched briefly on each of the key elements of Pennzoil's case. After mentioning the actual damages, he turned to "Texaco's wrongdoing" to hammer home the need for jurors who would consider awarding punitive damages.

The DeCrane notes and the Lynch notes alone were proof of actual knowledge on Texaco's part, he alleged.

JAMAIL: "The knowledge of the notes Mr. Terrell told you about? Those are Texaco's official notes. 'STOP THE TRAIN.' They are talking about the Pennzoil train. 'WE GOT 24 HOURS TO STOP THE SIGNING.'

"Is there anyone here who would not consider that as evidence of reckless disregard, callous, wanton, intentional evidence of Texaco's disregard for the rights of Pennzoil?

"We didn't make these notes. We got them properly under law."

His next series of questions asked if any panel members would not be able to assess punitive damages in a very large amount, no matter what the evidence was? Many of the hard-core dissenters had been disqualified for cause by the judge by the end of this fourth day.

Jamail's final harangue dwelt on a bizarre bit of evidentiary cloak and dagger. Getty Oil general counsel R. David Copely had been the secretary at the critical Getty board meeting of January 2–3. His handwritten minutes were then read into a dictaphone and transcribed by a secretary. After editing for spelling, punctuation, and whatever else, a final draft of the minutes was prepared. The notes filled sixty-five double-spaced, typewritten pages. These board minutes became known as the "Copley notes" and were an important trial exhibit.

Copley had filed the lawsuit for declaratory judgment in Delaware on January 6. Three days later, after traveling to Texaco headquarters in White Plains, Copley had destroyed the original handwritten notes. Jamail raised the specter of destruction of evidence, strongly insinuated that it was inspired by Texaco and therefore constituted further grounds for punitive damages.

"Your Honor, that concludes the voir dire for the Plaintiff," Jamail said. His presentation had nearly run to the end of the session, but Miller was going to get the ten minutes he had been granted that morning over Jamail's strenuous objections.

MILLER: "May it please Your Honor, ladies and gentlemen of the jury, these guys have filibustered me into the end of the week. His Honor is going to recess this case at noon today and we will not recommence until 1:30 on Monday.

"I have sat here since Tuesday listening to this tirade against my client hoping that eventually my time would come to talk to you. What I'm going to say to you today is going to be very brief and to the point, and it's going to be very difficult for me to do anything except to give you a few suggestions about how I would like for you to conduct yourselves during the weekend.

"A person who has the first say has a significant advantage, and whether they say the same thing over and over and over, I have to sit there and listen to it and I don't have the time or the right to reply. I'm afraid there may be somebody on the jury who thinks this is true.

"I'm sure there may be some of you who are saying 'How can Texaco ever win this case with this tale of alleged treachery and dishonesty and evidence destruction? How could Texaco possibly defend against that?' Well, we can if we have the chance.

"If you'll give us the chance, if you won't go home this day and let all of this sink in so that I can't do something about it when we come back here on Monday, that you are just going to shut your mind to our side of the case, you are going to disregard our rights to have our day in court and you are going to let your minds go poisoned with the greed of the Pennzoil Company and with the twisting and turning and ignoring of very important evidence that for their own purposes they chose not to say anything about during their four-day presentation, then I'm not going to get a fair shake."

Miller was having to start his case in a damage-control mode; Jamail had finished his voir dire presentation by again repeating his lengthy list of the "sins" of Texaco.

MILLER: "Now, I brought these guys from Texaco here so that you could see them. I want you to take a look at them. I want you to pay attention to the way they conduct themselves in this Court, and the way they respond to the questions that are asked of them. I want you to give them a chance to defend themselves, and I don't know whether in view of what has happened heretofore that there are people who have shut their minds and hearts to [Texaco's] side of the case.

"So I'm going to have to ask you, and I want to ask you straight out. Are we starting out behind after what has gone on? Am I starting out behind?

"I want to know the answer to that question, and I want you to think about it and say, 'Hey, Mr. Miller, you are too late. Dick, you just didn't get up in time. You should have objected and not let this tirade go on for four days.'

"Well, I would prefer that they had their right to have their say. I'll get my chance if you won't shut your minds, so let me check it out.

"Is there anybody on this side of the room [or] that side of the room who has made up their mind about this case?"

The only hand raised was that of a frequent visitor to the bench whose views had been repeatedly aired; Miller all but ignored him.

MILLER: "Incidentally, is there anybody going to hold it against Texaco, because I honest to God didn't think I was going to get to talk today, so I wore this green tie and I'm trapped with this tie, but I want to go ahead with it.

"Is there anybody who has got Texaco behind right now, that we've got something to overcome?

"Now, I haven't got time to question each of you. In the time I have practiced law, I have run into a few people, not many, but a few who believe the first story they hear.

"What I want you to do, ladies and gentlemen, is let's try this case in Court. The Pennzoil Corporation has had two of their public relations people over here since the commencement of this case and the time comes—"

Jamail's loud, clear voice cut through the courtroom as he jumped to his feet.

JAMAIL: "Excuse me, Mr. Miller. Your Honor, that is first not true and I—"

Judge Farris interrupted Jamail's outburst in open court and called the attorneys to the bench.

JAMAIL: "Your Honor, I quit early so that he would have an opportunity to say what he lied to the Court about this morning that he just wanted to ask them if they were going to go home and wait 'til they heard his side of the case. I didn't know he was going to come in here and make a closing jury argument and tirade Pennzoil. This is ridiculous. See, I can't make any agreement. You ruled this morning it would be limited to that, and this is so blatantly false about Pennzoil's people and irrelevant to anything this jury is about to do and I will object to this, Your Honor . . . "

Judge Farris cooled Jamail down and instructed Miller to observe the narrow guidelines they had agreed on in the early morning conference. He also reminded him that he had two minutes left.

MILLER: "What concerns me is that there may be articles in the newspaper or things on television that are not subject to cross-examination or to the test of law. . . . You know, I know enough of the way of how all of us are and we're interested in what we're doing and I know you are going to read that stuff. I don't know that I would ever want anybody to agree that they might not glance at it. I just don't want you to believe it. What I want you to do is decide the case according to what happens here in court.

             "Now, if there's anybody who wants to try this case in the newspaper, I can't use them. I'm going to get eight strikes and the first people I use those strikes on are people who are going to decide the case according to what they read in the paper."

The forceful tone of Miller's presentation also seemed to be his attempt to dent any cult of personality Jamail might be building among the panel.

MILLER: "Now, you don't know me and I don't know you, but the time will come when you are going to know me and I'm going to know you and then you will know my folks and you'll know whether we're a bunch of crooks and if they are, you'll find against them.

"What I want you to do, if you will, is when you come to court Monday, bring a pencil and paper with you and make some notes. . . . The only way we're going to lose this case is for somebody to say 'It's too hard to understand and I'm not willing to work hard enough to understand what this case is about.' That's the only way we could possibly lose.

"What I want are twelve jurors who will commit to work hard enough, to study this evidence hard enough, and to listen to this evidence close enough to know what went on.

"Now, is there anybody who is not willing to make that kind of a commitment to do his work or her work as a juror?"

JUDGE FARRIS: "Time."

MILLER: "See you all with your pads on Monday. Thank you very much."

It had been a tough week for Texaco. Since plaintiffs have the burden of proof, they always have the first and last word. In this case, the first word has been Jamail's masterful presentation of a complicated business case as a simple morality play.

Miller had to fight to even get ten minutes to close out the week, but his attempt to stem the tide against his client was not in vain. After hearing one side for so many days, simple human curiosity would make you want to hear the other side.

Monday, it will be Miller's turn. Joe Jamail has spun quite a yarn; it will be interesting to see just how Dick Miller goes about unraveling it.

12

The weekend had been a welcomed respite, and Judge Farris resumed the jury selection on Monday afternoon. Before the presentation could begin, however, an hour was consumed by ten or so panel members who lined up for a hearing at the bench. Many of the repeat visitors were finally able to demonstrate adequate bias to achieve the disqualification they sought; a few even had legitimate hardships that had developed since the voir dire began the week before.

Dick Miller had spoken for a few minutes on Friday, but he now began Texaco's portion of the voir dire in earnest.

MILLER: "The last thing my wife said to me this morning was 'Well, I hope you quit pouting if you get a chance to talk,' and I guess that's where I'm coming out on this. It's now my time on behalf of the Texaco people to address you and to talk to you about the case and what we believe the facts in the case will show.

"I join with Mr. Jamail in saying to you that what I say to you is not evidence. Now, I expect you to take my remarks with a degree of skepticism, just as I hope you have taken his remarks with that same degree of cynicism.

"I think you have to understand that in all of these cases somebody is going to be a winner and somebody is going to be a loser. There's no way to compromise some cases, and this case is one of those cases. So what we want is a verdict of the jury who have listened to the evidence and have weighed the evidence and who are willing to let the chips fall where they ought to fall."

With this necessary preamble out of the way, Miller began to warm to his subject.

MILLER: "Now, Mr. Jamail and Mr. Terrell have told you that this is the most important case that has ever been filed in the history of mankind. I take exception to that. I tried a case two months ago where a father was trying to get his children back. I consider that case considerably more important than I consider this case.

"This is a suit over money. Well, everybody needs a certain amount of money. You can't get along in the world if you haven't got a certain amount of money, but the idea that money turns this into the most important case that's ever been filed, I think, tells you something about the company that's bringing the case."

Recognizing the potential impact of Jamail's lengthy voir dire presentation, Miller quickly moved to defuse it.

MILLER: "I want to say to you at the outset that a good deal of information has not been passed on to you. I have no power over what Mr. Jamail selects to pass on to you. That's not up to me. That's up to him, but I frankly am astounded at some of the material he has not seen fit to show you.

"I'm going to tell you, as best I can, everything that's bad about this case from our point of view because I want you to hear it from me. I'm also going to tell you what's wrong with their case, and there's plenty wrong with it.

"This idea that the Texaco Company came in and butted into this deal, pushing its weight around and shoving its money around and taking advantage of this little bitty seven billion dollar company, to get a deal tortiously, to interfere with Pennzoil's alleged contract, is just so much sour grapes. That's what this case is, fourteen billion dollars' worth of sour grapes. I think you'll believe that after you hear the facts."

To counter Jamail's morality play recitation, Miller had to offer a plausible alternate scenario.

MILLER: "We didn't crash this party. We didn't butt into anything. The Getty Oil Company, its officers, and its directors through their representatives came to Texaco and asked us to make a bid on this company. Now, why did they ask us to bid? Very simple straightforward answer. The Pennzoil Company had prevailed upon Gordon Getty, who you will learn was not, is not and never will be a businessman, by making him a promise that he would head his father's company.

"And the board of directors of the Getty Oil Company was angry, outraged at the proposition that Pennzoil people had managed to talk Mr. Getty into relying first upon his emotional desire to succeed to his father's position, relying upon not the best motives of human nature but the worst.

"What they [the Getty directors] did not like in particular was to be in the grasp of a tender offer that had been instituted by Pennzoil on the 27th day of December in the greatest secrecy that can be imagined, scheduling their offer between Christmas and New Year's when they knew Getty would be the weakest and least able to defend itself.

"To put the board of directors of the Getty Oil Company in a vise and squeeze them, then make them commit to sell the stock for less than its worth. So there you have it. That's why the Getty Oil Company came to Texaco."

Miller briefly departed from his presentation to give human faces to his corporate client. The Texaco executives had been present throughout the first week of voir dire, and were still seated in a row behind their attorneys.

MILLER: "Before we go any further in talking about the case, I want you to know who's on my side, whose side I'm on.

"These three guys are down here. They're the three top executives of Texaco. They're down here for two reasons.

"First, they have been accused of very serious crimes of morality. Perhaps they're not crimes in the technical sense, but the accusations against them are very serious. They've been accused of fraud, of destroying evidence, which is a crime, of interfering with somebody else's deal, of lying and cheating and stealing, and for once, they'd like to face their accusers and their accusers' attorneys and answer questions.

"They're down here also because since they're the people who have been accused, they want you to have an opportunity to see them, make up your minds about them as executives and, more importantly, as people. So that's why they're here.

"First, I want to introduce John McKinley. He's the chairman of Texaco . . . John, would you stand up, please."

McKinley bore an uncanny resemblance to Lyndon Johnson, the beleaguered LBJ of the Vietnam war years.

MILLER: "This is John McKinley. He's born in Alabama, native son of Alabama. Been known to root for the University of Alabama, even when they played Texas. He has been chairman of Texaco for, I don't know, four or five years. Before that a longtime employee of Texaco, been with the company all his adult life. . . . He's the guy with the final say on what happens at Texaco.

"Also here today is Jim Kinnear, who is vice-chairman. . . . Jim is a graduate of the Naval Academy. He's been with Texaco ever since the war was over. He is the head of Texaco U.S.A., which has its offices here in Houston. I'm sure you all know that Texaco started out as the Texas Company, has long roots in this state.

"The last person here is Al DeCrane. Al is the president of Texaco Incorporated. He is a graduate of Notre Dame and of the Georgetown Law School and is the only one of the three who has a law degree.

"Those are the three people who are here on behalf of my clients."

There was an element of risk for Miller in having these men present, especially Chairman McKinley. When McKinley showed up for the beginning of jury selection, Jamail thought Miller had only brought him to show him off.

Having taken McKinley's deposition before the trial, Jamail felt there was a good chance Miller wouldn't risk putting him on the stand. Jamail grabbed Susan Roehm, one of the Baker & Botts attorneys working on the case. "Go down and get a subpoena," Jamail whispered to her. At the next break in the trial, a smiling Roehm approached McKinley with her hand out. McKinley extended his hand, apparently thinking she wanted to be introduced to him. Instead, she deposited the rude surprise in his palm. "Miller said it was a cheap shot," Jamail would later recall, "but is was Miller that fucked up."

John McKinley, chairman of Texaco, would be called as a witness for Pennzoil.

Miller continued his voir dire presentation with an explanation of the mechanics of a tender offer. He then briefly described the Getty family history that had left Gordon as sole trustee of the Sarah C. Getty Trust, in effect controlling 40 percent of the stock of Getty Oil. Miller began to depict the Gordon Getty that fit Texaco's defense.

MILLER: "Here is a description that comes from the proxy statement of the Getty Oil Company: 'Mr. Getty is chairman of the

L. S. B. Leakey Foundation, which is engaged in anthropological research. He has a B.S. in English literature from the
University of San Francisco in 1966. A consultant to the
Getty Oil Company in Saudi Arabia, in the Neutral Zone from
1958 to 1959 . . . consultant to the Getty Oil Company in 1965
through 1968; '73 through 1975, Trustee, San Francisco Con-
servatory of Music; member of the Board of Managing Directors of the Metropolitan Opera.'

"Mr. Getty is not a playboy. He is a serious man. He
knows next to nothing about business. His interests in life
are anthropology, archaeology, music, poetry, opera, to
which I think we all would say good for him.

"But there were people on the Getty board of directors
who did not feel that this background qualified him to run
an oil company. And indeed, as the evidence will show, there
were his relatives who felt the same way. A naive, inexperienced person, not trained to run a big business."

Next, Miller presented an abbreviated version of the conflict within
Getty Oil in 1983, described in Part I of this book, "Blood in the Water."

MILLER: "These disputes were reported widely in the press. In October 1983, the Pennzoil executives perceived this struggle
that was going on in this huge company. That company had
control of some very valuable reserves, oil and gas reserves,
and the Pennzoil Company coveted those reserves.

"So Pennzoil commenced to watch the Getty Oil Company. They commenced to gather clippings. They commenced to do research. They commenced to check up on the
directors who ran the company and on Gordon Getty.

"They found out Gordon Getty's intense desire to succeed
his father, a desire that came from his earlier failures and
his earlier inability to satisfy his now-dead daddy."

Now Miller's scenario began to take shape. Pennzoil's filing a tender
offer for 20 percent of the stock of Getty Oil became a sinister act as
told by counsel for Texaco.

MILLER: "When the Pennzoil people filed their tender offer documents in Washington, D.C., on December 27th, they told the
SEC they are making a tender offer because they wanted to
'participate constructively in the restructuring of the Getty
Oil Company.' I don't think you are going to believe that was

> a fair and honest statement of their purpose. Their true pur-
> pose was not to restructure the Getty Oil Company. It was
> to dismantle it, to get the resources represented by the public
> shares and bring them back to Houston at a price that was
> unfair to the public. They did not want to restructure any-
> thing. They wanted to tear it down and haul part of it off."

Miller's scorn for the lofty ambitions of the "purpose clause" of the
Pennzoil tender offer extended to the company and its officers.

He said that Liedtke and his family own or control over 350,000
shares of Pennzoil stock. If the damages sought by Pennzoil were paid,
it would mean over $100 million to Liedtke, according to Miller. He
contrasted this with the Getty Oil directors, none of whom owned a
significant number of shares. Miller was promoting the idea that these
directors were not motivated by financial self-interest.

After a brief recess, Miller resumed his version of the story.

MILLER: "I know how difficult it is to have to sit here and listen to
some guy holler on you on a subject that may be very
strange to you, and we certainly want you to know that we—
all of us, I'm not speaking just for myself—that we very
much appreciate what you are doing. I'm not going to ask
how many of you would stay here voluntarily.

"It's kind of like going to church when you are a kid.
Maybe you'd rather be somewhere else, but the point is here
we are and the best way to get out of it is to do a good job.

"Now, I was telling you before the recess about this
tender offer. I want to emphasize to you that there are two
kinds of tender offers. There are friendly tender offers and
there are hostile tender offers.

"The word 'hostile' in and of itself suggests a kind of economic war. People who conduct these tender offers are frequently known in the trade as raiders or sharks or
predators. . . . This whole process when you make a hostile
tender offer is always at a time when you think the target is
not going to be able to fight back, and you conduct your own
little economic Pearl Harbor."

Miller then used giant enlargements of some of the critical documents to give his version of Pennzoil's dealings with the Getty entities.
First, he showed the Memorandum of Agreement developed by Pennzoil, the Trust, and the Museum.

MILLER: "After they say who has developed the plan—that's all the
first sentence said—then they start talking about how fair

the price is. Nobody has said the price is unfair yet. Yet the first thing they do in the document is try to defend this price. 'So the guilty flee when no man pursueth.'"

After reviewing the document, he turned to the last page.

MILLER: "The last page of this plan called for the Getty Oil Company to sign it if the board approved it. They said, 'Well, wait a minute. Where's the signature?' There isn't any. The Getty Oil Company never signed the agreement, never."

The next document Miller put up on the easel was the "Dear Hugh" letter, which had been touched on by Jamail. Now, it was presented in detail by Miller, who read it aloud.

MILLER: "So Gordon Getty has signed a secret agreement with Pennzoil that if the board of directors of the Getty Oil Company would not vote for this plan, by George, we will get us some more directors. And if those directors won't support it, we will get another bunch. So finally, we now understand what is going on.
            "But I tell you, and the evidence will is going to show, that when the directors found out what Gordon and Pennzoil had in mind for them, it did not win them any friends.
            "You can see, however, that what had happened had created this distrust and suspicion, frustration, anger against Pennzoil for what Pennzoil was trying to do.
            "And to say to you ladies and gentlemen of the jury that the Getty Oil Company directors would have entered into a handshake with Pennzoil in these circumstances, that's a little rough, isn't it?"

In that context, Miller dismissed the 15-to-1 vote that concluded the Getty board meeting as the approval of "a price at which negotiations would commence."

MILLER: "Counsel for Pennzoil has said 'Hey, we love competition.' Yes, in a pig's eye. Everything they did from the beginning to the end, including making the claim that they have a contract, is so anticompetitive, because they were only interested in this fantastic bargain. They didn't want to pay a fair price."

Next on Miller's hit list is the January 4 Getty press release, as the afternoon wore on.

MILLER: "Folks, help is on the way because the Judge has said he's going to make me shut up at 4:15.

"I've got this one document I'd like you to look at before we recess, if you don't mind. This document is on the letterhead of the Getty Oil Company. It's dated January 4, 1984. It was issued before 8:00 o'clock on that day. There was a very, very compelling reason for the issuance of that press release. The Getty Oil Company had removed the Getty stock from trading on January the 3rd. And owners of the stock and people who wished to be owners of Getty stock were not permitted to bid on the stock on January 3rd. And that's something that is very important to people who have investments.

"They say, 'Hey, I can't buy or sell my stock, what's going on?' And in that circumstance, the SEC, not me, not Mr. Jamail, but the SEC requires you to respond."

Miller surveyed the text and called attention to two points: the press release was issued jointly by Getty Oil, the Trust, and the Museum, but not Pennzoil.

Second, the Getty entities announced that they had "agreed in principle with Pennzoil Company to a merger." Miller stated that "agreed in principle" is a "very important phrase in this case."

This analysis led Miller into his next line of argument.

MILLER: "When we first started this case, one of the things you heard was a comment I made to the press concerning whether you could or might or would want to enter into an oral agreement or what has been referred to by the Pennzoil people as a handshake."

The issue of a handshake was fair game, having been raised by Jamail. Miller isolated the Pennzoil handshakes; taken by themselves, they were open to ridicule.

MILLER: "They say, 'Well, we had a handshake.'

"We say, 'Well, who was the handshake with?'

"'Well, we shook hands with Gordon.' Right. Did you ever shake hands with the Museum? 'Well, no, but we waved to them.'

"Well, how about the Getty Oil Company? Did anybody ever shake hands with the Getty Oil Company? 'Hey, no, we never talked to them. We never talked to the Getty Oil Company.'

"So what is this handshake business that we've been hearing so much about. Did they ever shake hands with the Getty Oil Company? They didn't even speak to them, ever. So what does that [press release] say? They have 'agreed in principle.'

"Now, is that the same as an agreement? The answer to that is no.

"This phrase 'agreed in principle' in the context of what was going on in New York City means just the opposite. It means you have not agreed. You have not agreed. You have reached an agreement on a concept.

"And if 'agreed' is the same as 'agreed in principle,' why did he use the extra words?

"So it has the meaning that anybody who is involved in the takeover business knows about. And while Texaco has never, in the history of 80 years of its corporate existence, made a tender offer, Pennzoil has made seven. They made the first major tender offer ever made in this country. So they knew, *they knew*. They knew they didn't have a deal."

Miller was running short of time, but offered what he called "conclusive proof" that Pennzoil knew they didn't have a deal.

MILLER: "This is the first page of the document that's called Schedule 14(d)1. You see it up here at the top. Don't let that scare you. It's nothing but a government number for a document you have to file when you make a tender offer."

Any contacts, transactions, or negotiations involving the subject company must be reported to the SEC. Clearly, all three of these reportable activities occurred between January 1–3. Pennzoil complied by filing this amendment to its original tender offer declaration.

Filed on January 4, this amended 14(d)1 included the text of the Getty press release. Item 5 of the declaration requires Pennzoil to state its future plans for this tender offer.

This was Miller's "conclusive proof": Pennzoil stated in this document that the tender offer would be withdrawn "if a definitive merger agreement is executed."

MILLER: "Well, is 'if' I go to heaven the same as 'when' I go to heaven?"

With that document, the Texaco presentation ended for the day. Miller had taken Pennzoil's case and given it a serious lashing on his first day. How much damage had been done was harder to assess. Despite the extensive presentations by both sides, not one piece of evidence had been admitted. Not one witness had testified. As Jamail and Miller kept telling us, what the lawyers say is not evidence.

Jamail had been noticeably restrained, appearing unconcerned, even bored as Miller had had his say.

But Dick Miller had served notice to the panel that Texaco's defense would be prosecuted with great vigor and, it seemed, a great deal of animosity for the plaintiff, Pennzoil.

13

Tuesday, July 16, marked the one-week anniversary of this jury panel, still undergoing the voir dire extravaganza in the Pennzoil-Texaco case, but all signs pointed toward a jury actually being selected in the near future.

Miller resumed his presentation by covering some general points about contracts: what is necessary for parties to be bound, the right of a party to define what the essential terms are to them, and so on.

MILLER: "How many people have seen that old movie, *Boomtown*, with Spencer Tracy and Claudette Colbert and Clark Gable, where they flip for the oilfield? That's not this kind of deal. We don't want this case decided on the flip of a coin."

He also returned to the January 4 Getty press release, which contained another phrase Miller felt proved Pennzoil did not have a binding agreement. The agreement in principle referred to in the press release was "subject to execution of a definitive merger agreement."

MILLER: "I want to call your attention to the phrase 'subject to,' which I think everybody agrees means contingent upon . . . and what we have here is transactions that are subject to or contingent upon several other occurrences, 'Subject to the execution of a definitive merger agreement.'"

Another issue Miller raised was the session immediately following the conclusion of the board meeting on the evening of January 3. Lawyers, bankers, and public relations experts for Getty Oil, the Trust, the Museum, and Pennzoil gathered to thrash out the press release. It was a scene that many of the participants would describe as "bedlam" or "a madhouse," but it did produce that Getty press release that was approved by each of the three Getty entities.

Miller put out an enlargement of an earlier draft of a press release that was not used and suggested that it, too, indicated Pennzoil didn't have a deal.

His arguments were numerous, with varying degrees of persuasiveness. Overall, however, Miller's task was made substantially more difficult by the relatively technical nature of many of his defenses. Jamail, on the other hand, took his equally complicated case and cast it as a morality play, with definable heros, villains, and victims.

As his voir dire presentation was obviously winding down, Miller began to direct questions to the jury panel: Do you believe a party has the right to insist on a written agreement if that's what they want? Is there anybody here who has ever worked for an investment banking firm?

Miller then addressed Jamail's claim that Pennzoil had been the victim of a conspiracy of investment bankers, lawyers, and Texaco.

MILLER: "Now, there's some things a lawyer ought not be willing to do to win, and I put at the top of the list the attempt to create prejudice for race or creed or geography or anything like that.

"And I'm suggesting to you, ladies and gentlemen, when they talk about these New York lawyers in the tone of voice they used, that they're really suggesting more than that to you. And you know what I mean; I want to be certain that we don't have anything like that in this case. If they can win, let them win fair and square.

"And to say, you know, well, 'There's a conspiracy.' If there is, fine. Let them prove it. And if there was a conspiracy, you give them all the money in the world. If they got cheated, you will help them.

"But don't tell me that they ought to win because they're dealing with some New Yorkers in the same time frame they're hiring those same people themselves. It's going on."

The reference here is to the fact that Pennzoil had hired the Paul, Weiss law firm and Lazard Freres investment banking firm, both of New York, to assist in its bid for Getty.

MILLER: "Is there anybody who is going to decide this case on the basis of who is from where? I need to know that because if we can't look at a man and judge him on his merits and not his religion or his race or the place he comes from, then something is wrong with this case.

"Now, we're going to call some of these people from New York City to testify as witnesses in this case and I need to know if you're going to take a look at these people on the merits or you're going to judge them by where they're from or what church they go to. Am I dealing with that? If I am, I need to know so."

This was the persistent feeling Miller had that every time Jamail said "New York lawyer" or "New York investment banker," he actually meant *Jew* lawyer or *Jew* investment banker. If that was the concern, he could have framed the voir dire questions to deal pointedly with that specific prejudice. I was somewhat puzzled by the line of questioning at the time, but I would recall it vividly when the issue of anti-Semitism was raised after the trial. Throughout this case, however, it was absolutely a nonissue. I believe that will become apparent as this account of the trial progresses.

Another red herring Miller tossed into his voir dire concerned Thomas Barrow, who Terrell had announced would be Pennzoil's expert witness and would testify without pay.

Miller stated that Barrow was a long-time director on the board of Texas Commerce Bank, a client of Baker & Botts. Miller found this fact suspicious in the extreme.

MILLER: "As you all know, there is more than one way to receive compensation. It certainly could be in the form of money. It could be in the form of creating friendship and business opportunities. Compensation could take many forms."

This theory dwindled away without anything of substance being added. Miller then asked panel members about their prior service on juries. Finally, after a week of voir dire, there was a question that compelled me to raise my hand. I am Number 52.

NUMBER 52:   "I served on two criminal juries in Harris within an eighteen-month period about five years ago."

MILLER: "Have you ever served on a civil jury?"

NUMBER 52:   "Never."

MILLER: "You, of course, are familiar with the differences in the standards of proof in these cases?"

NUMBER 52:   "Since I've come here, I've been aware of it."

MILLER: "And you still understand that the business of proving something by a preponderance of the evidence is a substantial burden? I know it's not as large a burden as a criminal case, but it's still substantial. You know that?"

NUMBER 52: "I'm sure it will be further defined."

MILLER: "Absolutely."

NUMBER 52: "Because I'm not real sure what that means."

MILLER: "Well, it means by a preponderance of the evidence which means—I hate to put it in terms of numbers, but I suppose if you got technical—51% would get it. But 25% won't get it or 40% won't get it. Forty-nine percent won't get it, something like that. If the Court instructs you like that, do you feel like you can respond just like the law is?"

NUMBER 52: "Yes, sir."

Thus concluded my personal interrogation during jury selection. I felt reasonably secure that I wouldn't be chosen, because I was still quite a way down the line. If they picked twelve jurors and four alternates and each side had eight strikes, that was a total of thirty-two panel members. There were several more than that left before they got all the way over to Number 52.

Miller concluded his voir dire without another stemwinder speech, but Jamail rose for one last presentation.

JAMAIL: "Well, you have just seen the example of why we have the rule that what lawyers tell you is not evidence. I want you to remember what Mr. Miller said to you, I desperately want you to do that.

"He's made so many misstatements in this presentation and he can have equal time when I'm done. He can get up here and try to give you excuses as to why he did it.

"He says that the Pennzoil people did not put out a press release acknowledging this. That's just not so. . . .

"He talks about Gordon Getty as though he was some sort of an imbecile. Well, it was Gordon Getty's father that founded the company and that wanted his son to remain in control. Why would he leave him in charge of forty-plus percent of the stock?

"He makes him appear as some sort of a fool that *they* later took advantage of. They were real good to him. I think one of the first official acts [Texaco] did was to kick him off the board of his daddy's company almost immediately after they took it. Gives you insight into the arrogance and how they did this.

"For Mr. Miller or anybody at Texaco to say to you that an indemnity to guarantee the payment of damages against somebody in the amount of billions of dollars is routine is hogwash. Would anybody here do that?"

Jamail tarred Miller's presentation with a broad brush, moving from issue to issue.

JAMAIL: "He's asked this jury to say don't believe what you see, let me interpret it for you. Destroying the company notes, original notes, that's nothing. I'll tell you what's in them. Nothing has changed. Well, why can't we have them? Why did they do it after the lawsuit was filed? You've got a right to ask yourself these questions. Why did they do it after he left Texaco's headquarters in White Plains?

"Nobody from Pennzoil's side ever introduced any kind of prejudice, either racial, religious, or otherwise in this case until Mr. Miller brought it up.

"Pennzoil had advisors from New York who will testify. They had lawyers in New York who will testify.

"What's worrying them is the fact that Mr. Lipton is the man who, along with Mr. Boisi of the New York investment firm, saw a way to make more money for themselves. There was this motivation. That's why I took out after them very specifically. I made no condemnation and anybody who knows me knows I wouldn't do that."

Jamail addressed yet another point raised by Miller.

JAMAIL: "He talks to you about competition, and I want to talk to you about competition. Pennzoil believes in competition. That's the way it got to be a company. But we believe in fair, lawful competition.

"What Mr. Miller has, in effect, not so subtly got you to admit to is that you will accept, under the term 'competition,' any action that Texaco decided to engage in.

"Do you believe that it's right under the disguise of competition, after an agreement has been reached, that another party is free to go to one of the agreeing parties and undo it? That's what it's about. Is there anyone who believes it is proper under the American system of competition?

After a few more of these argumentative questions, Jamail concluded his voir dire. The jury panel was sent out on another break so the attorneys could make their strikes. With the panel members excused for hardship or disqualified for cause, the panel of one hundred had shrunk to less than half that number. Sixteen unlucky souls will make the final cut.

Judge Farris had told the attorneys that those selected would not be told which are jurors and which are alternates. He called this the "Hittner gambit," apparently devised by Judge David Hittner, whose courtroom is next door. The idea is for the alternate members not to know their true status, so they will all pay close attention to the case.

When the break was over, the remaining panel members filed back into the courtroom for the announcement. The judge called the session to order and instructed the court clerk to read the list. As their names were called, the selected jurors walked to the jury box at the front of the courtroom.

Much to my surprise and chagrin, my name was called. What's more, I had been counting the jurors as they were named and immediately realized that I was the thirteenth juror selected. I would be sentenced to many weeks in the courthouse, only to be sent home when the case finally went to the jury. My disappointment at being selected as an alternate was secondary to my utter dismay at the prospect of the ordeal that confronted me.

As I looked at the other members of the panel who made up the final sixteen, I was immediately struck by the thought, "What in the world do I have in common with this bunch?" Not a particularly kind sentiment, I realize. In the weeks and months ahead, I would often reflect on that first impression and realize how wrong I was.

Those citizens chosen to serve on this jury were a diverse lot. As I would come to know them, they were:

Fred Daniels, 41, was a letter carrier and Houston native. The father of four, he was an active member of his church and occasionally distributed religious tracts to the other jurors, court personnel, and even some of the attorneys. Once, Dick Miller accepted a proffered tract with a solemnity that bordered on the ridiculous.

Juanita Suarez, 53, another Houston native, had been a dormitory housekeeper at the University of Houston for over eighteen years. Her playful manner belied a quick mind. Her son was an attorney who would show up to take her to lunch when he was around the court-house. By the time I met him, I knew he would go far if he possessed the savvy of his hard-working mom.

Diana Steinman, 29, was a temporary office worker temporarily out of work as the trial commenced. Born and raised on Long Island, New York, she had come south after a brief marriage. Painfully shy, she told me that after the day's proceedings she would take long, soli-tary swims at the Jewish Community Center near her home. I would frequently catch a ride home with Diana after court. Although we were forbidden to talk about the case, there was little doubt that she grew increasingly unhappy as the trial progressed.

Richard Lawler, 30, was an army brat who had lived all over. He had come to Houston from Allentown, Pennsylvania, four years before. A heavy-equipment salesman for Briggs-Weaver, Rick lived with his wife and two children in far west Harris County.

Susan Fleming, also 30, was an accountant for a small steel com-pany. Another Houston native, this quiet single woman seemed most susceptible to job pressures. During trial breaks, she could be seen working on the company books with her pocket calculator.

Lillie Futch, 56, was the senior member of the jury. The long-time Houston resident worked in the group benefits section at the Harris County administration building across the street. A staunch Baptist with a sense of humor she kept firmly in check (most of the time), Lillie was privy to the gossip of the courthouse grapevine.

Velinda Allen, 25, was a clerk at St. Luke's Hospital. Velinda was generally good-natured, but occasionally seemed distracted by the tech-nical aspects of the proceedings, a trait she had in common with all the panel members at one time or another.

Laura Johnson, 51, was a surgical scrub nurse at Humana Hospital down at Clear Lake. Attorneys for one side had made an investigatory visit to her hospital during jury selection. Briefcase-toting men in three-piece suits set off shock waves in these days of malpractice suits so their presence was noted. This scenario caused some of us to won-der if we had been similarly scrutinized. After the trial, neither side would acknowledge sending investigators to Laura's hospital.

She wasn't consumed by the subject, merely curious. Laura took a lot of notes, prompting some Texaco lawyer to predict midway through the trial that she would be selected as jury foreman. Laura was a lovely person, but she was not one of the three members who would eventu-ally be elected to the post. (More on this after the trial.)

Israel Jackson, 45, was a warehouse foreman at an industrial photo-graphic supply firm where he had worked for twenty-four years. A large tatooed man fond of straw hats, the normally taciturn Israel was a great sports fan—not that any of Houston's teams were doing that

great in the second half of 1985; the trial managed to overlap the As-
tros', Rockets', and Oilers' seasons. The Rockets would make the NBA
finals in the spring, but basketball season never really starts in Texas
until football is over.

Ola Guy, 40, was among the most independent-minded of the jurors.
Never quite resigned to the regimentation of jury duty, she carried that
attitude into the deliberations. The professional housekeeper was yet
another element in the diverse mix that characterized this jury.

Theresa Ladig, 26, was an attractive divorcée with two young sons.
Employed as a clerical worker with Shell Oil, Theresa had been the
subject of much admiration by the young CPA sitting next to me during
voir dire. "Boy, I wouldn't mind being sequestered with her," he said
every day for a week. I offered to introduce him, sure that he would be
selected to the jury instead of me.

Shirley Wall, 55, was a former school teacher who had subsequently
gone into the computer business with her husband. A sharp mind and
quick wit made Shirley popular with her fellow jurors. She also followed Miller's suggestion and took prodigious notes throughout the
trial.

Linda Sonnier, 29, had been a ward clerk at St. Joseph's Hospital for
ten years. Linda was a single mother and, like my wife, a native of
Louisiana. An alternate juror, her calm common sense would be missed
in the deliberations.

Gilbert Starkweather, 51, had been a partner in a heat treating plant
for twenty-five years. He lived in Tomball in far north Harris County,
but his business was just seven minutes south of the courthouse. He
would invariably stroll into the courthouse one minute before the day's
proceedings began with his ever-present smile. "Zeb," as he was affec-
tionately known to the rest of us, was perhaps the best liked of all the
jurors. An alternate, he too was missed at deliberation time.

Douglas Sidey, 42, was the last juror and also an alternate. His job
in the oilfield supply division of Amoco Production required him to
work nights, so he did yeoman's duty throughout the nineteen-week
trial.

Doug and Linda were present at the courthouse when the verdict
was reached. Our shared experience would create a bond that could
not be imagined that day when we were first brought together as a
jury.

When writing these descriptions, I resolved not to use the word
"nice," but upon reflection I have to say that this was truly a nice
bunch of folks. Not that we always agreed on everything. . . .

*   *   *

Although jury selection had been held in the largest courtroom in the building, the trial itself would be held in Judge Farris's much smaller courtroom on the fifth floor.

After the oath was administered to the jury, we went to the fifth floor courtroom. The jury box there, of course, only had twelve seats. Four additional chairs had been placed at the end of the box to accommodate the four additional jurors. Conscious of my status as an alternate juror, I decided that if I were going to witness this trial, it would be from a seat inside the jury box proper. The last seat on the front row was vacant, so I grabbed it. I would later realize that taking this seat meant that when Jamail or Miller would look to the jury to make a point, they would frequently look me right in the eye from close range. I adopted a deadpan expression that allowed me to return their gaze without revealing my feelings.

That's how the jury was selected to try this case. As we sat there in the box that first afternoon, we had not heard any evidence at all. But we had been subjected to a week-long battle that had been waged for the hearts and minds of those who would take the seats we now held.

Did Pennzoil have a contract? I had no earthly idea. Jamail's presentation had been effective, if a bit confused at times. As Miller had pointed out, some people will believe the first story they hear. The story he told was not radically different from the story Jamail told, however; it just had a completely different meaning.

I won't say I was particularly eager to find out the answer to this riddle. The voir dire had promised too many days of dry testimony for any genuine eagerness. But I was somewhat intrigued by the prospect of seeing things I had never seen before.

Besides, as an alternate I could just sit back and enjoy the show. Right?

14

When we returned to the courthouse the next morning, the uncertainty of the long week of jury selection had been replaced with resignation to the fact that we were going to be here for a long time.

In a trial, it is in the opening statements that each side tells the jury what their case is about and what they believe the evidence will show.

After the incredibly detailed presentations made during voir dire, opening statements would seem to be an unnecessary redundancy. At the request of the parties, Judge Farris reserved the entire first day for opening statements. Redundancy would remain a prominent feature of the case for both the plaintiff and the defendant.

John Jeffers rose to make the opening statement for Pennzoil. This was the first time we heard from Jeffers, who had not addressed the panel during voir dire. Jeffers and Irv Terrell, both from Baker & Botts, were co-counsel with Jamail. Although a number of other attorneys were present for Pennzoil, these three would be the ones the jury would hear.

JEFFERS: "Good morning, ladies and gentlemen of the jury. I don't know about you, but I'm glad to be down in this courtroom. It's smaller [and] there's not as many of us now. I think if we hear better and see better, maybe we can communicate well.

"Mr. Miller in the last few days has talked a lot about the various law firms in the case, the Botts firm, the Jamail firm, and the Keeton, Brown firm and the other. But we have serious business to attend to here, so I would suggest that you just think of each of us at both counsel tables as

individual Houston lawyers who are going to try to bring
you the facts of this case, because that's all we are.

"Mr. Miller has more tricks in his mind than most anyone
you'll see, and it's going to be my job to try to point them
out. . . . He told you yesterday that he was going to tell you
everything bad about his case as well as the good. Well, I
tell you, he didn't tell you one thing bad about his case.

"He started out trying to get some kind of edge in this
case by trying to posture Pennzoil as a large, hungry predator out after a small prey.

"He characterized Pennzoil as a shark, a raider, a predator which perpetrates its own economic Pearl Harbor at a
time when Getty Oil Company was weak and defenseless,
he said. And he portrayed Texaco in this case in much the
same way, that Texaco was at a disadvantage because it
didn't get to go first in this trial.

"You know, somebody has to go first and somebody has
to go second. I think by the end of this long trial there's not
going to be any advantages or disadvantages in who went
first or second."

Jeffers reminded the jury that Miller had said that although Texaco
had its beginnings in Texas, "most of his witnesses are from New York
and he hoped there wouldn't be any prejudice invoked against New
York."

JEFFERS: "But we're not trying to invoke prejudice. We have a key
            witness in this case who is himself an eminent New York
            lawyer."

The reference here was to Arthur Liman, who would be a key witness indeed. Jeffers then turned to the relative size of the parties.

JEFFERS: "Pennzoil is a successful, strong Houston company. Last
            year its revenues were something over $2 billion.

            "In 1983, which was its last full year of operation, Getty
            Oil Company had almost $12 billion in revenues when it
            was supposedly weak and defenseless. Last year, Texaco
            with Getty having been added into it had, I think, forty-
            seven billion in revenues.

            "So let's not talk about Pennzoil as a large and hungry
            predator after small prey. Let's just get on with the facts of
            the case."

The background of the Pennzoil Company was his next topic, ending with an account of how it had developed around 300 million barrels of proved oil and gas reserves, "the lifeblood of an oil company." He compared this with Getty's 2.3 billion barrels of proved reserves.

JEFFERS: "Mr. Miller told you a lot about J. Paul Getty. I don't think he knew him. I guess he read something about him. I don't know whether all the things that were written about him are true. Mr. Miller said he was a man who spent his whole life in quest for a dollar, had no care for his family. I don't know about that, but I do know that he built one magnificent oil company in his lifetime."

Jeffers waxed poetic about the many and varied assets of Getty Oil. He referred to large deposits of something called diatomite, from which it may soon become commercially practical to extract oil.

JEFFERS: "They tell me that this diatomite deposit that Getty has in California is a hard substance near the surface that, on a hot day, literally turns into oil. That's the kind of rich assets this company had."

This paean to the Getty assets was followed by an abbreviated account of the turmoil within Getty that led to the Pennzoil tender offer in December 1983. The negotiations on January 1 that produced the Memorandum of Agreement were recounted, as well as the Getty Oil board meeting of January 2–3.

Jeffers responded to Miller's insinuations the previous day about the January 4 Getty press release.

JEFFERS: "Mr. Miller said, 'Well, the problem is Pennzoil didn't sponsor this press release. It's the Getty Oil press release.' I don't know why he said that. The evidence is going to show, through some of *his* witnesses, [that] Mr. Goodrum, a lawyer for Pennzoil, approved this press release before it was issued.

"But, you know, that's not the key. Here is Pennzoil's press release; the same identical press release issued the same day."

The meaning of the phrase "agreement in principle" used in the press release had been a major point of Miller's voir dire presentation, and the methodical Jeffers spent some time responding to that argument and others.

That is what transformed these "opening statements" into something altogether different from their usual function in courtroom procedure. The extent to which the case had been argued on voir dire had set a pattern and tone that would see infrequent deviation throughout nineteen weeks of trial. That is the reason the voir dire is so exhaustively detailed in this book; its significance to the trial itself cannot be emphasized enough.

JEFFERS: "What is Mr. Miller, Texaco really, saying then? Well, he said three different things about this [phrase] 'agreement in principle.'

"First, he said an agreement in principle means you don't have any agreement. An agreement in principle, he says, is a nonagreement. That just means that everything that they spell out in [the press release], 4/7ths–3/7ths, the price, everything, that's all no agreement.

"Well, does that make any sense to you?

"Then he said at a second place what an agreement in principle is an agreement on a concept of some sort, an approach.

"Then, finally, he said that an agreement in principle is just an invitation to the world that you are open for bids.

"In three different places, he said those three different things."

This exposition by Jeffers was leading to a critical point of Pennzoil's case.

JEFFERS: "There are investment bankers who are going to testify at this trial that in their circles and by the way they practice their trade, an agreement in principle is an invitation to the world that you are open for bids. And we think that one of the things that has to be done in this case, that ought to be done in this case is to bring that circle of people square with the law.

"That may be their tactics, but it's not right and it's not the law."

Like a boxer who steadily builds up points and at the end you realize he has won the fight, Jeffers built his case jab by jab in this opening statement. He finished with a flourish that, while not exactly thrilling, encapsulated what he had talked about for two hours.

JEFFERS: "Ladies and gentlemen of the jury, this courtroom moves to center stage now. The men from Texaco who have been here—Mr. McKinley, Mr. Kinnear, Mr. DeCrane—sit at the pinnacle of one of the largest corporations in the world.

"But jurors like yourselves judge the rich and the powerful, the small and the weak, and that is what you are here for.

"People from Texaco will tell you that what they did was right. But if they were wrong, who is to tell them they were wrong if not this jury and this Court at this time?

"We have a government of laws and not of men. We ask your patience and your indulgence as we tell our story. Thank you very much."

After lunch, it was Texaco's turn. Miller's partner, Richard Keeton, rose to present its opening statement.

KEETON: "Ladies and gentlemen of the jury, my name is Richard Keeton. I am yet another face and another name for you to learn. I think during the course of these next few weeks, we will all learn to know each other.

"You probably can already tell from my accent that I am not from New York. I am from Houston, and I do represent Texaco."

Keeton's slow Texas drawl did not conceal his probing intellect, and he didn't really fool anyone when he would exaggerate it to ridicule an answer or a witness. It was an effective counterpoint, however, to Miller's softer Oklahoma drawl that ranged from fire-and-brimstone preacher to "just plain folks" congressional candidate.

KEETON: "I am going to try to avoid some of the repetition that has already crept into these already-long proceedings, so that we can get on with the evidence in this case.

"The purpose, however, of my presentation is to try to give you a framework, if you will, of what Texaco believes the evidence will show and a framework for understanding the evidence in Texaco's answers to the charges that have been made by Pennzoil . . . and there are answers.

"In this afternoon session, I am certain I will not give you every answer to every charge in some definitive manner. There is simply not time, and it would serve no useful purpose. . . . However, I do want to hit the high spots."

These "high spots" largely amounted to a recitation of the points raised by Miller during Texaco's voir dire. He reminded the jury that they were not being asked to arbitrate the long-running dispute between Gordon Getty and the Getty Oil management that had triggered the Pennzoil tender offer; he correctly stated that these actions were not at issue in this case against Texaco. He raised the issue, however, to address the motives of Pennzoil.

KEETON: "Pennzoil executives Mr. Kerr and Mr. Liedtke have been described to you as pretty good card players in a business sense, shrewd persons. They are. They're shrewd. I don't know how good their card playing is, but they certainly sought to take advantage of a situation where they saw persons fragmented by discord and disharmony."

Casting the negotiations of the first three days of January 1984 in the most negative light possible, Keeton described how Pennzoil manipulated Gordon, all but blackmailed the Getty board, and strong-armed its way to what Miller had said was "a nonagreement."

The "Dear Hugh" letter was pointed to as the paramount example of the cupidity and cunning of Pennzoil. The January 4 press release was again compared to an earlier draft from the hectic drafting session that took place after the board meeting concluded with a 15-to-1 vote on the evening of January 3. This line of argument focused on the fact that the earlier draft spoke of an "agreement" while the final press release called it an "agreement in principle." If the legal definition of those two terms was the point on which this case would turn, why hadn't the issue already been settled?

Some of the other companies contacted by Getty Oil's investment banker Geoffrey Boisi on January 3 did in fact investigate the possibility of making a bid for Getty, a fact that Keeton emphasized.

KEETON: "Well, there are sure a lot of sophisticated people in companies that don't think the curtain is falling when they read that 'agreement in principle' because this press release has been out.

            "Now, it doesn't say 'invitation to bid,' but what this is is something that says, 'There is going to be, in all likelihood, not certainty, but in all likelihood, some sort of transaction is going to happen. If you want to get in, if you want something, put in your bid.'"

            "Now, that doesn't say that. I'm not going to fool you. It doesn't say that. I'm talking about to knowledgeable people. That may include some of you. That's what it means. It means we're working and there's a tender offer going on, so there's time pressure. Nothing immoral about competition except in the eyes of two sharp card players who may think they've got a pigeon in the game."

This little speech illustrated a central difficulty with that aspect of the Texaco defense. It relied on an interpretation of words that didn't mean what they seemed to say, except to "knowledgeable people." The fact these same cognoscenti may have had an overriding financial incentive in that particular definition tended to cast further doubt on an already murky word game. In addition, it had been suggested that other people, equally knowledgeable, didn't share this understanding.

The repeated references to the "two sharp card players" struck a dissonant note in this reasoning, but shed no further light on the point at issue. Texaco's attempt to cast Pennzoil as the villain in the case began during its voir dire and continued through final arguments.

In contrast, Keeton described Texaco's entry into the fray. He said the solicitation from Boisi was enough to make Texaco take a serious look at Getty Oil. After the January 4 press release announcing Getty's agreement in principle with Pennzoil, Texaco retained an investment banker. Morgan Stanley, its first choice, had gone with Chevron only hours before it was contacted by Texaco. First Boston's Wasserstein and Perella had been supplying DeCrane with information about the progress of the Pennzoil-Getty negotiations for two days and were finally retained by Texaco.

McKinley called a meeting of Texaco's board of directors for the next day, January 5.

KEETON: "And the Texaco board authorized its executives to make an offer to purchase Getty. They authorized that, so long as it was with the agreement of all three of these parties. . . . No matter what anybody may have said, the evidence will clearly establish that Texaco went into this [saying] 'If everybody doesn't agree, we are gone.'

"I told you before, nothing illegal about trying to do it the other way. I'm telling you how Texaco did it.

"On the afternoon of the 5th, the board meeting ended, and then they made appointments to see the various parties representing the Getty entities.

"There was first a meeting where they went and met with the lawyers and the investment banker for the Museum."

Keeton offered no explanation as to why the Museum was first on the list. Since Texaco said it wouldn't do the deal unless all three parties consented, the order in which they were approached was irrelevant to the scenario described by Keeton.

Next, the Texaco contingent went to the Pierre Hotel for the critical meeting with Gordon Getty. Keeton listed the lawyers and advisors present with Gordon and described how Larry Tisch joined the meeting. Then Marty Lipton arrived. What was the lawyer for the Museum doing there?

KEETON: "It wasn't a matter of 'Mr. Getty, if you don't do this, we're
           going to squeeze you out because we're going to buy the
           Museum and we're going to make a tender offer for every-
           body else, and we're going to leave you with these pieces of
           paper.
                "There will be no evidence that Texaco ever intended to
           perform in that fashion or threatened anyone in that way.
           None."

After the meeting at the Pierre ended with Gordon signing a handwritten letter saying that he intended to sell his stock to Texaco, the scene shifted back to the Wachtell, Lipton office. The drafting of the documents for the purchase of the Museum shares was completed in the early morning hours of January 6. The Museum agreement included the indemnity that specifically mentioned the Pennzoil agreement, which Keeton contended had been thoroughly investigated.

KEETON: "They [Texaco] have both their general counsel and their
           outside [law] firm meeting and looking at these things,
           everyone saying you're running no risk and people telling
           them that. And it's not just the Museum. It's Mr. Tisch for
           the company. It's Mr. Petersen for the company. It's Mr.
           Boisi for the company. It's Gordon Getty for himself. It's
           Mr. Siegel.
                "Everyone was saying you're free to go. They sign a de-
           finitive agreement. It's not got blanks. It's signed."

The Getty board would meet later that day to accept the Texaco proposal formally, but the decisive battle of the Six Day War was waged in Gordon Getty's hotel suite that night, the same suite in the Pierre Hotel where the Pennzoil-Getty agreement had its genesis a few days earlier.

KEETON: "Now, the only thing that you all are really concerned with
           out of this whole panoply of stories and facts and evidence
           [is] . . . was there a contract?"

The intricacies of what makes it a contract are reduced to this question:

KEETON: "Do all the parties intend at this time to be legally bound? 'Am I willing for the curtain to fall and that's it?' Everybody's got to have an intent to be bound."

Keeton concluded his opening statement by addressing the charge that Texaco had "tortiously interfered" with the Pennzoil-Getty contract. He said "A tort is a civil wrong."

KEETON: "You are going to be asked 'Did Texaco tortiously interfere with some contract?'
"I don't think you are going to find there is a contract. But in order to find that Texaco has done anything wrong, they are going to have to tortiously interfered, and they are going to have to have known. Known, not suspect. Known, not should have have known. Known.
"Now, I am not saying for a minute that just because they say 'We didn't know,' you ought to accept that. I would insult your intelligence if I took that position."

In truth, however, Texaco and Pennzoil would both insult the intelligence of the jury at times with tedious repetition. Keeton concluded with another statement that illustrated the daunting task facing the Texaco defense team.

KEETON: "But I submit to you that when you look at the array of information Texaco had available to it on the 5th and 6th and 7th, when it was entering these contracts, there is absolutely no way you can reach the conclusion that Texaco knew there was a contract, if there was one.
"Now, lawyers are often criticized and it's difficult for lawyers to say 'There is no contract, but if there was, they didn't know about it,' because laypersons often think, 'Well, that's tricky lawyer talk.' . . . It's like saying 'I didn't kill the man because I've never been in Houston, but if I was in Houston, it was self-defense.' Those two don't work, but that's not what we have here.
"You are going to actually have evidence that Texaco didn't have. . . . The whole curtain is drawn back but Texaco acted responsibly. They made responsible inquiry of responsible people, people they were entitled to listen to.

They asked their own lawyers. They looked at those things they had available.

"You, with more evidence, are going to reach the same conclusion. There was no binding contract to be interfered with, but in that very unlikely event you should reach a contrary conclusion, there is no way you could reach any conclusion but that Texaco acted responsibly, not hitting the sand like an ostrich. . . . Texaco did not know if there was [a contract], so there was no tortious interference.

"Stand back and say 'What happened here?' Shareholders of Getty received a premium price. Have we stolen something from Pennzoil? No. Texaco has competed. They competed fairly.

"When the evidence is in, I think everything I just said will be the same conclusion that each of you will have.

"I thank you for your attention. It has been a long time to sit. It's been a long day for you and a long week and a half. We'll begin the evidence tomorrow.

"I hope I have not intruded on your time. It is important that you carefully attend to things and not let some cheap promotionalism carry you along. Thank you."

At this point, I will offer an observation about those opening statements:

Neither Jeffers nor Keeton broke much new ground because of the extensive map of the case charted by Jamail and Miller during voir dire. Near the end of his presentation, Keeton mentioned the possibility that the evidence would show Pennzoil indeed had a contract. Although Keeton had said this was "very unlikely" to happen, the whole concept would be unceremoniously dumped by Miller. The defense he would offer on Texaco's behalf could be summed up in two words: "No contract."

We headed home with the knowledge that the testimony would begin the next day, testimony from someone actually involved in the case. The prospect didn't exactly leave me salivating like Pavlov's dog.

PART VI

CRITICAL DOCUMENTS I

Evidence of a Contract

This brief document itemized the agreement reached by Pennzoil, the Museum, and the Trust. It required the approval of the Getty Oil board, to whom it was presented at the beginning of the January 2 board meeting.

The note at the top right is from Lipton to his associate, Patricia Vlahakis. Tom refers to Thomas Woodhouse, one of Gordon's attorneys.

"PAV please check this very carefully to be sure its OK. If satisfactory to you give the change to Tom"

Item 1 spells out how Pennzoil will increase its tender offer to $110 a share for a total of 24 million shares.

Item 2(a) calls for the Getty Oil to purchase the Museum shares (over 9 million) for the same price of $110 a share.

On the next page, 2(a) contains another provision calling for the Museum to receive any higher price subsequently paid for any Getty shares.

MEMORANDUM OF AGREEMENT

January 2, 1984

The following plan (the "Plan") has been developed and approved by (i) Gordon P. Getty, as Trustee (the "Trustee") of the Sarah C. Getty Trust dated December 31, 1934 (the "Trust"), which Trustee owns 31,805,800 shares (40.2% of the total outstanding shares) of Common Stock, without par value, of Getty Oil Company (the "Company"), which shares as well as all other outstanding shares of such Common Stock are hereinafter referred to as the "Shares", (ii) The J. Paul Getty Museum (the "Museum"), which Museum owns 9,320,340 Shares (11.8% of the total outstanding Shares) and (iii) Pennzoil Company ("Pennzoil"), which owns 593,900 Shares through a subsidiary, Holdings Incorporated, a Delaware corporation (the "Purchaser"). The Plan is intended to assure that the public shareholders of the Company and the Museum will receive $110 per Share for all their Shares, a price which is approximately 40% above the price at which the Company's Shares were trading before Pennzoil's subsidiary announced its Offer (hereinafter described) and 10% more than the price which Pennzoil's subsidiary offered in its Offer for 20% of the Shares. The Trustee recommends that the Board of Directors of the Company approve the Plan. The Museum desires that the Plan be considered by the Board of Directors and has executed the Plan for that purpose.

1. **Pennzoil agreement.** Subject to the approval of the Plan by the Board of Directors of the Company as provided in paragraph 6 hereof, Pennzoil agrees to cause the Purchaser promptly to amend its Offer to Purchase dated December 28, 1983 (the "Offer") for up to 16,000,000 Shares so as:

    (a) to increase the Offer price to $110 per Share, net to the Seller in cash and

    (b) to increase the number of Shares subject to the Offer to 23,406,100 (being 24,000,000 Shares less 593,900 Shares now owned by the Purchaser).

2. **Company agreement.** Subject to approval of the Plan by the Board of Directors of the Company as provided in paragraph 6 hereof, the Company agrees:

    (a) to purchase forthwith all 9,320,340 Shares owned by the Museum at a purchase price of $110 per

G 1200

167

The handwritten interdelineations here are by Lipton, and initialed by Gordon Getty ("GPG") in the right-hand margin.

Gordon has already signed this document and a counterpart also signed by Liedtke for Pennzoil.

Item 2(b) spells out how the company will buy all the remaining shares of stock, again for $110.

Item 2(c) has the Company grant Pennzoil an option to buy 8 million shares of unissued stock from the company treasury.

Item 3 has the Museum consent to these provisions.

Item 4 is the agreement between Pennzoil and the Trust, who will become owners of the new Getty Oil.

Item 4(a) establishes the ratio of this ownership at 4/7ths for the Trust and 3/7ths for Pennzoil.

Share (subject to adjustment before or after closing in the event of any increase in the Offer price or in the event any higher price is paid by any person ~~other than the Company~~ who hereafter acquires 10 percent or more of the outstanding Shares) payable either (at the election of the Company) in cash or by means of a promissory note of the Company, dated as of the closing date, payable to the order of the Museum, due on or before thirty days from the date of issuance, bearing interest at a rate equivalent to the prime rate as in effect at Citibank, N.A. ~~(the "Company Note"~~ and backed by an irrevocable letter of credit (THE "COMPANY NOTE")

> (b) to proceed promptly upon completion of the Offer by the Purchaser with a cash merger transaction whereby all remaining holders of Shares (other than the Trustee and Pennzoil and its subsidiaries) will receive $110 per Share in cash, and

> (c) in consideration of Pennzoil's agreement provided for in paragraph 1 hereof and in order to provide additional assurance that the Plan will be consummated in accordance with its terms, to grant to Pennzoil hereby the option, exercisable at Pennzoil's election at any time on or before the later of consummation of the Offer referred to in paragraph 1 and the purchase referred to in (a) of this paragraph 2, to purchase from the Company up to 8,000,000 Shares of Common Stock of the Company held in the treasury of the Company at a purchase price of $110 per share in cash.

3. <u>Museum agreement</u>. Subject to approval of the Plan by the Board of Directors of the Company as provided in paragraph 6 hereof, the Museum agrees to sell to the Company forthwith all 9,320,340 Shares owned by the Museum at a purchase price of $110 per Share (subject to adjustment before or after closing ~~in the event of any increase in the~~ Offer price) payable either (at the election of the Company) in cash or by means of the Company Note referred to in paragraph 2(c).

[margin handwriting: AS PROVIDED IN PARAGRAF 2(a)]

4. <u>Trustee and Pennzoil agreement</u>. The Trustee and Pennzoil hereby agree with each other as follows:

> (a) <u>Ratio of Ownership of Shares</u>. The Trustee may increase its holdings to up to 32,000,000 Shares and Pennzoil may increase its holdings to up to 24,000,000 Shares of the approximately 79,132,000 outstanding Shares. Neither the Trustee nor Pennzoil will acquire in excess of such respective amounts without the prior

Item 4(b) is the so-called "divorce clause," which provides that if Pennz-oil and the Trust cannot agree on a plan for restructuring after one year, the assets will be divided along that same 4/7–3/7 ratio.

Item 4(c) provides for the partners to appoint members to the board of the new Company; 4 for the Trust; 3 for Pennzoil.

It goes on to designate Gordon Getty as Chairman of the board; J. Hugh Liedtke as president and CEO, with Pennzoil president Baine Kerr as chairman of the executive committee.

written agreement of the other, it being the agreement between the Trustee and Pennzoil to maintain a relative Share ratio of 4 (for the Trustee) to 3 (for Pennzoil). In connection with the Offer in the event that more than 23,406,100 Shares are duly tendered to the Purchaser, the Purchaser may (if it chooses) purchase any excess over 23,406,000; provided, however, (i) the Purchaser agrees to sell any such excess Shares to the Company (and the Company shall agree to purchase) forthwith at $110 per Share and (ii) pending consummation of such sale to the Company the Purchaser shall grant to the Trustee the irrevocable proxy to vote such excess Shares.

(b) Restructuring plan. Upon completion of the transactions provided for in paragraphs 1, 2 and 3 hereof, the Trustee and Pennzoil shall endeavor in good faith to agree upon a plan for the restructuring of the Company. In the event that for any reason the Trustee and Pennzoil are unable to agree upon a mutually acceptable plan on or before December 31, 1984, then the Trustee and Pennzoil hereby agree to cause the Company to adopt a plan of complete liquidation of the Company pursuant to which (i) any assets which are mutually agreed to be sold shall be sold and the net proceeds therefrom shall be used to reduce liabilities of the Company and (ii) individual interests in all remaining assets and liabilities shall be distributed to the shareholders pro rata in accordance with their actual ownership interest in the Company. In connection with the plan of distribution, Pennzoil agrees (if requested by the Trustee) that it will enter into customary joint operating agreements to operate any properties so distributed and otherwise to agree to provide operating management for any business and operations requested by the Trustee on customary terms and conditions.

(c) Board of Directors and Management. Upon completion of the transactions provided for in paragraphs 1, 2 and 3 hereof, the Trustee and Pennzoil agree that the Board of Directors of the Company shall be composed of approximately fourteen Directors who shall be mutually agreeable to the Trustee and Pennzoil (which Directors may include certain present Directors) and who shall be nominated by the Trustee and Pennzoil, respectively, in the ratio of 4 to 3. The Trustee and Pennzoil agree that the senior management of the Company shall include Gordon P. Getty as Chairman of the Board, J. Hugh

Item 4(e) calls upon the Trust and Pennzoil to coordinate any public announcements with the Company regarding this change in ownership.

Item 6 clearly dictates the necessity of procuring the Getty Board's consent to the plan.

The first signature is that of Gordon Getty.

Under the next signature (that of Museum president Harold William's), Lipton has printed this language:

> UPON CONDITION THAT IF NOT APPROVED AT THE JAN. 2 BOARD MEETING REFERRED TO ABOVE, THE MUSEUM WILL NOT BE BOUND IN ANY WAY BY THIS PLAN AND WILL HAVE NO LIABILITY OR OBLIGATION TO ANYONE HEREUNDER.

This writing was apparently inserted to emphasize that the Museum would not be bound to this agreement if not approved at the Getty board meeting.

> I agreed to
> handle this
> way instead
> of changing 6

The writing at the lower left is another note from Lipton, also presumably to his associate Vlahakis.

Gordon Getty and J. Hugh Liedtke both signed a counterpart of this document.

Liedtke as President and Chief Executive Officer and
Baine P. Kerr as Chairman of the Executive Committee.

   (d) Access to information. Pennzoil, the Trustee
and their representatives will have access to all
information concerning the Company necessary or pertinent to accomplish the transactions contemplated by the
Plan.

   (e) Press releases. The Trustee and Pennzoil (and
the Company upon approval of the Plan) will coordinate
any press releases or public announcements concerning
the Plan and any transactions contemplated hereby.

   5. Compliance with regulatory requirements. The
Plan shall be implemented in compliance with applicable regulatory requirements.

   6. Approval by the Board of Directors. This Plan
is subject to approval by the Board of Directors of the
Company at the meeting of the Board being held on January 2,
1984 and will expire if not approved by the Board. Upon such
approval, the Company shall execute three or more counterparts of the "Joinder by the Company" attached to the Plan
and deliver one such counterpart to each of the Trustee, the
Museum and Pennzoil.

   IN WITNESS WHEREOF, this Plan, or a counterpart
hereof, has been signed by the following officials thereunto
duly authorized this January 2, 1984.

                               Gordon P. Getty
                               Gordon P. Getty, as Trustee
                               of the Sarah C. Getty Trust

                    The J. Paul Getty Musuem

                    By _Harold Williams_   _JPG_
                       Harold Williams, President   UPON CONDIT
                                              THAT IF NOT APPROVED AT THE CALL A BOAR
                                              MEETING REFERRED TO ABOVE, THE MUSE
                       Pennzoil Company   WILL NOT BE BOUND IN
                                               ANY WAY BY THIS
                                               PLAN AND WILL
                                               HAVE NO LIABILI
                    By _____   OR
                       J. Hugh Liedtke, Chairman of OBLIGATIC
                       the Board and Chief Executive TO AN (O)
                       Officer   HEREUNDE
                                       CX1003

I agreed to
handle this
way instead
of changing 6

The last page of the Memorandum of Agreement had only one section, labeled "Joinder by the Company."

The deal among the Trust, the Museum and Pennzoil, who were the principal signatories to this document, required the approval of the Getty Oil board of directors. The board's assent was to be indicated by an authorized signature on this page.

During the trial, Texaco attorneys alleged the absence of the signature here indicated that the board had not approved the document.

Pennzoil witness Arthur Liman, however, testified that it would have been "impossible" for the company to have authorized such a signature since the price contained in the agreement had been increased during the board meeting. After the board meeting, attorneys for Pennzoil and the Getty entities decided to proceed immediately to the drafting of the final merger document, rather than revise the Memorandum of Agreement.

Joinder by the Company

    The foregoing Plan has been approved by the Board of
Directors.

                         Getty Oil Company

January 2, 1984   By _____

The press release issued by the Getty entities on January 4 contains terms from the Memorandum of Agreement and tended to cast doubt on the Texaco claim that all the Getty board approved was the price.

This release was approved by attorneys for the Trust, the Museum, and Getty Oil before it was issued. Pennzoil also released an identical notice through their own publicists the same day.

As spelled out in the first paragraph, the press release is jointly issued by Getty Oil, the Museum and the Trust. It is on the letterhead of Getty Oil.

Paragraph Two does give the price. The "$5 per share within five years" had a present cash value of $2.50, which made the real purchase price to $112.50 a share.

CONTACT: JACK LEONE
(213) 739-2231

FOR IMMEDIATE RELEASE
JANUARY 4, 1984

LOS ANGELES -- Getty Oil Company, The J. Paul Getty Museum and Gordon P. Getty, as Trustee of the Sarah C. Getty Trust, announced today that they have agreed in principle with Pennzoil Company to a merger of Getty Oil and a newly formed entity owned by Pennzoil and the Trustee.

In connection with the transaction, the shareholders of Getty Oil other than Pennzoil and the Trustee will receive $110 per share cash plus the right to receive a deferred cash consideration in a formula amount. The deferred consideration will be equal to a pro rata share of the net after-tax proceeds, in excess of $1 billion, from the disposition of ERC Corporation, the Getty Oil reinsurance subsidiary, and will be paid upon the disposition. In any event, under the formula, each shareholder will receive at least $5 per share within five years.

Prior to the merger, Pennzoil will contribute approximately $2.6 billion in cash and the Trustee and Pennzoil will contribute the Getty Oil shares owned by them to the new entity. Upon execution of a definitive merger agreement, the December 28, 1983, tender offer by a Pennzoil subsidiary for shares of Getty Oil stock will be withdrawn. .

(more)

G.O.C. - 000002

On Page Two, the first paragraph describes the option Getty Oil granted to Pennzoil to purchase 8 million shares of treasury stock.

The second paragraph contains a phrase Texaco emphasized throughout the trial: "The transaction is subject to execution of a definitive merger agreement. . . . "

The third paragraph spells out two more terms from the Memorandum of Agreement that have nothing to do with price: The ratio of ownership in the new Getty Oil to be owned by the Trust and Pennzoil, and the so-called "divorce clause" that allowed the partners to divide the assets if they cannot agree on a plan for restructuring after one year.

The agreement in principle also provides that Getty Oil will grant to Pennzoil an option to purchase eight million treasury shares for $110 per share.

The transaction is subject to execution of a definitive merger agreement, approval by the stockholders of Getty Oil and completion of various governmental filing and waiting period requirements.

Following consummation of the merger, the Trust will own 4/7ths of the outstanding common stock of Getty Oil and Pennzoil will own 3/7ths. The Trust and Pennzoil have also agreed in principle that following consummation of the merger they will endeavor in good faith to agree upon a plan for restructuring Getty Oil on or before December 31, 1984, and that if they are unable to reach such an agreement then they will cause a division of assets of the company.

* * * * * * *

G.O.C.- 000003

Getty Oil general counsel R. David Copley took handwritten notes to prepare the minutes of the marathon two-day board meeting on January 2–3, 1984. Copley read his notes into a tape recorder and the notes were then typed. The handwritten original notes were discarded.

Pennzoil made a great deal out of what they called "destruction of evidence," and referred to this document as the "ersatz," "sanitized" Copley notes. Ironically, the typewritten notes contained passages that seemed to confirm Pennzoil's claim of board approval of the Pennzoil–Getty deal. Pages 63–66 reproduced here document the climactic vote.

Paragraph 2 on page sixty-three has the Chairman (Petersen) call on Lipton to repeat the Pennzoil proposal. Although many of the details concern price, the citing of the 24 million shares Pennzoil was to buy is consistent with the Memorandum of Agreement, as is the notation about the Company purchase of the Museum shares.

The minute details of the "stub," the premium to be paid for the sale of ERC, are also discussed. Lipton's explanation continues on the next page (64) of the Copley Notes.

Mr. Higgins stated that Salomon could give a fairness opinion to the Museum based on the proposal outlined. Mr. Boisi stated that he was not prepared to comment as to whether or not the proposal was fair. Mr. Lipton noted that the proposal would only apply to 48,000,000 shares. The Chairman thereupon recessed the meeting at 3:15 p.m. (E.S.T.) January 3, 1984. The meeting was reconvened at 6:15 p.m. (E.S.T.) January 3, 1984 with Director Mitchell present.

The Chairman requested that Mr. Lipton repeat the proposal. Mr. Lipton stated that the proposal from Pennzoil to the Trustee was that Pennzoil will pay $110 per share cash for all shares other than the shares of the Trust. Mr. Lipton stated that he thought the transaction would be a cash merger but it had been discussed as a tender the previous day. The transaction would involve the purchase by Pennzoil of 24,000,000 shares less the number of shares of the Company's common stock presently held by Pennzoil, and the purchase by the Company of the Museum's shares followed by a cash merger to take out the balance of the shares. In addition an interest would be given to all the shareholders other than the Trust from the proceeds of the sale of ERC in excess of $1 billion after tax and expenses. If less than $3 per share had been paid to the shareholders 5 years after the agreement, a minimum of $3 per

-63-

The formula on the premium for ERC worked out to a present value of $112.50 a share.

share would be paid to the shareholders at that date. Mr.
Lipton stated that he thought Pennzoil would agree to increase
the $3 amount to $5 per share. Mr. Lipton went on to state
that this would give a current value of approximatley $112.50
per share on the basis of the $5 increment. Mr. Lipton said
this proposal was acceptable to the Museum and Salomon would
give a fairness letter to the Museum. Mr. Lipton said he had
been advised that the Trust would agree to the transaction
with either the $3 or $5 increment on top of the $110 per
share.

Mr. Winokur stated that he thought the transaction
had to be done as a merger and this would involve some delay
in the shareholders receiving their cash payment. The delay
was stated to be in the order of possibly several months and
Mr. Winokur inquired if this had been discussed with Pennzoil.
Mr. Siegel stated that Pennzoil would prefer a tender offer.
Mr. Winokur asked if there had been any discussion of compensation for the shareholders for the delay in receiving their
funds as a result of a merger, to which Mr. Woodhouse responded
"No." Mr. Lipton then stated that he thought Pennzoil would
want the Museum and the Trust "locked up."

Mr. Stuart then inquired as to what happened if a

-64-

The second full paragraph recounts the misgivings of some of the directors, particularly Chauncey Medberry.

The last paragraph on this page records the motion of Museum president Harold Williams "that the Board accept the Pennzoil proposal."

shareholder's stock is not turned in on a merger. "Does the
shareholder have appraisal rights." Mr. Lipton responded "Yes"
under Delaware law.

The Chairman then inquired if there were any further
comments or any further discussion.

In response to the Chairman's question, Mr. Medberry
stated "I have concerns. I have a different opinion as to
values. Major shareholders make it impossible for us as
Directors to perform our duty as well as we might. I think
more is to be obtained for the shares." Mr. Stuart commented
that most of us feel that the real value of the Company under
normal conditions is greater and our investment bankers have
so stated and under the circumstances we cannot obtain it.
Mr. Taubman then told a story about the Michigan boy that dated
the girl with a pimple on her nose and said after he married
her he shouldn't complain about the pimple. Mr. Taubman went
on to state that if shareholders bought stock in a company with
two large shareholders, they should be aware of the problem
and it was like dating the girl with the pimple. Mr. Stuart
responded, "In this case, she didn't used to have the pimple."

Mr. Williams then moved that the Board accept the
Pennzoil proposal provided that the amount being paid relating
to ERC be $5 per share. The motion was seconded by Mr.

-65-

The last paragraph from the previous page carries over to page 66 and records the Board vote on Harold Williams' motion to accept the Pennzoil proposal: "The vote was taken with all directors voting 'For' other than Mr. Medberry, who voted 'No.' This is the 15-to-1 vote the jury found made the Pennzoil-Getty deal binding.

Paragraph four describes a motion by director Henry Wendt to enhance those termination and pension adjustments that are generally known as "golden parachutes." The notes refer specifically to those executives "likely to be affected with the change in management."

On the heels of that 15-to-1 vote for the Pennzoil proposal, Wendt's motion is itself suggestive.

Mitchell. The vote was taken with all directors voting "For" other than Mr. Medberry who voted "No."

The Chairman then inquired as to whether or not there were any other items to be brought before the Board.

In response, Mr. Stuart stated that he wanted the record to reflect the splendid job done by the Chairman in handling the meeting under difficult conditions. Mr. Stuart moved that the Board commend the Chairman for the conduct of the meeting. The motion was seconded by Mr. Taubman and the vote was unanimously "For."

Mr. Wendt moved that the Company indemnify the Chairman, the President and the General Counsel for actions taken during the last eighteen months in carrying out their duties, including actions relating to dealings with the Trust and the Museum. The motion was seconded by Mr. Tisch and the vote was taken with all Directors voting "For."

Mr. Wendt then made a motion recommending that the Chairman of the Executive Compensation Committee recommend termination and pension adjustment for key executives of the Company whose careers were likely to be affected with the change in management. There was a second by Mr. Boothby. Prior to taking a vote the motion was amended by providing

THE "DEAR HUGH" LETTER

This document, ostensibly from Gordon Getty to Hugh Liedtke, was actually prepared by Pennzoil attorneys.

Many jurors saw the fact that Gordon signed a letter promising to support the Plan spelled out in the Memorandum of Agreement as another indication of his intent to be bound to the agreement with Pennzoil.

The only qualification Gordon inserted was that he would support the Plan "subject only to my fiduciary obligations." This exception was repeatedly cited by Texaco attorneys during the trial.

Pennzoil president Baine P. Kerr signed the document for Pennzoil in the place of chairman Liedtke, who was unavailable at the time.

Kerr's initials ("BPK") as well as those of Gordon ("GPG") flank each of the additions.

January 2, 1984

Dear Hugh:

This is to confirm the understanding between us relating to the Plan dated January 2, 1984 that is being presented to the Board of Getty Oil Company today.

Subject only to my fiduciary obligations:
I agree that I will support that Plan before the   *GPG*
Board and will oppose any alternative proposals or other ar-   *BPK*
rangements submitted to the Board that do not provide for your
participation in Getty Oil Company on the same basis as outlined
Subject only to my fiduciary obligations:
in the Plan. If the Board does not approve the Plan today, I
will execute a Consent to remove that board and to replace the   *GPG*
directors with directors who will support the best interests of   *BPK*
Subject only to my fiduciary
the shareholders, as reflected in the Plan. I will also use my
best efforts to urge the J. Paul Getty Museum to execute a Consent to the same effect.

For your part, you agree that Penzoil Company will not
participate with any third party in any plan or arrangement that *GPG*
me as Trustee of   *BPK*
does not provide for the participation by the Sarah C. Getty
Trust dated December 31, 1934 on the same basis as provided in
the January 2, 1984 Plan unless, despite your efforts, the Plan
is not approved by the Getty Board this week.

                        Very truly yours,

                        Gordon P. Getty

                        Gordon P. Getty, as Trustee of the
                        Sarah C. Getty Trust

APPROVED: PENNZOIL COMPANY

By: *Baine P. Kerr*
    ~~J. Hugh Liedtke, Chairman of the~~
    ~~Board and Chief Executive Officer~~
    BAINE P. KERR, *President*

            EXHIBIT A

THE SIEGEL AFFIDAVIT

This affidavit was filed by Martin Siegel, Gordon Getty's investment banker with the firm of Kidder, Peabody on January 5, 1984 in an attempt to have a California court lift a restraining order obtained by Gordon's niece, Claire Getty. The restraining order prohibited Gordon from executing the definitive merger document that formalized the Pennzoil agreement approved by the Getty board the previous evening.

The Siegel Affidavit became a significant exhibit at the trial because it was filed mere hours before Texaco made their move to acquire all of Getty Oil. Much of the language contained in the affidavit seems to confirm Pennzoil's contention that it had reached a binding agreement with the Getty entities when the Getty board voted to approve the deal. Siegel was present for much of the two-day board meeting, and was present in Gordon's suite at the Pierre Hotel when Texaco came calling on January 5, 1984.

On Page Three, Item #5 states that the Getty Oil board "approved a corporate reorganization transaction (the "Transaction)" among Pennzoil and the Getty entities, and goes on to give the ratio of ownership in the new Getty Oil called for in the Memorandum of Agreement.

Item #6 on that same page states that "the principal terms of the Transaction are set forth in the Getty press release of January 4, 1984," which is also reproduced in this book. As we have seen, that press release also contains many of the terms of the Memorandum of Agreement.

Item #8, which begins on the bottom of Page Four and continues on the top of Page Five, also speaks of "the Transaction now agreed upon."

On Page Six, Martin Siegel signed the affidavit, which was sworn before a New York Notary Public and transmitted to California that same day.

Pennzoil confronted virtually every Texaco witness with this affidavit, to great effect. Siegel himself refused to disavow the document, testifying in his video deposition that the affidavit was "true and correct."

Taken by itself, the Siegel Affidavit did not cinch the contract issue for Pennzoil, but it certainly contributed greatly in establishing their deal with the Getty entities by a preponderance of the evidence, which is the standard of proof in civil cases.

MOSES LASKY
CHARLES B. COHLER
LASKY, HAAS, COHLER & MUNTER
Professional Corporation
505 Sansome Street
San Francisco, California 94111
Telephone: (415) 788-2700

Attorneys for Gordon P. Getty,
as sole trustee of the Sarah C.
Getty Trust dated December 31, 1934

IN THE SUPERIOR COURT OF THE STATE OF CALIFORNIA

IN AND FOR THE COUNTY OF LOS ANGELES

In the Matter of	No. P. 685566
THE DECLARATION OF TRUST OF SARAH C. GETTY, DATED DECEMBER 31, 1934.	AFFIDAVIT OF MARTIN A. SIEGEL Hearing date: January 5, 1984 Hearing time: 4:00 P.M. Department 40

STATE OF NEW YORK)
) ss.:
COUNTY OF)

MARTIN A. SIEGEL, being duly sworn, deposes and says:

1. I am a vice president and director of Kidder, Peabody & Co., Incorporated ("Kidder, Peabody"), which has been retained as an investment banker to render financial advice to Gordon P. Getty as trustee (the "Trustee") of the Sarah C. Getty Trust dated December 31, 1934 regarding the investment of the Trust in shares of the Getty Oil Company ("Getty Oil") and various financial alternatives to such investment. I have personally acted on behalf of Kidder, Peabody in this engagement for the Trustee since October 5, 1983.

2. I am a director of Kidder, Peabody, responsible for its mergers and acquisitions activities. During the past five years Kidder, Peabody has been involved in more than 450 merger related assignments. In 1983 alone, Kidder, Peabody was involved in more than 80 merger transactions.

3. On December 28, 1983, a wholly-owned subsidiary of Pennzoil Company ("Pennzoil") initiated a tender offer to purchase at least 16 million shares of Getty Oil common stock (20% of its outstanding common stock) for $100 cash per share (the "Pennzoil Offer"). Since then I have been advising the Trustee with respect to the alternatives realistically available to him. In the course of advising the Trustee I have, among other things, evaluated possible alternative transactions and the relative risk and certainty involved and evaluated the likely financial effect of alternative transactions if they could be achieved.

-2-

4. On January 2 and 3, 1984, I attended a special meeting of the board of directors (the "special directors' meeting") of Getty Oil. I was present as an adviser to the Trustee pursuant to the Trustee's engagement of Kidder, Peabody referred to in paragraph 1, above.

5. During the special directors' meeting the Getty Oil board of directors approved a corporate reorganization transaction (the "Transaction") among Getty Oil, Pennzoil, the J. Paul Getty Museum (the "Museum") and the Trustee as a result of which the Trustee will own approximately 57% and Pennzoil approximately 43% in Getty Oil. Pennzoil, the Museum, and the Trustee have all agreed in principle to the Transaction approved by the Getty Oil board of directors.

6. The principal terms of the Transaction are set forth in the press release issued by Getty Oil, the Museum and the Trustee, a copy of which is attached hereto as Exhibit 1.

7. As of the time of the Getty Oil special directors' meeting, I had been engaged for several days in discussions with the Trustee analyzing the realistic alternatives concerning the Trustee's interest in Getty Oil. In reviewing the alternatives available to the Trustee in the context of the Pennzoil Offer, a central concern was the

-3-

unresolved and continuing risk that the Trustee could be left as a minority shareholder in a Getty Oil controlled by another shareholder. This possibility posed a high degree of risk that the value of the Trustee's Getty Oil stock would suffer a substantial loss when compared to its pre-transaction value. I was also concerned that the existence of the Pennzoil Offer created an uncertain and unstable situation which could have a deleterious effect on the operations of Getty Oil and could adversely affect the value of the Trustee's stock. I therefore concluded that the most advisable course of action from the Trustee's point of view was to enter into a transaction as promptly as possible that would put an end to the volatility and uncertainty of the existing situation and would also assure that the Trustee not be left in a position as a minority shareholder. Circumstances required that the Trustee take an active role in pursuing such a transaction since failure to do so would virtually assure that Pennzoil or some other third party could unilaterally impose a result on the Trustee that would leave the Trustee and the Trustee's stock in this vulnerable position.

8. Accordingly, among all of the alternatives available to the Trustee, my evaluation convinced me, and I so advised the Trustee, that in my opinion the Transaction

-4-

nificant decrease in shareholder values, including the value of the Trustee's stock.

Sworn to before me this
5th day of January, 1984

[signature]
Notary Public

[signature]
Martin A. Siegel

---

now agreed upon created a far greater certainty of being consummated without taking major and unnecessary risks of diminution in realizable value. None of the other available alternatives for the Trustee's consideration offered a similar degree of certainty for achieving majority control of Getty Oil on comparable financial terms.

9. In providing financial advice to the Trustee with respect to the Trustee's ownership of Getty Oil stock, Kidder, Peabody also reviewed the specific terms of the transaction as well as certain other relevant information and has rendered an opinion to the Trustee that the financial terms of the Transaction are fair to the Trustee from a financial point of view. (A copy of Kidder, Peabody's opinion is attached hereto as Exhibit 2.)

10. Any delay in execution of the agreement for the Transaction, even of a very few days, has the potential for substantial harm to the Trustee. It is important to the stability of Getty Oil that there not be created greater uncertainty as to the future ownership of the company, lest that increase the difficulty of maintaining operations and retaining key employees. The increased turmoil and continuing uncertainty that would result from any delay in the execution of the Agreement would present significant risks to Getty Oil and the substantial possibility of a sig-

-5-

193

THE GARBER NOTE

Getty Oil treasurer Steadman Garber was in California when he made this note of his conversation with another high-level Getty employee (believed to be general counsel R. David Copley). It is a report on the outcome of the January 2–3 Getty board meeting made prior to the public announcement of the Getty-Pennzoil deal.

Agreement

—T + M + P
—Board Agreed but for Chauncey that deal should be done
—PZO to purc. 24m @ 100 + 5
—Co will buy Museum at same price
—Getty to buy 24m shs less shs acquired by PZO in a cash out merger
—PZO to take Museum then up to 20

This note is telling because this report telephoned to California shortly after the Getty board approved the deal contains a number of terms of the Memorandum of Agreement.

"T + M + P" obviously refers to the Trust, the Museum, and Pennzoil, which had in fact joined forces to make the deal the board approved.

Except for Chauncey Medberry, the lone holdout in the climactic vote, the Getty board agreed "that deal should be done" with that 15–1 vote.

The number of shares to be purchased by Pennzoil and repurchased by the Company would result in the 4/7–3/7 ratio called for in the Memorandum of Agreement.

Agreement

- T + M + P
- Board Agreed but for Chancey
  that deal should be done

  - PZO to purch 24 nn ello+5
  - Co will buy museum shs at
    same price
  -

- Getty to buy 24 nn shs less
  shs owned by PZO
  in a cash out merger

- PZO to take Museum then up
  to 20.

PART VII

BURDEN OF PROOF

Pennzoil's Case

15

During the week-long jury selection, the top executives from both firms had been present. For Pennzoil, that was Hugh Liedtke and Baine Kerr.

Seated together in the courtroom, Liedtke and Kerr were a study in contrasts. Not a large man to begin with, Kerr was dwarfed by the hulking Liedtke, whose ruddy visage remained grim throughout the long proceedings. Kerr, on the other hand, seemed calm almost to the point of detachment, as if he was aware of the intricacies of the legal process and resigned to letting the system work.

Baine Kerr had been a lawyer for forty years, many of them spent with the firm of Baker & Botts, who was representing Pennzoil in this case. He had become general counsel of Pennzoil in 1968, then president in 1977. By the time he testified in the trial in 1985, he had been retired for a few months. Kerr retained his seat on the Pennzoil board and, more important, the confidence of chairman Liedtke.

Joe Jamail had delivered a blistering presentation of Pennzoil's case on voir dire, followed by a calm, deliberate opening statement from John Jeffers.

To begin the complicated task of presenting the evidence, Jamail tapped Jeffers. Pennzoil's first witness would be Baine Kerr.

There was not what you'd call a ripple of excitement as Kerr took the witness stand, where he would remain for the next seven days of the trial. Whatever circuslike atmosphere had existed downstairs in the larger courtroom, with one hundred prospective jurors and half the lawyers in town straining to get a look, it had not survived the week. The low-key opening statements of the day before had further depressed any expectations held by the jurors.

The task confronting Jeffers and Kerr was a formidable one. Despite the vivid picture of the case painted by Jamail, he continually reminded us that what the lawyers say is not evidence. The evidence comes from the witness stand. The jury is the sole judge of the credibility of the witnesses, and must determine what weight to give their testimony in the deliberations at the end of the trial.

So too it would be in this case, although by the time testimony began, we had already heard detailed presentations from both sides and seen many of the documents that would be admitted as evidence. Officially, we knew nothing; we had heard no evidence.

Jeffers began with the usual questions about the background of the witness. When his story turned to the company he had helped Hugh Liedtke to build, Kerr's narration became almost inspirational, lacking only thundering orchestration and shimmering visuals depicting the oilfield equivalent of amber waves of grain.

When the questions turned to the Getty Oil Company, Kerr's tone leveled off considerably.

Prior to Pennzoil's tender offer, it had followed the events that had spilled the blood in the water at Getty Oil. As a publicly owned company, Getty was required to make certain filings with the Securities & Exchange Commission. These filings then became a matter of public record. The internecine warfare of 1983 had been largely confined to company circles, although there had been rumors up and down Wall Street.

When the Standstill Agreement, that ostensible peace treaty among the company, the Museum, and the Trust, had been signed in November, it signaled that there was substance to those rumors. If there was not at least the threat of war, why was a treaty required?

At any rate, this document was followed on December 5 by the Consent of Majority Stockholders, called "the Consent," when the Museum joined with the Trust to impose new bylaws on the board after the notorious back-door incident.

This was the background of Pennzoil's tender offer for twenty percent of the Getty stock in late December, 1983.

Of course, the internal dispute among the Getty entities was not at issue in this lawsuit. Texaco tried to have the entire matter excluded from the trial, but Pennzoil opened a narrow door by citing information obtained in the Pennzoil study of Getty in the last three months of 1983.

Jeffers handed Kerr the documents one by one. The Consent contained an account of the back-door incident at the November 11 board meeting. Jeffers had him read it into the record.

JEFFERS: "Would you just read the next paragraph?"

KERR: "The next paragraph is also headed 'Item.'
        "'The trustee's personal absence from the November 11 board meeting was procured by fraudulent means for the period during which the matter of the

proposed intervention was discussed. These means included: (1) inducing the Trustee to leave the meeting while the proposed intervention was discussed without first telling him that it was to be discussed; and (2) using the back-door maneuver of secretly introducing into the room, through another entrance, counsel to the corporation to advise on the proposed intervention. The result was a rump meeting of the board, of a kind unquestionably prohibited by the Standstill Agreement.'''

JEFFERS: "And the final paragraph on this page?"

KERR: "It's also headed Item. 'In a subsequent telephone conversation, counsel to the corporation admitted in so many words that the purpose of these maneuvers was to 'snooker' the Trustee.'''

JUDGE FARRIS: "Would you spell that for the court reporter?"

KERR: "S - N - double O - K - E - R."

Since Kerr was not present at that meeting, he could not testify about what happened there. He could and did testify about how he obtained the document, what he understood it to mean when he read it, and how Pennzoil's subsequent actions were influenced by it.

This is how the Getty struggle was admitted into evidence, through the testimony of Pennzoil's president. The real dilemma this posed for Texaco was that it had engineered the Getty deal with the same individuals who had "snookered" Gordon at the board meeting. Under the Pennzoil plan, Gordon would have become chairman of Getty Oil, a prospect that had to have dismayed those who plotted to undermine his authority as Trustee less than two months before.

For Pennzoil, Jeffers was assigned the task of getting this into the record. The task was complicated by Dick Miller, who objected vigorously to every misstep by Jeffers or Kerr. Under the hearsay rule, witnesses are generally prohibited from testifying about anything that they do not know from their own personal knowledge. There are few exceptions to this rule.

Judge Farris correctly sustained almost all the hearsay objections Miller raised. Jeffers would stop, shift gears, and try another approach. To let Kerr know the direction he was then taking, Jeffers would sometimes insert a clue into the question. If the clue was too big, Miller would object that Jeffers was "leading the witness," as of course, he was.

In a way, Jeffers plodding, methodical style was more suited to this kind of painful extraction than that of the more excitable Jamail or

Terrell. Jeffers also recognized when Miller's objections had merit; he would retreat and try again. When he felt he was on firm legal ground, however, Jeffers guarded his turf like a dog with a bone.

Over in the jury box, however, it made for slow going. The clashes between the lawyers were slightly more stimulating than listening to Kerr read legal documents, but whenever it got hot, Judge Farris would call the attorneys up to the bench, where they would argue in whispers as the court reporter listened in.

A number of times during these bench conferences, Judge Farris would suddenly bellow, "Mr. Bailiff, take the jury out." We would be exiled to the hall, and spent many long hours on the benches outside the courtroom, subject to instant recall to the jury box.

Carl Shaw, the long-suffering bailiff, had a lot of responsibilities. In addition to catering to Judge Farris, Carl was also the chief counselor and nanny for sixteen juror-campers at the 151st District Day Camp.

Frequently, our hallway sojourns would end with Carl telling us to go on break or to lunch.

Jeffers continued the direct examination of Kerr, who really was an all-purpose witness. As Pennzoil president, he could be said to have participated in all aspects of the case, even if it was limited to reading documents about what happened before and after the Pennzoil-Getty deal.

Sometimes there would be a problem with the document itself. This exchange is reflective of the kinds of minutiae that bogged down Kerr's testimony:

KERR:	"This is signed by Gordon P. Getty as Trustee of the Sarah C. Getty Trust."
JUDGE FARRIS:	"Is that Sarah P. or Sarah C. ?"
KERR:	"Sarah C., Your Honor."
JUDGE FARRIS:	"The exhibit list says P. Which is it?"
JEFFERS:	"That's a mistake, Your Honor. It's Sarah C. The exhibit list is mistaken. Sorry."

Despite this mania for precision, the exposition continued. Jeffers asked Kerr questions that would place the documents critical to Pennzoil's case before the jury and, more important, in the record.

There were other obstacles to be overcome. No witness was permitted to give legal opinions in court. The applicable law would come from the judge in his charge to the jury at the conclusion of the case.

This presented problems throughout the trial. To deal with it, many questions were prefaced with "Based on your experience as a businessman . . . ". The witness would then proceed to deliver what

was really a legal opinion in thin disguise. These legal opinions did not
have great influence with the jury; we had long since learned that no
matter what a lawyer said, there was another lawyer nearby who
would swear that exactly the opposite was true. I mention it here only
because Kerr was the first of seven lawyers to take the stand.

The testimony moved on to the actual dealings between Pennzoil and
the Getty entities. Kerr had been with Liedtke at the Pierre on January
1 when they had struck the deal reflected in the Memorandum of
Agreement. That document was introduced into evidence and Jeffers
led Kerr on a page-by-page review of its contents.

Speaking "as a businessman," Kerr explained each section of the
Memorandum of Agreement, how it would be implemented, and what
it meant to Pennzoil.

The "Dear Hugh" letter of January 2 was introduced. Kerr had actually signed that letter for Pennzoil in Liedtke's absence. In the context
of Getty Oil management's rough treatment of Gordon that Kerr had
earlier reviewed, the "Dear Hugh" letter seemed less a hardball tactic
than Pennzoil's assisting Gordon in the manly art of corporate self-
defense.

After the Getty board had approved a deal with Pennzoil at $112.50,
Kerr had instructed his lawyers to begin preparation of the final
merger document. Jeffers tried to ask him a question about that.

JEFFERS:	"What did you, Mr. Kerr, have in mind that those doc-uments would be?"
KERR:	"I had in mind—"
MILLER:	"Excuse me, Mr. Kerr. The question is not what the witness had in mind. It's what instructions he gave to his agents and his lawyers and I object."
JUDGE FARRIS:	"Wouldn't he have to have something in mind before he gave instructions?"
MILLER:	"Why, certainly. I would like for him to tell us what instructions he gave. Then we'll know what was in his mind."
JUDGE FARRIS:	"Ask the question, Mr. Jeffers."

The examination of Kerr was riddled with objections from Miller,
which could have been viewed as obstructionist if it were not apparent
that there was merit to most of the objections. This was confirmed by
Judge Farris, who sustained almost all of Miller's objections.

The flow of documents continued in chronological order. The January 4 Getty press release was next. Pennzoil had issued its own version of the release that same day, utilizing the same text.

The Getty press release was an important piece of evidence for Pennzoil. While Texaco contended that the Getty directors had approved the price only at the end of the January 2–3 board meeting, the subsequent Getty press release contained many of the terms spelled out in the Memorandum of Agreement.

Jeffers had Kerr compare the terms, using the giant enlargements of both documents. One matter the press release did not specifically address was the question of who would purchase the Museum shares.

Phrasing a question on this topic that Miller would not object to proved to be a daunting task for Jeffers.

JEFFERS: "What did you understand, by the time of the press release, the situation to be with respect to the Museum shares?"

MILLER: "Your Honor, that question by its nature asks the witness to give hearsay evidence and what he understands is not relevant to any issue. If he can testify as to the facts, we would have no objection. 'His understanding,' as that question is phrased, gives the witness complete freedom to say anything. We wish to interpose an objection to the present form of the question, Your Honor."

JUDGE FARRIS: "Mr. Jeffers, are you going to withdraw the question or are you going to rephrase it?"

JEFFERS: "I'm going to rephrase it."

JUDGE FARRIS: "All right."

JEFFERS: "Based on the press release itself, Mr. Kerr, what was going to happen to the Museum shares?"

MILLER: "May it please the Court. The press release is in evidence and the jury is perfectly capable of interpreting the press release. The witness' interpretation of a document he did not prepare and did not participate in in any way whatsoever is not relevant. We wish to object."

JEFFERS: "He's the president of Pennzoil and Pennzoil issued the press release. I don't know what could be a fairer question than to ask him what, in this press release, is going to happen to the Museum shares."

MILLER: "May we approach the bench?"

This argument was joined by the other attorneys; its ended with Judge Farris instructing Jeffers to withdraw that line of questioning.

To say it was tough sledding for Jeffers during the Kerr examination would be something of an understatement; Jeffers himself would later say the experience was "like marching to Russia." It was no picnic for the jury either, but the knowledge of those documents being introduced would remain useful throughout the trial.

Jeffers persisted with questions about the Museum shares.

JEFFERS: "So it would be true then that whether Pennzoil or Getty Oil Company bought the Museum shares, the price would be the same?"

MILLER: "May it please the Court. Counsel has asked these leading questions over and over and over and has forced me to get up and object and I do wish to object again.

"The witness is an experienced lawyer. The lawyer who is asking the questions is an experienced lawyer and they shouldn't put me in the position of having to do that."

JAMAIL: "Your Honor, may we approach the bench?"

Jamail had had enough. He had sat passively while the examination of Kerr had progressed in fits and starts, but Miller had touched a nerve. A fierce argument in whispers was held twenty feet from the closest juror. Down at the far end of the jury box, I couldn't hear a word. The transcript revealed the following exchange:

JAMAIL: "This is the very thing they want to get up and make some speech about, forcing him to object. Nobody is forcing him to object. That's sidebar . . . but to stand up and say that we're forcing him to do anything is just wrong and I'm asking you to instruct him and ask the jury to disregard his remarks about being forced to do anything.

"If he wants to make an objection, he knows how to make it without making the cheap shots."

MILLER: "Since we started this afternoon, I have had to make at least 25 or 30 objections to leading questions, every one of which has been sustained, and it does make me look bad in the eyes of the jury . . ."

JEFFERS: "The last question I asked was just summing up his testimony so I could move from from one point to the other, because we're talking about a matter that's not going to be easy for the jury to understand."

MILLER: "That's the problem right there. If I can say so, this business of summarizing the witness' testimony after

he has given it is itself objectionable and it simply pro-
longs the trial."

JUDGE FARRIS: "Try harder to rephrase your questions so they can
be remotely succinct. And, Mr. Miller, if you have an
objection to make, make it and don't make a speech."

MILLER: "All right, sir."

JAMAIL: "I'm asking the court to ask the jury to disregard his
remarks. If not, I'm going to get up and make sidebar
remarks also and you can hold me in contempt."

JUDGE FARRIS: "If you want to put a spotlight on it."

JAMAIL: "I want to put a spotlight on it."

TERRELL: "What Mr. Miller is doing, he's just making a speech
to the jury. The jury can hear him back here. He's making a grandstand ploy."

BROWN: "That is something that Mr. Terrell has interjected
into this. We've all been standing before the bench and
we have all been speaking in a voice so that the jury
cannot hear our comments but, You Honor, we don't
phrase the questions to which we have to object,
opposing counsel does that.
    "We didn't choose this fight; he thrust it upon us."

JAMAIL: "Judge, there's a way to do it, and that's not the way
to do it, so I'm asking for the instruction and we're
entitled to it."

JUDGE FARRIS: "All right, gentlemen."

Jamail had come to Jeffers's aid in his own unique way. When Terrell tried to insert himself into the argument, he drew a sharp response
from Bob Brown, Miller's partner. Judge Farris would give Jamail the
instruction he requested.

JUDGE FARRIS: "Members of the jury, at this point the Plaintiff is putting on its case and it is the duty of the attorney for
the Plaintiff, who is now examining this witness, to
ask the questions, and the attorney for the other side
to, if he feels there is an objection to be made, make
the objection.
    "They're both doing what they're supposed to be do-
ing. However, they're not supposed to be making
speeches. So, if you hear an objection and you also
hear a speech with the objection, ignore the speech.

"If you hear a response to the objection and it includes a speech, ignore *that* speech.

"In other words, you're supposed to be hearing questions and answers and objections and not speeches. We do that before we start with the evidence and after we get through. Continue."

This pronouncement by Judge Farris was typical of the way he conducted the trial. He would sometimes literally halt the proceedings to explain or ask an attorney to explain some arcane point to the jury. During the testimony of Baine Kerr, the mechanics of a tender offer and the circumstances surrounding the taking of a deposition were carefully explained. Although the individual jurors had little experience in matters of high finance, the detailed explanations and concrete examples provided a nineteen-week cram course that allowed us to evaluate the evidence with a reasonable degree of certainty.

Another legal term that was defined was interrogatory.

JUDGE FARRIS: "Mr. Jeffers, explain to the jurors what interrogatories are."

JEFFERS: "Ladies and gentlemen of the jury, prior to the time that the case comes to trial, the lawyers both do a lot of what's called discovery, and that is, they gather evidence that's to be presented at the trial. . . . [An interrogatory] is where, instead of asking individual questions, you send written questions to the company itself. In this case, Plaintiff sent written questions to Texaco for Texaco to answer for the corporation. In other words, not for any one person to answer but for the corporation to answer.

"Texaco could do the same thing to Pennzoil. It's just one of the means of gathering information and the rules request that when the party that's given the written questions answer, they have to do so under oath.

"And so we sent in 1984, more than a year ago, fourteen questions to Texaco to be answered under oath in writing and I'm offering now the question and answer to the second written question. They call them interrogatories, but they're just written questions.

"May I proceed, Your Honor?"

JUDGE FARRIS: "You may."

JEFFERS: "Question Interogatory No. 2 states as follows: 'Please identify each Texaco individual who saw on January

5, 1984, the article entitled 'Pennzoil's Alliance with Trust Leads to Accord for Getty Oil Company' that appeared in *The Wall Street Journal*, Eastern edition, issue of that date. A copy of such article is attached hereto as Exhibit G. Your identification of each such person shall include his name and address, his employer at that time, his position at that time, the time, place and occasion when he saw the article, and whether and with whom he discussed the article on January 5, 1984. You may limit your answer to the Texaco board of directors and to the Texaco employees, agents, and representatives working on matters related to Getty Oil or Pennzoil.'

"Answer: 'No one within the identified group remembers seeing this article on January 5, 1985.' This answer is sworn to by Mr. Robert E. Fuller of Texaco and submitted to the Court by Richard B. Miller."

This, of course, became known as the one day in the history of Texaco that nobody read *The Wall Street Journal*. Throughout the rest of the trial, every Texaco witness would be asked if he usually read *The Wall Street Journal*; most answered that they did. They were then asked if they had read it January 5. All answered that they had not. The article referred to repeatedly described the Pennzoil-Getty deal as "an agreement."

Pennzoil said the article, combined with the press release, served as notice to the business community of its deal with Getty.

Texaco, on the other hand, said that on the day its board authorized the largest corporate merger in history, not one director, officer, or employee working on the deal read the page 3 story (with a headline on page 1) in the "paper of record of American business" about the same company they had targeted. I suppose this was possible; however, I don't think one person in the jury box believed it. Shortly afterward, Jeffers concluded his direct examination of Kerr and passed the witness.

Miller walked up and showed Kerr a document to begin the cross-examination.

MILLER: "Mr. Kerr, here is Plaintiff's Exhibit 3, the Getty press release of January 4, 1984. Did you ever talk with Mr. Glanville about that press release?"

KERR: "I don't recall Mr. Glanville—any conversation with Mr. Glanville. I might have. He might have been in the room somewhere, I don't know."

MILLER: "Do you ever recall discussing the press release with Mr. Ward Woods?"

KERR: "No."

MILLER: "Do you ever recall discussing the press release with Mr. Liman?"

KERR: "At what point, Mr. Miller? I'm sorry. I don't understand the time frame."

MILLER: "There is no time frame. At any time."

KERR: "Well, I think we talked with all of them after we saw it and they saw it."

MILLER: "I just got through asking you if you had talked to Mr. Glanville about the press release and do I understand you never did talk to him about it?"

KERR: "I assumed you meant prior to the issuance of it."

MILLER: "Well, the question was 'Have you ever talked with Mr. Glanville about that press release?'"

KERR: "I recall no specific conversation with him."

MILLER: "All right. Have you ever talked with Mr. Ward Woods about that press release?"

KERR: "I recall no specific conversation with him."

MILLER: "Have you ever talked to Mr. Liman about the press release?"

KERR: "I think we did talk about it, yes."

MILLER: "When was that?"

KERR: "I think it was on the morning of the 4th."

MILLER: "Was anybody present with you and Mr. Liman when you talked to him about the press release, if you did?"

KERR: "If I talked to him, it would have been with a group of people, that's right, in the room."

MILLER: "You don't remember talking to Mr. Liman about the press release, do you?"

KERR: "That's true, either way."

James Glanville and Ward Woods are investment bankers with Lazard Freres, the firm that had represented Pennzoil in the Getty deal. Arthur Liman is an attorney with the New York firm of Paul, Weiss that was retained by Lazard. Miller's opening broadside was typical of the cross-examination that would consume the next four days.

Like most of the principal figures in the case, Kerr had given a lengthy deposition. Miller sought to point out inconsistencies between his deposition and trial testimony. In fact, there *were* some inconsistencies. Kerr tried to explain them, which was not easy to do in responding to narrow, yes-or-no questions.

Judge Farris had something of a fetish for yes-or-no questions, which Miller would skillfully exploit. Farris insisted the witness first give a yes-or-no answer; then he would be permitted to explain his answer.

Jeffers, who had suffered through Miller's constant interruptions during direct, repaid the favor in kind on cross with series of objections. These included entreaties to the judge that Miller had not let Kerr answer the question or had mischaracterized this or another witness's testimony.

Miller recalled Kerr's testimony concerning Pennzoil's analysis of the turmoil at Getty Oil.

> MILLER: "As you analyzed the situation, you were going to become a fourth player whose role would be uncertain until the role was identified?"
>
> KERR: "That's right."
>
> MILLER: "And that's what an old card player like yourself would want to do; is that right?"
>
> KERR: "I'm not much of a card player. I may be old, but I'm not much of a card player."
>
> MILLER: "That's not what I understood Mr. Jeffers to tell the ladies and gentlemen of the jury on his opening statement. I understood him to say that you and Mr. Liedtke were very good card players."
>
> JEFFERS: "That's in the business sense, Your Honor. I object."
>
> JUDGE FARRIS: "In the business sense. We're merely talking about semantics here."

The impatience of the judge convinced Miller to give up on this line of questioning and switch to a different topic.

At another point, Kerr tried to recall a meeting during the negotiations with Gordon on January 1.

> MILLER: "Are you telling me something about a meeting that you attended?"
>
> KERR: "I'm telling you of a meeting that occurred, to the best

of my knowledge and belief, and that was reported to me."

MILLER: "You are telling me things that people have repeated to you?"

KERR: "Yes."

MILLER: "I don't want to be deprived of my right to cross-examine witnesses, Mr. Kerr, by being told hearsay evidence. Now, if you will just answer my question, sir—"

JEFFERS: "I object. That's argumentative sidebar. I request that we proceed by question and answer."

MILLER: "May I address that question, Your Honor?"

JUDGE FARRIS: "Yes."

MILLER: "The witness is a lawyer with forty years of experience. He knows that hearsay evidence is improper and he has repeated it."

JUDGE FARRIS: "Mr. Miller, I want you to make response without making a jury argument."

MILLER: "All right, sir. My objection is that the witness is repeating hearsay information that he knows is hearsay and we object."

JEFFERS: "We don't agree that it's hearsay at all, but my objection was that we proceed by question and answer and not by sidebar."

JUDGE FARRIS: "Mr. Miller, the next time that you feel the witness is, as you say, answering by using hearsay evidence, approach the bench."

MILLER: "Yes, sir, I will."

JUDGE FARRIS: "Do not address the witness in that manner again in the presence of the jury."

MILLER: "Mr. Kerr, if you will listen to my question, I will try to make it as precise as possible and give you a full opportunity to answer."

"When you left the Pierre on the night of the 1st, had you and Gordon Getty essentially agreed to the planned Memorandum of Agreement as it is in front of you?"

KERR: "That is correct."

And the cross-examination was off and running again. A short while later, it was Kerr who felt the sting of a rebuke from the Judge Farris. Miller cut off the end of an answer, bringing Jeffers out of his chair to object.

JEFFERS: "I don't think the last part of his answer was picked up."

JUDGE FARRIS: "Did you finish, Mr. Kerr?"

KERR: "I've forgotten where I was."

MILLER: "I can ask . . ."

KERR: "No, let me get her to read it back to me so I can finish it."

JUDGE FARRIS: "Let me point out to you that *I'm* the one that tells the court reporter whether to read back."

KERR: "I'm sorry."

JUDGE FARRIS: "Proceed."

If Judge Farris seems like something of a tyrant here, I guess that is a fair assessment. Keep in mind, however, the difficulty of the task he faced—running an emotional trial with a covey of high-powered attorneys locked in a bitter, high-stakes struggle.

Miller later focused on the "Dear Hugh" letter, making it clear he did not accept the benign description Kerr had given of the document, which he said was "never intended to be used as a club." This caused Kerr to erupt in what was, for him, an extraordinary outburst.

MILLER: "This letter may not have been intended to be used as a club, but the possibility of removing [the directors] was intended to be used as a club, though. That's right, isn't it?"

KERR: "Mr. Miller, this had been mentioned and known to everybody as a possibility for some time."

"I mean, how much longer was Mr. Getty going to put up with this stuff? How long was he going to let them snooker him, as they said. I-I think that at some point he's got to exercise his rights."

MILLER: "And you told him that?"

KERR: "I told him I was surprised that he hadn't."

MILLER: "[You] said, 'Why don't you assert yourself? You own 40% of the stock'?"

KERR: "He couldn't do it himself, but, I mean, I couldn't understand why they would continue to have the company do
what they did. The company going out and finding one of
the brothers or half-brothers who was a person of, according
to all these newspaper reports and articles you have, was a
dope addict living as a recluse in London who wouldn't even
pay for the medical expenses of his son who had become a
total vegetable. That's the man they got to authorize the lawsuit and Gordon was a good enough person where he, with
no legal obligation to do so, paid all these medical expenses."

Miller had provoked this response from Kerr, but if he was waiting for
him to blurt out something damaging to Pennzoil's position, it never
happened.

The portion of Jamail's voir dire presentation about handshakes was
read to Kerr, with much greater effect.

MILLER: "Then on page 1344 he said, 'The question was asked of Texaco's lawyer 'What's a handshake worth these days out in
the oil patch and is that a binding agreement?'
"'And the Pennzoil Company thinks it's worth $14 billion
and is suing Texaco to get that money.' Now, you never shook
hands with anybody at the Getty Oil Company, did you?"

KERR: "Well, only with Gordon Getty, I guess . . . I didn't shake
hands with Mr. Harold Williams. I didn't shake hands with
Mr. Petersen; I *would* have shaken hands with him."

MILLER: "Well, of course."

KERR: "But I was never given that opportunity."

Miller had caught Kerr between rhetoric of Jamail's voir dire presentation and the facts of the case. After reviewing with Kerr the list
of people with whom neither he nor Liedtke shook hands, Miller moved
in for the kill.

MILLER: "So Pennzoil didn't shake hands with the Getty Oil Company, did it?"

KERR: "I don't know who the Getty Oil Company is exactly. I mean,
there's several different entities involved with Mr. Petersen
or Mr. Copley or Mr. Lipton or Mr. Williams. Personally, we
didn't shake hands with them."

MILLER: "You shook hands with Gordon Getty?"

KERR: "That's right."

MILLER: "All right."

KERR: "I don't mean to imply, though, that we didn't have an agreement with them. I'm just saying that we didn't physically shake hands."

His last attempt at damage control notwithstanding, Baine Kerr's sixth day on the stand had been Texaco's best of the trial so far. Although at this point the jury still clung to hopes of a six- to ten-week trial, the fact was both sides would have many good and bad days during what would be almost twenty weeks of trial.

Kerr's testimony concluded the next day. Miller raised the subject of the Schedule 14(d)1 Pennzoil filed with the SEC after the Getty board vote. Since Pennzoil had filed a tender offer, it was required to disclose any contacts with the target company. The filing, prepared by Pennzoil lawyers, said the tender offer would be withdrawn if a definitive merger agreement was signed. Miller asserted this meant Pennzoil knew it didn't have a binding agreement.

MILLER: "Then it's obvious to you, is it not, that the lawyers who were preparing this . . . said we're not going to withdraw the tender offer until they sign this agreement?"

KERR: "Mr. Miller, I think you're straining at gnats now to try to read into that some significance."

MILLER: "Well, were you present when I questioned the jury panel and asked them if they would take into account the difference between 'if I go to heaven' and 'when I go to heaven'? Were you there?"

KERR: "I don't remember that. That's an interesting thought, but I just don't recall it."

After four grueling days of cross-examination, Kerr's underplayed amusement at that last question left little doubt the place he was thinking of for Miller was not heaven.

The cross-examination ended with Miller going at the "Christmas surprise" angle again, trying to get Kerr to admit the tender offer had been timed to catch Getty Oil with its defenses down between Christmas and New Year's Day. Kerr answered his questions but refused to get drawn into his scenario, saying "Oh, Mr. Miller, I'm not going to argue about when Christmas comes."

The prolonged testimony of this witness let the jury know that the trial was for real, and that we would be here for a long, long time. There wasn't any real backlash against Kerr, although his answers to many of Miller's questions that he "could not recall" or "did not recollect" a particular conversation or event managed to arouse the ire of one juror.

When we were sitting outside during a break, Kerr shuffled past on his way back to the courtroom. Theresa leaned over and said, "No wonder they got rid of him; he can't remember a thing." I laughed and winced at the same time at her cruel (and funny) remark. I'm sure she was just teasing.

After his marathon voir dire performance, Joe Jamail had taken a back seat to John Jeffers. He remained a constant presence in the courtroom, however, and it now seems there was method to his madness. As he related to me after the trial, "Many people thought we ought to start out with Liedtke to make that big impression with the jury. I said bullshit. If I haven't impressed them enough when the voir dire's over, then we've got trouble anyway. We've got to start with the glue and the facts man and the dull bullshit. But it has to be done for the record.

"Not only that, for appeal purposes you have to build your record. It's like building a house."

Jamail selected both the lawyer and the witness to lead off his case: "I decided it would be Kerr. Baine is mild-mannered, but I knew him to be tough inside, just like a piece of iron. Then I made the decision that I'd start with Jeffers. They had heard me and would have heard Miller, the real antagonists. John was so courtly and gentlemanly, hard to understand at times, but I knew Miller wouldn't be able to hesitate from doing what he does and that's jumping on Baine Kerr. I wanted that contrast to be so great right at the beginning. You know, [Miller] finally got so exasperated he was abusive to Kerr."

With Kerr as the Facts Man, Jeffers was able to introduce the major elements ("the glue") of Pennzoil's case. Jamail had other reasons for this course of action: "I wanted that jury to want to hear from me again," he said, "and I knew we were going to videos for a long time. I didn't want you to get tired of me."

Not having to introduce the video depositions was fortunate for Jamail; at least he stayed and watched them. For the jury, of course, there was no choice in the matter.

16

When the jury returned the following Tuesday, the courtroom had been transformed into an electronic battleground. Video monitors were deployed in all directions, connected by long strands of black coaxial cable that seemed to be everywhere.

Robert Ball, a video technician with longish hair and a bushy mustache, would operate the system on cues from Jeffers or Terrell. Although neatly attired in a tan suit, Ball struck a discordant visual note amid the buttoned-down attorneys we were used to seeing.

Video depositions are a curious tool available to a trial lawyer, occupying a realm somewhere between live witnesses and the reading of transcribed depositions. Pennzoil relied on video to bring the players of the Getty drama into the courtroom. As with live testimony, you can see and hear the witness on the screen, and observe his or her mannerisms and emotions, if any. On the other hand, the video depositions played for the jury had been severely edited to remove anything that was remotely objectionable from a legal point of view.

Since attorneys for both sides question the witness in a deposition, they are both allowed to designate certain portions to be played, subject to the aforementioned editing.

As we would discover in this trial, the result is almost always a stalemate. In an adversarial proceeding, the opposition can usually come up with something to mitigate any admission in this rigidly controlled format.

In questioning a live witness, a lawyer can go right to the edge of what is objectionable and sometimes cross over, creating an image that will linger in the jury's mind no matter how strenuous the judge's instruction to disregard it. Such a maneuver is always eliminated in an edited video deposition, and properly so. The problem is that the elimination of spontaneous exchanges all but precludes the chance of unexpected revelations. In other words, expect these folks to stick to their story.

The judge is not present when depositions are taken. An attorney can flat out refuse to let his client answer a question. The remedy for the questioner is to return to the judge and request an order compelling an answer, a messy procedure that is less frequently invoked. The result is battling lawyers at the deposition and a generally bland-leading-the-bland offering to the jury.

For Pennzoil, the necessity of describing what happened in the Getty deal meant it had to rely on adverse witnesses. Under reciprocal jurisdiction, witnesses in California and New York can be compelled by a local court to give sworn testimony to a Texas court; they cannot be compelled to appear. Texaco had granted indemnities that seemed to oblige the Getty entities to assist in the defense. Pennzoil chose not to wait and see who Texaco brought to court, and for good reason; if it didn't prove its case, Texaco wouldn't have to call a single witness.

Getty Oil chairman Sidney Petersen was the first witness whose video deposition was played for the jury. The patrician Petersen had fought his way to the top of Getty Oil, only to see the company where he had worked all his adult life dissolve beneath him. Coming up through the financial side of the business, Petersen was most decidedly not an oilman. His vision of the future of Getty Oil proved to be apocalyptic when it collided with the ambitions of the 40 percent stockholder, Gordon Getty.

Faced with the prospect of giving up his hard-earned chairmanship to the upstart Gordon, Petersen had gladly acquiesced to the Texaco deal. His dispirited defense of that deal in this video testimony, however, underscored how difficult it was for him to be enthusiastic about the dismantling of the Getty Oil Company.

Petersen did testify that the 15-to-1 board vote on the Pennzoil deal concerned the price only. He was then confronted with a copy of a *Fortune* magazine article that quoted him as saying of the Pennzoil deal, "We thought there was a better deal out there, but it was a bird-in-the-handish situation. We approved the deal, but we didn't favor it."

Petersen said he didn't remember saying those words, but stopped short of totally disavowing the quote. The article was introduced as an exhibit in the trial. The only parts admitted into evidence were the heading (e.g., "MERGERS/by Peter McNulty"), the date, and the quote itself. Not one additional word of the article was shown to the jury. After the trial, I went to the library and looked up that issue of *Fortune*. The title of the article, "How Texaco Outfoxed Gordon Getty," was especially ironic since Gordon Getty took his $4 billion and his indemnity from Texaco and walked away.

The second video deposition offered was that of Peter McNulty, author of the *Fortune* article. In his brief deposition, McNulty described the circumstances of his interview with Petersen and said that he had no verbatim recollection of Petersen actually saying the "bird-in-the-hand" quote. He also indicated that did not diminish his belief that the quote was correct. In response to a question, he said Petersen had never called him to correct the quote, a fact confirmed by Petersen.

Getty Oil general counsel Ralph D. Copley had been secretary of the January 2–3 board meeting. His sixty-five double-spaced typewritten pages of notes were to be the minutes of the meeting, but they were never formally approved by the Getty board. Before his deposition was played, portions of the "Copley notes," as the would-be minutes were known, were read into the record by Pennzoil attorneys.

Texaco objected to the partial offer of the notes, and full copies were later distributed. This would be the most frequently consulted piece of evidence in the trial, especially page 63. Whenever an attorney would ask a witness to look at page 63, I could recite the passage verbatim.

Copley had also approved the January 4 press release on behalf of Getty Oil and testified that it was "true and correct." He also said the board voted on price only, although this seemed inconsistent with the press release.

On the subject of his notes, Pennzoil had determined from a previous deposition taken from Copley that he had destroyed the handwritten originals of the notes the day after he returned from Texaco headquarters in White Plains. By that time, Getty Oil had already filed the suit for a declaratory judgment against Pennzoil in Delaware, which led Pennzoil to claim that evidence had been destroyed. The inference was that Copley left White Plains with orders from his new bosses at Texaco to purge the notes. This charge would become largely irrelevant in the long run, however; Pennzoil actually used what it called the "edited, sanitized, ersatz" Copley notes to prove its case.

The attack on the typewritten notes was actually a preemptive strike by Pennzoil, because the notes did not explicitly say "Pennzoil-Getty-Binding Contract." On balance, however, they were more favorable to Pennzoil.

The next video deposition was from Gordon Getty's lawyer, Charles Cohler. Cohler had been present in Gordon's suite at the Pierre when Liedtke and Kerr came calling on January 1 and had attended the January 2–3 board meeting.

He had flown to California when Claire Getty had found a judge who enjoined Gordon from signing any definitive documents. Cohler had called the Pierre on the night of January 5 to report that he had filed the affidavit by Martin Siegel with the California court. John McKinley and the Texaco contingent were at the Pierre meeting with Gordon at the time. It was immediately apparent to Cohler that the Siegel affidavit's contention that "there is a transaction now agreed upon" between Pennzoil and the Getty entities was fraught with peril in light of this new turn of events. In addition, Cohler had filed an affidavit of his own, stating the same thing.

He couldn't do much about the Siegel affidavit, but Cohler quickly amended his own affidavit, scrawling in some handwritten qualifiers that made the "done deal" of the original affidavit more tentative. What the California court made of Cohler's before-and-after affidavits was not recorded. To obtain the approval of Gordon's relatives, Texaco president Al DeCrane would fly to California, eventually raising the price to $128 a share, a little move that cost Texaco another quarter-billion dollars.

In conjunction with the playing of Cohler's deposition, Pennzoil introduced both the original affidavit and the amended version as evidence.

As with most trial exhibits, copies were made and handed to the jury. Looking at the original affidavit, it was obvious that Cohler was telling the California court that the Trust, at least, believed it had a deal with Pennzoil. The implications of the amended version of the affidavit were not lost on the jury.

Pennzoil capped a week of video depositions with one from Geoffey Boisi, "the eighteen million-dollar man" from Goldman, Sachs. Getty Oil's investment banker would be called to testify for Texaco later in the trial, so most of my memories are from his personal appearance. Some things from his video deposition stand out, however. Boisi, a mergers and acquisitions specialist, couched many of his descriptions of the events of the Six Day War in that curious lexicon common to his trade.

Seated at his side, visible throughout the video, was an attorney representing Goldman, Sachs. This attorney played a role in the deposition, asking for clarification of many questions. His presence appeared to muzzle Boisi somewhat, who didn't seem all that glad to be there in the first place.

Boisi reluctantly admitted his role in the back-door incident at the November 11 board meeting and warily described shopping the Pennzoil-Getty deal.

The following exchange gives some idea of the flavor of his testimony:

TERRELL: "You were candid with Perella and Wasserstein when you talked with them on the 4th about where things stood with Pennzoil? That's my question."

BOISI: "Yeah. You keep using the word 'candid,' and, yes, I was candid with them because I told them what I thought, but—"

TERRELL: "That's all I'm asking about."

BOISI: "I just have this feeling that there is another meaning to that word and I'm not picking it up."

As I said, the most vivid impressions I have of Boisi's story come from his live testimony nearly two months after this video was played. Through this week of video depositions, Pennzoil was able to introduce a number of documents into the record, documents that bolstered its case. It also gave us a look at many of the players on the Getty side of the drama.

The video equipment was cleared away during a break, and a tall man with a shock of graying hair strode to the witness stand. Pennzoil's next witness would be the man who had been its emissary to the Getty board—Arthur Liman.

17

With over three hundred lawyers, Paul, Weiss, Rifkind, Wharton & Garrison is one of New York's biggest institutional law firms. One of its best (and best known) attorneys is senior partner Arthur Liman.

When Lazard Freres became Pennzoil's investment banker for the Getty tender offer, it retained the Paul, Weiss firm, securing the services of Liman. Although he would later gain national fame as Senate counsel in the Iran-Contra hearings in 1987, Arthur Liman had been a big fish in the New York pond for a long time.

In 1971, he headed the investigation of the bloody Attica prison riots as general counsel of a commission appointed by Governor Nelson Rockefeller. The worst prison uprising in this century generated significant controversy, especially in New York; forty persons, including many hostages, were killed when the state police retook the prison.

In the 1980's, Liman headed a commission that investigated allegations of a cover-up of police killings by the medical examiner's office in New York City. Mayor Ed Koch, who appointed him, would repeatedly invoke Liman's name as a talisman to assure suspicious citizens of the integrity of the commission investigating this highly charged issue.

Liman was a politically well-connected corporate attorney as well as a highly skilled trial lawyer, a potent combination in any locale. In the Pennzoil-Getty deal, he had worn his corporate hat, with no idea that he would end up as a key witness in a landmark lawsuit twenty months later. But in August 1985, he was in a Houston courtroom, the second live witness to come before this jury.

Irv Terrell handled the direct examination for Pennzoil. After describing his background, Liman recounted how he had participated in the preparation of the tender offer.

The testimony covered a lot of what was virgin territory to the jury. On December 30, Liman and Glanville had a meeting at the Lazard offices with Gordon Getty's investment banker, Martin Siegel of Kidder,

Peabody, who was accompanied by Gordon's lawyers, Charles Cohler and Thomas Woodhouse.

In this session, the preliminary groundwork was laid for Liedtke's January 1 meeting with Gordon. Many of the terms that would end up in the Memorandum of Agreement were thrashed out by Liman and Siegel on December 30.

Bob Brown, Miller's partner, handled this witness for Texaco, and he objected repeatedly to Liman's account of his discussions with Siegel as hearsay or, worse, double hearsay.

Terrell responded that the negotiations qualified for an exception to the hearsay rule because they spoke to the "operative facts" of how the deal came about. Judge Farris overruled most of Brown's objections.

When he continued to object to nearly every question Terrell asked, the judge said Brown could have a running objection to the entire line of questioning, "with the same ruling. Overruled."

Liman described how, as the negotiations progressed, Siegel began to press for a meeting with Liedtke to iron out the details before Liedtke met with Gordon.

TERRELL: "Mr. Liman, what did Mr. Liedtke say in response to the proposals conveyed to him through you by Mr. Siegel?"

LIMAN: "He said he wanted to negotiate directly with Gordon Getty and he didn't want to deal with agents."

Siegel the "agent" would not get his private negotiating session with Liedtke, but the Liman-Siegel discussions had definitely set the stage for the January 1 meeting at the Pierre.

In a sense, you *could* say that Liedtke and Gordon struck a deal two hours after meeting each other for the first time, but the preliminary spadework of Siegel and Liman had given the potential deal a framework before that January 1 meeting ever began.

Unlike Siegel, Liman did not go to the Pierre that afternoon. Instead, he attended Martin Lipton's New Year's Day party. Liman had called the Museum attorney after his meeting with Siegel two days earlier; the Museum would be an integral part of any transaction.

After the meeting at the Pierre concluded, Siegel also showed up at Lipton's party. This was when Liman and Lipton both learned for the first time that Gordon and Liedtke had reached an understanding, subject to the agreement of the Museum.

After the party, Liman testified, he went to the Pennzoil suite at the Waldorf Towers, where representatives of the Trust and the Museum had gathered with the Pennzoil contingent. Out of that session came the draft of the Memorandum of Agreement, reflecting the deal structured by Pennzoil, the Trust, and the Museum.

Bob Brown had continued to object to Liman's hearsay account of the inception of the deal, but it made an intriguing story; Liman, the seasoned trial lawyer, was a captivating storyteller. He would turn to the jury and give his answers to the questions in a forthright, direct manner. The urbane Liman did not affect an overt folksiness; no detectable hint of condescension crept into his testimony.

Before the questions turned to the Getty board meeting of January 2–3, Terrell asked Liman for his impression of Gordon Getty. This brought Brown to his feet yet again, and Judge Farris sent the jury home from the day.

When testimony resumed the next morning, Terrell framed his question this way:

TERRELL: "Did you form an opinion of Mr. Getty as far as his intelligence and ability to act in this period of time, during the January 2 and 3 board meeting?"

LIMAN: "I did."

TERRELL: "What was that?"

LIMAN: "My opinion of Gordon Getty is that Gordon Getty knew what he wanted; that Gordon Getty understood the strength of his bargaining position; that Gordon Getty was approaching the decisions that were being made in a very businesslike way; that Gordon Getty was exceptionally methodical; that Gordon Getty was very careful about not being taken advantage of as a person of great wealth and was concerned that people might try to manipulate him and he was on his guard about that; and that Gordon Getty went to lengths that I had never seen before in selecting his advisors to make sure that he was getting the best advice he could obtain."

This incredibly detailed recounting served as Pennzoil's answer to the Texaco contention that Liedtke had taken advantage of a naive, inexperienced fool.

The questions then turned to January 2 and the Intercontinental Hotel, where the Getty board meeting began at 6:00 P.M. Lipton told Liman that Harold Williams had signed the Memorandum of Agreement for the Museum just as the meeting was beginning.

Gordon and his advisors had taken a suite on the same floor as the board room; here Liman would await news of any developments.

During an early recess, Liman said he was informed that the board had rejected the Memorandum of Agreement. Gordon and Williams, signatories to the document, had voted for it, as had Larry Tisch and the other three new directors placed on the board in the wake of the December 5 Consent.

Liman testified that Lipton, Siegel, Tisch, and Boisi all told him that the board wanted an additional $10 a share over the $110 provided for in the Memorandum of Agreement. Goldman, Sachs had valued ERC, a reinsurance company Getty Oil had purchased several years earlier, at $1.2–1.5 billion and the board wanted a premium for it.

Terrell had Liman read from the Copley notes of the board meeting to describe the action inside the board room. Boisi devised a plan for a $10 debenture to cover the desired premium for ERC. This was detailed in a handwritten note that was delivered to Glanville, the Pennzoil investment banker.

About 3:00 A.M., the board meeting recessed until the next afternoon.

Hugh Liedtke was unhappy about the additional $10 the Getty board wanted. Liman testified that Liedtke felt that he had already gone up from $100 to $110 and he was bidding against himself. Pennzoil's response to the Boisi proposal was an offer to pay either $110 a share or $90 a share plus the value of the ERC.

As the meeting reconvened, Liman was back at the Intercontinental with the counterproposal.

LIMAN: "I handed it to Mr. Lipton. He looked at it and said something like, 'It won't sell. It's too cute. I won't present it to the board.'"

Liman testified that Lipton had told Liedtke that the Museum would not let the board stand in the way of the deal, that he would persuade the Museum to join with the Trust and essentially make good on the threat of the "Dear Hugh" letter.

When Lipton told Liman that Pennzoil needed to offer a "sweetener" on top of the $110, the news wasn't well received at the Waldorf Towers. Liedtke went along, however; the "sweetener" would be in the form of a note or "stub" that guaranteed the stockholders would receive a pro-rata share of all proceeds over $1 billion from the sale of ERC. The minimum guaranteed payment would be $3 in five years.

Liman recalled the conversation where Liedtke approved the $3 stub:

TERRELL: "And you presented this to Mr. Liedtke. What was Mr. Liedtke's response?"

LIMAN: "And [Liedtke] said, 'You are authorized to present that to Lipton and tell him that if this isn't good enough, then I expect him to live up to what he said before,' which is to get the Museum to recommend removing the directors by consent."

TERRELL: "All right, sir. Did you go back to Mr. Lipton and report to him your instructions from Mr. Liedtke?"

LIMAN: "It wasn't hard to get back, he was probably ten feet from me."

TERRELL: "Did you tell him about the $3 stub?"

LIMAN: "I told him that and he did some arithmetic and he said, 'You know, if you discount $3 at the present value, it's less than $2. Why don't you get them to make it a guarantee of $5?'"

TERRELL: "And what did you say to Mr. Lipton?"

LIMAN: "I said, in effect, 'Martin, I value my life. I'm not going back to Hugh Liedtke and asking him to bump this again. If you come back and the board has approved the whole plan and the only thing it wants more is to move the $3 guarantee to $5 so that we have a firm deal . . . I undertake to recommend and sell it to Mr. Liedtke but I'm not going back and asking him to keep raising it without having a firm deal with you.'"

TERRELL: "At that point, Mr. Liman, you were out on a limb at $5 . . ."

LIMAN: "Mr. Terrell, I knew that once I told him that I would recommend and sell $5, that there was never any way Mr. Liedtke was going to get this deal at $3 but I also felt that Marty . . . Marty said he would try to do it at 5. I also felt that this was the best shot that my client had at getting the deal that day."

Liman testified that Lipton recorded the details of the $5 stub in a note, which was in evidence in this trial. Terrell had Liman read it to the jury. This "Lipton note" joined the parade of handwritten counter-offers from the board meeting, including Boisi's $110 plus $10 and Pennzoil's $90–110 that Lipton had said was "too cute."

The $5 stub that Liman "had gone out on a limb" for was going to come up for a vote.

TERRELL: "At some point, were you advised of a resumption of the formal board meeting itself?"

LIMAN: "Yes. There was a lot of milling around, and then they were going back in and meeting and I was trying to find out,

well, 'Are they going to accept it? Did we have a deal?' And
I felt like a hostage; 'Can I go home? Can we get it over?'
and that was the kind of discussion I was having, and then,
all of a sudden, everyone was back in the board room, I was
alone with my partner . . ."

To continue the story back in the board room, Terrell had Liman
read from page 63 of the Copley notes: "Mr. Williams then moved that
the board accept the Pennzoil proposal provided that the amount being
paid relating to ERC be $5 per share. The motion was seconded by Mr.
Mitchell. The vote was taken with all directors voting "For" other than
Mr. Medberry, who voted "No."
    Outside the boardroom, of course, Liman knew none of this. Then
the doors opened, and Siegel, Lipton, Cohler, Boisi, Winokur, and a
host of other people flooded into Gordon's suite.
    Liman testified that Siegel and Lipton told him the board voted 15
to 1 to accept the proposal if the guarantee on the stub was increased
from $3 to $5. Liman was also told the board had voted to indemnify
the management for its actions in dealing with Gordon Getty, the Museum, and Pennzoil. In addition, they had voted to settle the contracts
of top management, to award them "golden parachutes."
    After learning these details, Liman placed a call to Liedtke.

LIMAN: "And he said, 'Do we have a deal at this juncture?'
            "I said, 'That's it.'
            "He said, 'You can tell them yes.'
            "And so I said to Marty Lipton, 'My client accepts.'
            "Then Martin said, 'We have to go back in session.'
            "And I said, 'Why do you have to go back in session? I
            thought you told me that the board had voted this and it
            was done.'
            "He said, 'No, they had voted the counterproposal. Now
            I have to communicate to them that you have accepted the
            counterproposal.'
            "And he and everyone else then went back into the board
            room. He said, 'You can come and wait outside. This is only
            going to take two seconds,' and then he went in."

TERRELL: "Then what happened?"

LIMAN: "Then, within a few minutes, Marty Lipton and Marty Sie-
            gel came out, the door opened and they said 'Congratula-
            tions, Arthur. You got a deal.'"

TERRELL: "Then what did you do?"

LIMAN: "They said 'Good work,' and I said, 'Really, you're the guys that were doing all the work. You were in there on the board.'

"I said, 'I'd like to ask for permission to go into the board meeting and to shake hands with all of the directors because they've been at it for so many hours and I'd like to just shake hands with all of them.' And they said that would be okay. So, I was finally allowed into the board room."

TERRELL: "How may people, roughly, would you say were in the board room when you went in?"

LIMAN: "Everyone was sort of milling around. I guess there were probably twenty people, because there were directors and lawyers, bankers . . . I went around the room and I introduced myself and I shook hands with everyone I could; I said, 'Congratulations.' They said, 'Congratulations to you.'

"I said, 'Thank you for putting in all these hours. I know that you were doing your best.'"

TERRELL: "Were the men that you shook hands with, Mr. Liman, at that point, were they hostile towards you?"

LIMAN: "No, they were tired. So was I."

Liman's account of the machinations that produced the Pennzoil-Getty agreement was the jury's first detailed look at the dynamics of this brand of high-powered deal making.

After the board meeting, Liman played host to representatives of Getty Oil, the Trust, and the Museum at the Paul, Weiss offices to draft a press release announcing the deal; the Memorandum of Agreement called for the parties to coordinate such announcements. The January 4 Getty press release was the product of this session; Pennzoil also released an identical press release on their own letterhead.

Terrell moved on to the dark side of Liman's story—the entrance of Texaco. He showed Liman the January 6 Texaco press release announcing the purchase of the Museum shares.

TERRELL: "Mr. Liman, when did you first find out about Texaco?"

LIMAN: "I received a telephone call from Glanville at Lazard, who read me an announcement . . ."

TERRELL: "What was your reaction upon having this announcement read to you?"

Brown again rose to object, saying Liman's reaction was immaterial. Judge Farris overruled the objection, and Liman answered the question.

LIMAN: "Disbelief."

Liman described how he had called Martin Siegel after hearing the news from Glanville.

TERRELL: "What did you say to Mr. Siegel?"

LIMAN: "I said, 'Marty, I just had this announcement about Texaco read to me. What happened?'"

TERRELL: "And what did Mr. Siegel say to you?"

BROWN: "For Texaco, Your Honor, we object on the grounds that's hearsay."

JUDGE FARRIS: "It's overruled."

LIMAN: "He says that Marty Lipton and Texaco had come to him and the Trustee late the night before, that Texaco had sewn up the Museum and Getty Oil Company before, that he had no choice but to go along. Otherwise, Gordon Getty would have been left with his 40% and a minority position and he said, 'Arthur, I'm sorry."

TERRELL: "What did you say to him, Mr. Liman?"

LIMAN: "I said, 'Marty, I'm not going to be the messenger this time to convey what you say. You better call Hugh Liedtke and tell him yourself and not slink off.'"

TERRELL: "Mr. Liman, to your knowledge, did Mr. Siegel call Mr. Liedtke or did he slink off?"

LIMAN: "You will have to ask Mr. Liedtke."

TERRELL: "When you had this read to you on the morning of January 6, at any point that day did you call Martin Lipton like you called Marty Siegel?"

LIMAN: "No."

TERRELL: "Why not?"

LIMAN: "I was too upset."

TERRELL: "At that point, had the Museum already sold its stock to Texaco, according to this press release?"

LIMAN: "According not only to this press release, but what Marty Siegel had told me."

With that, Terrell concluded the direct examination of Arthur Liman and passed the witness to Texaco.

Bob Brown rose to confront the witness who had delivered this dramatic account of the Six Day War.

BROWN: "Mr. Liman, I'm going to try to speak clearly, but I do have a little bit of a cold. If you don't understand me, if I am not speaking loud enough for you, please let me know."

And also, if anybody on the jury can't hear me, I wish they would let me know also."

JUDGE FARRIS: "This jury will, Mr. Brown."

In fact, while the members of the jury had very little say in how the proceedings were conducted, we zealously guarded against anything that interfered with following our sworn oaths. That included inaudible questions and, as would later be demonstrated, confusing answers.

Meanwhile, Brown began the cross-examination of Liman.

BROWN: "Mr. Liman, I am going to try to give you a very precise question . . . that you can answer yes or no.

"Of course, I invite you to explain any answer you need to . . ."

Brown started by asking questions that established Liman's role in representing Pennzoil, to let the jury know this fascinating testimony had actually come from a Pennzoil lawyer. He then turned the interrogation to the subject of Martin Lipton, who had been prominently featured in Liman's recitation of the Pennzoil-Getty dealings.

BROWN: "Is Mr. Lipton, in your judgment, an honest gentleman?"

LIMAN: "Yes."

BROWN: "Is he, in your judgment, a trustworthy gentleman?"

LIMAN: "I've always found Martin Lipton someone that I could trust, putting aside the dispute that was involved here."

BROWN: "Have you always know him to be a man of his word?"

LIMAN: "Yes. Again, putting aside the dispute we're in here."

Having had Liman vouch for Lipton, Brown asked him about another potential Texaco witness.

BROWN: "Would you tell the ladies and gentlemen of the jury what you know about Mr. Tisch?"

LIMAN: "Mr. Tisch is the chief executive officer of the Loews Corporation. He is a fine human being. He has been a friend of mine and my family for a long time."

BROWN: "If he comes before this jury and testifies, can this jury, in your opinion, believe everything he says?"

TERRELL: "Your Honor, I object to that. How can the witness answer it when he doesn't know what he's going to say?"

JAMAIL: "Or what he's going to be asked?"

TERRELL: "I object to it."

JUDGE FARRIS: "Sustained."

BROWN: "Do you know of any reason why Mr. Tisch would come into this courtroom and speak anything other than the truth?"

LIMAN: "I have no doubt that Larry Tisch will tell this jury the events as he remembers them, just as I've tried to do. I have no doubt about that."

The blank space provided for the Getty Oil Company signature on the Memorandum of Agreement had been the focus of attention by Texaco since the start of voir dire. When Brown questioned Liman about that subject, it produced an effect he had not anticipated.

BROWN: "My question, sir, is did Getty Oil Company ever sign this document?"

LIMAN: "It did not and, if I may add, it could not have."

BROWN: "When the board meeting was over on the 3rd, did you speak to anyone at the Getty Oil Company and say, 'I want an executed counterpart of this document for my client, Pennzoil'?"

LIMAN: "The answer is no, and I'd like to explain."

BROWN: "When the board meeting was over—"

TERRELL: "Excuse me, Your Honor. May the witness explain, Your Honor?"

JUDGE FARRIS: "He may."

LIMAN: "Mr. Brown, the proposal that you have here was supplemented and sweetened with the stub, and since the stub had been the subject of approval and then a counterproposal by Getty at $5, which we accepted . . . Getty couldn't have signed this unless we had this retyped to include the stub . . . "

An exasperated Brown had wound up on the receiving end of a lecture on deal-making. His next question revealed his pique.

BROWN: "All right, sir. Is that your total and full and complete and
           final explanation of why this original Memorandum of
           Agreement is not signed?"

LIMAN: "Yes, because it wasn't complete since it didn't have the
           stub."

Looking for another crack in Liman's defense of the Pennzoil deal,
Brown brought up a matter relating to the mechanics of the transaction not specifically covered in the Memorandum of Agreement.

BROWN: "I thought you had told me earlier that every term in
                the Memorandum of Agreement was material and essential. Did I misunderstand you?"

TERRELL: "That's an absolute misstatement of this witness's testimony and I object to it.
                     "It's argumentative and also not true. I object to it
                as argumentative."

BROWN: "That's why we need the clarification, Your Honor.
                This is cross-examination."

JUDGE FARRIS: "Yes, provided you are quoting the witness correctly."

BROWN: "I want to understand the witness. That's why I want
                the clarification."

JUDGE FARRIS: "Why don't you ask him to clarify it then. Otherwise,
                we have to go through the lengthy routine of having
                the court reporter go back through her notes and find
                the exact quotes."

LIMAN: "I understood his question, Your Honor. May I answer
                it?"

JUDGE FARRIS: "Yes, if you understood it."

LIMAN: "You said . . . you understood me to say that every
                essential   term   was   in   the   Memorandum   of
                Agreement . . . I said every essential term was in the
                Memorandum of Agreement, but there were also
                terms that I would consider boilerplate and not essential, so that it's just the opposite of what I said.
                     "All essential terms were in, but there was, in addition, some additional language that I would not consider to be essential terms of a transaction.

"I don't think it was an essential term that the parties agreed to a press release."

JUDGE FARRIS: "Mr. Brown, are we really talking about the juxtaposition of the word 'essential'?

"Couldn't we shorten this somehow? If that's what we're talking about, maybe it wouldn't be so long."

BROWN: "Maybe we can clarify it this way."

JUDGE FARRIS: "All right."

BROWN: "When you and Mr. Goodrum wrote this Memorandum of Agreement, why didn't you just put in the essential terms?"

LIMAN: "First place, Mr. Goodrum wrote it. And, second, if you ask a lawyer why did he not limit something to its basest terms, that's asking a lawyer, particularly a corporate lawyer, to act out of character.

"I've never seen a document that didn't have more words than it needed and I've often had clients complain that lawyers get paid by the word."

JUDGE FARRIS: "I heard that."

It seemed as if every time Brown had found an opening, Liman would turn it around and make his questioner seem almost absurd.

Taking nothing away from Bob Brown, in retrospect it is obvious to me that the best lawyer in this debate was on the witness stand. It didn't hurt that the facts generally appeared to favor the witness.

For Texaco, this underscored its rancor at Pennzoil's heavy reliance on testimony from a witness who was, in reality, its own lawyer. It was Pennzoil's good fortune that it had an articulate advocate like Liman present at so many crucial battles of the Six Day War.

Brown returned again to the Getty board meeting.

BROWN: "I want to discuss with you now the character of this board of directors meeting that you were attending, not actually in the meeting but in attendance in the adjoining space."

LIMAN: "We understand where I was. I wouldn't call it attending, Mr. Brown, but I understand it. So does the jury."

BROWN: "Was that board meeting, to your knowledge, accompanied by frequent heated discussions?"

LIMAN: "Well, I wasn't in there, but Gordon Getty reported to me that it got heated at some points."

BROWN: "Did he report to you where it got heated or specific points that discussion got heated?"

LIMAN: "The answer to that is yes."

BROWN: "Can you recollect what points?"

LIMAN: "My recollection is that he said some directors on the board made a personal attack on him because he believed that the proposal was in the best interests of the stockholders . . . and that they had attacked him personally and that it was very, very hard for him to take because this was the same board that snuck a lawyer in the side door and voted to sue him.

"And that I have a very vivid recollection of the way Gordon Getty felt when these people—I mean, I wasn't there, but he felt [they] had double dealt on him, all of a sudden attacking him."

JUDGE FARRIS: "Did you say 'snuck'?"

LIMAN: "I said snuck. If that was—"

JUDGE FARRIS: "I thought that it was only Texans that used the word."

LIMAN: "Judge, if I stayed down here any longer, I think I may become eligible for citizenship."

The point Brown had been trying to make lay in ruins in the wake of the erudite Liman and his knack for sensing just the right thing to say.

Brown tried to make the best out of what had turned into a bad situation, gamely pressing his point.

BROWN: "Well, would it be accurate that the board meeting was not just a friendly handshake kind of a board meeting?"

LIMAN: "That's the way I understood it."

Brown was on safer ground when he had Liman review successive drafts of what was called the Transaction Agreement, the final merger document between Pennzoil and Getty. Three drafts of this document, totaling dozens of pages, were passed to the jury while Brown quizzed Liman on the differences between them. Although there were no startling revelations (or revelations of any kind, for that matter), the jury suffered in silence while this review dragged on. Any points Brown had earlier managed to score on Liman were totally obscured by this protracted review.

Toward the end of cross-examination, Brown turned again to the subject of Liman's friend, Martin Lipton.

BROWN: "I think that yesterday or the day before when we were talking about Mr. Lipton . . . I had asked you if he were an honest
lawyer. Do you remember that?"

LIMAN: "Yes. He is an honest lawyer."

BROWN: "Now you seem to have some reservation about that. Did I
detect something that wasn't there?"

LIMAN: "No, you didn't detect it. I think you asked me a question
that was calling for me to—at least, the way I interpret it—
to express an opinion on what he did in this case. And I expressed myself by saying 'except for the issues in this case.'

"But in terms of Martin Lipton's integrity, there's no reservation whatsoever. He's a friend and a thoroughly decent
person."

BROWN: "Your reservation about Mr. Lipton, were you trying to say
that he has done something that's dishonest in this
proceeding?"

LIMAN: "No . . . I was not saying that he had done anything dishonest and I didn't attempt to suggest that. If you want to know
what I was suggesting, I'll say it very bluntly."

BROWN: "I was trying to find out what your reservation is in this
case."

LIMAN: "I think that Marty Lipton was my friend, still is my friend,
and my opinion of him and respect for him is undiminished.

"Martin Lipton was presented by Texaco with an offer he
couldn't refuse. What do I mean by that?

"His client was a seller. He was offered a substantial premium over what we offered.

"Everybody who is a party to a contract has a right to
either perform or to pay damages. You could always breach
and pay damages, and he was offered in this case an indemnity so his client wouldn't have to pay damages . . . His client
was getting more money and instead of having any kind of
risk of damages for walking out of the contract, the buyer,
Texaco, was going to pick that up.

"I don't see how Martin, given the fact that he was trying
to get as much as he could for the Museum, could have
turned down the offer that was made to him.

"And, when I ultimately saw the document he drew with
Texaco, he specifically refused to represent that he didn't
have a contract with Pennzoil. He excluded that and got an
indemnity from them, so I think that Martin put his interest
in getting as much for his client above any kind of obligation

that his client had to Pennzoil and said Pennzoil can resolve that with Texaco since they gave the indemnity.

"That's how I feel. It does not in any way reflect on Martin's integrity. I think he did an extraordinary job for his client in getting them the highest price *and* the indemnity, an extraordinary job."

With this one answer, Liman had managed to articulate the very essence of Pennzoil's case against Texaco and, at the same time, preserve the mantle of friendship and loyalty he had draped around Lipton. It was a balancing act that bordered on the artistic, and demonstrated how cross-examination can be fraught with peril for even the most skillful questioner.

If Bob Brown recognized the damage Liman was inflicting on the Texaco defense, he gave no clue.

BROWN: "What I am trying to get to, Mr. Liman, is this: You've known Mr. Lipton for twenty-five years. Do you think that if Mr. Lipton came to testify to this jury about the facts, that he would tell them something that is not true?"

LIMAN: "He would make every effort, in my opinion, to tell the facts as he recalls them. That's my belief about Martin Lipton."

BROWN: "Do you think that Mr. Lipton, based on your knowledge of his reputation, would involve himself in a conspiracy to cheat Pennzoil out of anything?"

LIMAN: "He wouldn't do that, but he would have done what I just said, in order to further the interests of his client, getting the indemnity."

The interrogation of Liman was winding down, but the issue of Rule 10(b)13 of the Securities & Exchange Commission was raised by Brown.

Rule 10(b)13 is designed to insure that, in a tender offer, all stockholders receive equal treatment. The question of who would buy the Museum shares under the Pennzoil-Getty deal ran into the provisions of the rule. Since the company would be subjected to a major tax penalty for buying the Museum shares, Pennzoil would have to buy them. The tender offer of December 28 was still in effect, so a 10(b)13 violation was a real possibility.

Previous testimony had established that Liman had served as an assistant U.S. attorney in Manhattan, specializing in securities law. He testified that since all stockholders would receive the same price as the Museum, the spirit of 10(b)13 would be met. Therefore, he said, it would be a simple matter to obtain either an exemption or what is called a "no action" letter from the SEC.

In addition, Museum president Harold Williams had served as chairman of the SEC during the Carter administration, another indication that there was no suspicious intent present.

Even Bob Brown appeared marginally satisfied with Liman's explanation. The 10(b)13 issue would not die here, however; months after the trial it would be waved like a bloody shirt by those who were dissatisfied with the outcome of the proceeding. For the jury, however, it was another nonissue.

The testimony of Arthur Liman occurred relatively early in the trial, but it still made a definite impression on the jury.

If we had any doubts about its significance, they were quickly laid to rest. For the remaining three months of the trial, Texaco attorneys went to great lengths trying to discredit Liman. Their efforts served only to remind the jury repeatedly just how convincing Arthur Liman had been.

18

It was the second month of an emotional trial, and tempers occasionally grew short.

Consider, if you will, the natural superstition and the paranoia of trial lawyers, who must ultimately rely on the vagaries of twelve unpredictable citizens empaneled as a jury. In this situation, counsel might read anything into the slightest occurrence.

Much of the conflict took place out of the jury's presence, but the tension was ever present.

Joe Jamail was unrelenting throughout the entire trial. He approached Texaco the same way he had approached the Japanese in World War II. During Arthur Liman's week-long testimony, the trial transcript reveals this exchange, which occurred one afternoon after the jury had been dismissed:

JAMAIL: "Before everybody leaves, we have another problem. Mr. Miller is sitting back over there by the jury and I don't mind that too much, except . . . he is engaging in some sort of conversations with some other jurors as they come through the gate."

In fact, Dick Miller would often seek out a seat in the sparsely populated spectator section when one of his co-counsel was handling a witness or playing a video deposition, claiming he had difficulty viewing documents projected on the screen. Comments were sometimes exchanged when members of the jury had to traverse the cramped quarters of the smallish courtroom.

Although Jamail had temporarily handed the reins of his case to Jeffers and Terrell, he still kept a watchful eye over every aspect of the proceedings. There was no doubt who was in charge of Pennzoil's case, and the division of labor meant Jamail was temporarily free to concentrate on psychological warfare. As usual, his target was Miller.

JAMAIL: "I think that if they get real close to him, they will learn to hate him. But I think that he ought to be–this conversation and this 'watch the gate' sort of thing with the jurors is no more than trying to curry favor with the jury, and I think perhaps he ought to move up to the counsel table, Your Honor."

JUDGE FARRIS: "Mr. Miller?"

MILLER: "I don't believe I ever saw a lawyer so goosey."

JAMAIL: "I am not goosey. I just think that if you are going to try this honestly, you ought to be up here."

MILLER: "You are goosey. I got caught in the door coming through that because one of the jurors had his feet in the way.

"I said, 'I am going to have to watch this door.' It wasn't said to the jurors; I said it in front of everybody that heard me."

JAMAIL: "I take it back. If he wants to try to curry favor with them, he can sit over there."

MILLER: "Thank you very much."

KRONZER: "Let's play by the pool house rules, Judge. One side of the rail."

JAMAIL: "I think it's demeaning, and if that's what he thinks he has got to do, the Court ought to be aware of it."

MILLER: "Is it an objection or not? I can't tell whether there is an objection or not."

JUDGE FARRIS: "[Miller] told me that he had some eye problem, and I said he could sit on the other side."

JAMAIL: "How does that cure his eye problem? What does that have to do with his eye problem?"

JUDGE FARRIS: "Perhaps out there it's a little darker."

MILLER: "Well, do you object or not?"

JAMAIL: "I am having mixed emotions about it. If they get close enough to him, they will hate him. I think he probably ought to sit up here at the counsel table, unless he is ashamed of these people he is representing."

JUDGE FARRIS: "Not as long as he tells me he has an eye problem."

JAMAIL: "Can we move him away from the jury and the Court ask him not to talk to the jury?"

JUDGE FARRIS: "Only the last part."

MILLER:	"I haven't said a word. I have said what I said and it wasn't to the jurors. It was to everybody in hearing, and I said it loud enough for even Mr. Jamail to hear it."
JUDGE FARRIS:	"Well, the next time, watch your leg."
MILLER:	"I will, of course, and get Mr. Jamail a hearing aid."
JAMAIL:	"Judge, how close can I get to this jury? Can I get over there and sit with them?"
MILLER:	"It's okay with me."
JUDGE FARRIS:	"I don't think you can get any closer than the chair at the end of the table."
JAMAIL:	"I may move over there tomorrow."

Many trial lawyers are indeed, by their very natures, a superstitious bunch. That fact, combined with Jamail's war of nerves, produced many clashes like the one just outlined. The comment of Jim Kronzer was typical of the contributions he made to the Pennzoil team. His collection of folksy aphorisms was almost as extensive as his mental index of legal precedents.

In the meantime, there was a lawsuit to try.

After the testimony of Arthur Liman, the courtroom was again outfitted for video depositions. This time, it would be two senior executives of Texaco whose testimony would become part of Pennzoil's case.

Sitting in the darkened room watching videos that were dull by any standards, it was often difficult to stay awake.

The essential component of this round of video depositions was to have admitted into evidence Texaco's handwritten notes about its move on Getty. The first deposition played was that of deputy controller Patrick Lynch, whose notes contained the statements "STOP THE TRAIN" and "STOP THE SIGNING."

A curious footnote to Lynch's testimony is that he was present at the meetings where these notes were taken only because his boss was out of town on vacation. To say that he didn't volunteer any information in this deposition would be an understatement. You had to sympathize with Lynch to a certain degree; his answers were no doubt closely monitored by his superiors at Texaco.

The testimony did establish Lynch as the author of the notes, which were duly admitted into evidence. Their contents, however, were generally credited to Wasserstein and Perella of First Boston. Although not formally retained by Texaco until the evening of January 4, First Boston had been a primary source of intelligence about the Pennzoil-Getty dealings.

The second video deposition was from Texaco President Alfred De-Crane. His handwritten notes covered four pages of a yellow legal pad and would later be called the "smoking gun," proving Texaco's knowledge of and interference with the Pennzoil-Getty deal.

Although noticeably lacking in fireworks, DeCrane's lengthy video deposition had some small interest for the jury. On the screen, you could see DeCrane sitting at the head of a long conference table, his blue suit coat buttoned. To his left sat the court reporter. Dick Miller and Irv Terrell were in shirtsleeves on opposite sides of the table, their voices easily recognizable to the jurors who had been listening to them for weeks.

Terrell kept peppering DeCrane with questions about a wide range of subjects. DeCrane would pause, then give a minimal, precise answer in a flat, unemotional tone of voice.

Even if it hadn't been disclosed in the early questions that he was an attorney, DeCrane's tendency to invoke a lack of knowledge or memory when he didn't want to answer a particular question was itself revealing.

Terrell led DeCrane through the notes practically on a line-by-line basis. The wealth of detail provided to Texaco by Wasserstein and Perella is reflected in DeCrane's notes.

DeCrane would quibble over minor details, but could not deter the exposition Terrell's questions were designed to produce. This exchange is typical of the review of the notes:

TERRELL: "The next item says, 'Alert,' and is it underlined twice?"

DeCRANE: "I do not know. It is underlined."

TERRELL: "It says, 'Colon, isn't an ordinary merger, dash, putting the Museum stock in an irrevocable escrow agreement which will be a lock-out for anyone, dash, only have an oral agreement.'
            "Did I read that correctly?"

DeCRANE: "Yes."

TERRELL: "Whose word is the word 'alert'? Is that the word that you chose to write down or is that the word that was used by Mr. Wasserstein or Mr. Perella?"

DeCRANE: "I do not remember, but I would expect that most of this is a report, that somebody used that word in talking about that situation."

TERRELL: "What did you think was meant by 'only have an oral agreement'?"

DeCRANE: "That there was nothing final, nothing conclusive. That there was a preliminary agreement in principle."

TERRELL: "That's what you understood an oral agreement to mean?"

DeCRANE: "Yes."

Nearly three months later, DeCrane would take the witness stand, the last live witness presented by Texaco. Terrell would again raise the subject of oral agreements with the Texaco president. In a face-to-face confrontation not subject to editing, the results would be considerably more dramatic.

Meanwhile, the video depositions of Lynch and DeCrane had taken the better part of three days. In truth, it had been a long three days, but for the remainder of the trial, the Lynch and DeCrane notes were before the jury.

Texaco's explanations of its actions in the Getty acquisition kept running into the strategy outlined in these notes. There was no particular cure for this conflict, so Texaco devised one: this was not the firm's strategy!

TERRELL: "Was this meeting that we've gone over on the night of January 4, reflected by your notes, a strategy meeting by Texaco with its representatives, First Boston and Skadden, Arps, to determine how to go about acquiring Getty Oil Company?"

DeCRANE: "It was a meeting to review the situation, to get the thoughts from the First Boston people, and then to determine what needed to be done next, so it was not a strategy meeting essentially."

TERRELL: "However, it was a strategy meeting in part because First Boston was making recommendations to Texaco on how to proceed?"

DeCRANE: "First Boston was making recommendations."

TERRELL: "On how to proceed to acquire Getty Oil Company?"

DeCRANE: "Yes. They were making recommendations. We were not sitting down and exchanging views on strategy, so that I would say that it's not necessarily a strategy session as such."

Many of the apparently damaging references were in Lynch's notes. Terrell questioned DeCrane about one passage of the Lynch notes.

TERRELL: "There is a reference in notes that he took on January 5 in a meeting which I believe you were in attendance.

"At the bottom of page 13, where it says, 'If there is a tender offer and Gordon doesn't tender, then he could wind up with paper,' do you remember hearing any discussion on January 5 where this, in substance, was discussed or said?"

DECRANE: "Not those specific words."

TERRELL: "Do you recall anything close to those words in reference to this subject of, 'If there is a tender offer and Gordon doesn't tender, then he could wind up with paper'?"

DECRANE: "At some point, I believe we touched on this before, in talking about the Trust situation and the restrictions on the ability of the Trustee to move or sell or on his freedom of action with the body of the Trust.

"There were discussions of tender offers versus mergers and what could happen, what the result might be under a tender offer if it were a two-part tender offer."

TERRELL: "Whereby Mr. Gordon Getty might end up with paper?"

DECRANE: "No, it was a discussion of how the tender offers proceed and if it related to an interpretation of the Trust agreement . . . "

TERRELL: "Do you recall any plan discussed on the 4th or the 5th within Texaco or with First Boston about which Gordon Getty might be compelled to sell to Texaco because otherwise he might have been left in a minority position and would end up only with paper instead of cash?"

DECRANE: "That was not a Texaco proposal."

TERRELL: "Well, do you recall any discussion of that as possible strategy?"

DECRANE: "No, I don't recall that as a strategy that was proposed by Texaco."

TERRELL: "Or proposed by First Boston to Texaco?"

DECRANE: "No, I don't recall that."

The importance of these notes as evidence of Texaco's thinking would grow as the case progressed. The implications of "leaving Gordon with paper" flew in the face of repeated contentions that Texaco actions in the acquisition were entirely friendly.

DeCrane could never distance Texaco from his notes and those of Lynch, who had faithfully recorded far too much detail for the defense to explain away. Pennzoil skillfully exploited these notes, but the statements they contained were so explicit that even a first-year law student could not have failed to grasp their significance. It certainly wasn't lost on anyone sitting in the jury box.

19

Pennzoil alternated between live witnesses and video depositions throughout the presentation of its case. This resulted in an occasional scrambling of the order of the evidence. But it was not a particularly glaring problem, since many of the witnesses testified about all aspects of the case anyway.

Following the video depositions of Lynch and DeCrane, Pennzoil began to establish its case for the $7.53 billion in damages it claimed it would take to restore the bargain it had lost when Getty Oil was acquired by Texaco.

The first witness on damages was Dr. Ron Lewis, a Pennzoil vice-president in charge of the offshore division. Called late in the afternoon, Dr. Lewis had obviously not expected to take the stand that day. He was the winner going away of my unofficial Ugliest Necktie of the Trial contest, sporting a monstrous green affair with what appeared to be different-sized paisleys with the tails cut off. My dad was an engineer at NASA for many years, so the engineers' fashion ethic was very familiar to me.

I imagined Dr. Lewis arriving home that evening and innocently telling his horrified wife, "Guess what, honey? I went and testified at the trial today."

When he appeared in court again the next morning, Dr. Lewis was attired in an immaculate blue suit and sober tie. As a petroleum engineer, he would testify about Pennzoil's historical finding and development costs for a barrel of oil.

As Irv Terrell had outlined in his voir dire presentation, Pennzoil's damage claim was based on the cost to find and develop slightly more that 1 billion barrels of oil, 43 percent of Getty Oil's proved reserves. Dr. Lewis had prepared a series of charts and graphs showing much more detail than anyone would ever want to know about finding and development costs. Under extensive questioning by Terrell, he described a reporting requirement of the Securities & Exchange Commission for oil and gas companies, where they have to file annually what is called a Form 10-K. These documents then become a matter of public record, providing a source of information on which to base investment decisions concerning these companies.

Dr. Lewis also noted that this was how companies routinely kept up with the competition. He produced a copy of a document put out by the accounting firm Arthur, Andersen that compiled that 10-K data on oil and gas reserves for 375 public companies.

Using a pocket calculator, Dr. Lewis was able to show how the figures on his charts had been derived from the data. In this way, Dr. Lewis was able to testify not only about Pennzoil's reserves and finding and development costs but also those of Texaco.

Pennzoil had charged that Texaco's declining reserves and high finding and development costs were a powerful motive for its interference in the Getty deal. This motive was so pervasive, Pennzoil charged, that it still represented a tremendous bargain to Texaco even if substantial damages were awarded in this trial.

Dr. Lewis was cross-examined by Richard Keeton for Texaco. The testimony had been largely dry and technical, and Keeton's objectives on cross were relatively modest. He wanted to have Dr. Lewis admit that his testimony was, by its very nature, imprecise.

Keeton asked him to extract some additional figures from the exhibits, saying, "Dr. Lewis, did you bring your little calculator with you again today?" The witness held his own, though he did concede some points. Keeton noted that calculating finding and development costs over a limited period of time can give a misleading picture. Twice, he had Dr. Lewis read a sentence from the Arthur, Andersen report that said: "Even a four-year period is too short to make reasonable assumptions of finding costs per barrel. Such an analysis could lead to erroneous conclusions about the performance of specific companies even on a relative basis."

Keeton had not really scuttled the testimony, but had started to shift the momentum a little. When he cut off one of Dr. Lewis's answers, that gave Terrell an opening:

TERRELL: "Excuse me, Your Honor. I let Mr. Keeton cut him off a number of times and I would beg the court's indulgence to let the witness finish his answer and not talk over each other."

JUDGE FARRIS: "Doctor, have you finished your answer?"

LEWIS: "Well, I was going to say that when I made the comment that the industry was experiencing a decrease in

$3/7$ OF GETTY'S ASSETS =

1.008 BILLION BARRELS OF PROVED ENERGY RESERVES

PENNZOIL'S AVERAGE HISTORIC
FINDING & DEVELOPMENT COSTS

$$= \$10\frac{87}{}\text{ PER BARREL}$$

PENNZOIL'S MAXIMUM NET COST
PER BARREL FOR GETTY'S RESERVES

$$= \$3\frac{40}{}\text{ PER BARREL}$$

DIFFERENTIAL

$$= \$7\frac{47}{}\text{ PER BARREL}$$

LOSS TO PENNZOIL DUE TO TEXACO'S INTERFERENCE

1.008 BILLION BARRELS × $7.47 PER BARREL

= $7.53 BILLION

Pennzoil's replacement model of damages is graphically depicted in this trial exhibit.

the cost to drill wells and to do their normal business,
we've done the same thing. We've done that also."

JUDGE FARRIS: "Mr. Keeton, would you be sure when you ask your
next question that you are absolutely certain that the
witness has answered fully."

KEETON: "Yes, Your Honor, and I apologize. I mean no impolite-
ness to Dr. Lewis, of course."

JUDGE FARRIS: "I've got to be careful. You know, the lawyers in the
last evaluation said I was mean not only to the lawyers
but also to the witnesses. I'm not going to be mean to
this witness."

KEETON: "Judge, I didn't even vote in that."

JUDGE FARRIS: "That was for the witness's benefit."

KEETON: "All right. I didn't think you meant me."

The recent Bar Association poll had obviously singed the Judge's
feelings a little, but he still thought being mean to attorneys meant he
was doing his job. In fact, it was obvious that Judge Farris relished the
give and take of the courtroom and kept the attorneys in check most
of the time.

Keeton concluded his cross-examination of Dr. Lewis on a rather
desultory note after one additional question. Apparently he would
leave the task of refuting this testimony to Texaco's damage witnesses.

In keeping with the mixture of live and video witnesses, Pennzoil
next played the deposition of Martin Siegel, Gordon's investment
banker.

Like Boisi, Siegel had an attorney at his side throughout the deposition. Pennzoil's purpose with the Siegel deposition seemed to be to establish that Gordon had received expert advice throughout his dealings
with Liedtke. This was to counter the Texaco claim that Pennzoil had
taken advantage of a naive, inexperienced Gordon Getty.

A graduate of the Harvard Business School, Siegel was the star of
mergers and acquisitions for Kidder, Peabody when that firm was retained to advise the Trust.

In 1982, Siegel's bold move in Martin Marietta's defense against a
takeover attempt by Bendix had seemed headed for disaster. Siegel had
his client respond to the Bendix takeover attempt by launching a coun-
termove to take over Bendix. United Technologies stepped in at the last
minute, rescuing Martin Marietta and cementing Siegel's reputation at
the same time. This so-called "Pac Man" defense had been the center-

piece of what had been the most controversial corporate takeover battle ever—until the Pennzoil-Getty-Texaco affair.

Once again seated at center stage was Martin Siegel. On the video, the jurors could recognize the voices of Terrell and Miller.

Responding first to questions by Terrell, Siegel recounted how he learned of the Pennzoil tender offer when he was vacationing in the Virgin Islands. The scene in Gordon Getty's suite at the Pierre on January 1 was again described. Siegel added one detail to the story that had been told by Baine Kerr, who had also been in attendance.

When discussing what price should be offered in the Memorandum of Agreement, Siegel recalled telling Liedtke that it would take $110 a share; he quoted Liedtke's response as, "How about $105?"

Siegel also added to our knowledge of the exotic language of his trade by referring to the $5 debenture for ERC as a "boot." Boisi and Liman had called it a "stub" or a "sweetener."

On the second day of Siegel's video deposition, John Jeffers had the bailiff leave the lights on while the first few minutes of the video deposition were played. In response to a question from Terrell, Siegel told how he had prepared an affidavit for a court proceeding in California at the request of Trust attorneys.

Jeffers stopped the tape and offered the January 5 affidavit of Martin Siegel as Plaintiff's Exhibit 19. Judge Farris admitted it over the objections of Miller, and copies were passed to the jury. Jeffers read the affidavit into the record, including Siegel's comments that the Pennzoil-Getty deal was "the transaction now agreed upon."

Back on the video, Siegel testified that the affidavit was accurate when he signed it, and he still believed it to be accurate. The alternative to this testimony would have been for Siegel to admit he had perjured himself on January 5, 1984.

Under questioning by Miller, Siegel did introduce enough equivocation to almost effect the stalemate that was the result of most of these video depositions. But Pennzoil had placed that affidavit in the record, and the jury would grow very familiar indeed with its contents.

Miller's subsequent questions led Siegel through the manner in which Texaco dealt with the Trust. He got Siegel to comment on the favorable reputation of John McKinley and how his actions in dealing with the Trust had not altered that reputation in Siegel's view.

Martin Siegel would not be called as a Texaco witness during its defense. The name of Ivan Boesky did not surface during the Siegel video deposition. Both these facts would arouse great interest in my evaluation of the case after the trial.

Based on the evidence presented in court, however, there were two conclusions to be drawn from Siegel's deposition. First, the Trust had indeed been adequately represented during the Six Day War, and second, Siegel, like Cohler, had definitely intended to tell the California court on January 5 that the Trust, the Museum, and Getty Oil had a deal with Pennzoil.

Following the Siegel deposition, Pennzoil called Clifton Fridge to the stand to resume the testimony on damages. While Dr. Lewis had testified as to Pennzoil's historical finding and development costs, Fridge would be called on to calculate the price Pennzoil would have paid for 43 percent of Getty Oil.

This was a complicated equation that involved analyzing the many nonoil and gas assets of Getty Oil and assigning a value to them, taking the cumulative debt of Getty Oil into account, and arriving at a number that reflected the adjusted price that Pennzoil would have paid.

According to the theory of damages advanced by the plaintiff, this historical finding and development cost per barrel, minus the price it would have paid for each barrel of Getty Oil, equaled the true measure of damages . . . when you multiplied it by 1.008 billion. That had been called the "conservative" estimate of 43 percent of the proved reserves.

Like Dr. Lewis, Fridge came equipped with a series of charts and tables. His testimony was dry and unexciting, as would be expected of a CPA who had risen up through the ranks to the upper levels of the Pennzoil financial side.

The valuation of many of the Getty assets was simplified by the fact that they had already been sold by Texaco, so their value was set by the marketplace.

Fridge was an unassuming sort of guy who lived in Pearland, Texas, a small country town south of Houston that had become a bedroom community for commuters as the city expanded. He concluded his direct examination, and suddenly, his world turned upside down.

When Dick Miller rose to conduct the cross-examination for Texaco, Fridge had no doubt been prepared to defend his projections and valuations. But Miller didn't ask him a single question about his direct testimony. Instead, it was "What did Mr. Liedtke say?" What did Mr. Liedtke do?" in the preparation of the tender offer for Getty Oil.

Miller had pressed Kerr for the better part of a week, but had scarcely dented the old lawyer. Fridge was a different matter entirely. While it would seem that the subject of his testimony should arouse no great controversy, the opening to Liedtke was too much for Miller to pass up.

In fact, Fridge *had* been involved in some of the preliminary planning sessions, where the Pennzoil financial people were instructed to submit Getty Oil to some rudimentary fiscal analysis. Instructed to compose a hypothetical study that looked at an acquisition of Getty at

$90 a share, Fridge and company did so. The resulting study was known as a pro forma, sort of a financial "what if" look at a possible deal.

Fridge described the pro formas as relatively simple, "back of an envelope" calculations, and admitted another bit of information: he had done a pro forma that showed that even at $120 a share, Getty was still a good deal.

Miller's eyes lit up like a little boy who had discovered a shiny new bike under the tree on Christmas morning. Fridge tried to back away from the subject, saying it was of no real significance, but Miller wouldn't hear of it. Wheeling out the blackboard, Miller had Fridge reconstruct the $120 pro forma while Miller recorded it in three different colors of chalk.

The hapless Fridge was forced to recount again and again the displeasure Liedtke had shown when told about the $120 pro forma. "I don't want to see that $120 pro forma walking around," ordered Liedtke. If word had leaked out that Pennzoil's own people thought $120 was a good price, the negotiating would be impossible. At this point, Liedtke seemed to be working his way up to the $100 a share eventually provided for in the tender offer. If he thought it would go at $90, it would have been $90.

The $120 pro forma had been duly "discarded," in Fridge's words. "You mean destroyed," Miller shot back. Fridge refused to concede the point.

With all the attention Pennzoil had focused on the "destruction of evidence" (the original Copley notes), here was a document of Pennzoil's that had been made to disappear. We would hear about it for the rest of the trial.

Miller seized this victory with relish. Although Pennzoil's case had not exactly been setting the world on fire, there had been little to cheer about over at the Texaco table either.

Miller had photographs made of the "reconstructed" $120 pro forma he had drawn on the blackboard. Each juror received an 8 × 10" glossy color print for our by-now bulging exhibit folders. The 8 × 10s didn't really do it justice, however; the next day, they passed out 11 × 14" prints.

By this time, I thought Clifton Fridge might head home to Pearland and never venture into the big city again. Miller had not been cruel, particularly, but he had been unrelenting. After all, that was his job.

Fridge had been placed in an untenable situation, strapped to a log and run into a legal buzz saw. His confusion was apparent, not that it generated much of a backlash of sympathy from the jury. We had been subjected to too many weeks of dull, technical testimony to turn up our noses at a little blood letting.

Several weeks after the trial ended, I had a chance to chat with Fridge. Almost four months had passed since his encounter with Miller; Fridge swallowed and managed a smile as I recalled the ferocity of the cross-examination.

On the balance, however, Fridge had done his job on direct examination, getting information vital to Pennzoil's case on damages into the record. Miller's attack had not been on Fridge personally, although it must have been pretty hot in that witness chair. It really served as a preview of coming attractions; Dick Miller was sharpening his sword for Hugh Liedtke.

This and other clashes would soon relegate the Fridge cross-examination to footnote status. I can remember it like it was yesterday, however.

With the exception of the latter part of Kerr's testimony, the evidentiary phase of the trial had been marked with what seemed to the jury like an overabundance of good manners. The guts-and-glory promises of the voir dire had gone largely unfulfilled. The pitched battles were being waged largely out of the presence of the jury, but soon, this too would change.

20

The sole expert witness called by Pennzoil was Thomas Barrow, recently retired as vice-chairman of Standard Oil Company of Ohio (Sohio). As I had duly noted on my juror information card, my wife was employed at Sohio as a contract geological drafter when the trial commenced. I've stated that I had never heard of Liedtke or McKinley before showing up for jury duty; Barrow's name was equally unfamiliar to me.

Most corporate chieftans have not been generally well known in years past; the exploits of a Ross Perot or Lee Iacocca might make headlines, but can you name the men who occupy the top spots of their competitors?

As Irv Terrell had outlined during his voir dire presentation, Pennzoil had originally lined up a man named Pendleton Thomas, former chairman of the board of Sinclair Oil Company and B. F. Goodrich as its expert witness. Thomas had died unexpectedly, prompting Pennzoil to seek out another witness.

Under questioning by John Jeffers, Barrow, a big, distinguished-looking man with a full head of white hair, described how he had been approached by Jeffers and Terrell abut giving expert testimony in two areas: contract formation and damages.

> JEFFERS: "What conditions did you advance as to appearing as an expert witness for Pennzoil?"
>
> BARROW: "First, on the condition of damages, that I would have a completely free reign to review the material available to determine what, in my opinion, was not only the best answer but the method or methods to do that."
>
> "And, secondly, that my identity be kept confiden-

tial until June 1st, at which time I was retiring from Sohio."

JEFFERS: "What arrangement did you make with Pennzoil for this testimony you are giving?"

BARROW: "None."

JEFFERS: "And why was that?"

BARROW: "First, I was not interested in becoming a professional expert witness and have no intention of doing this kind of thing again."

"Second, that I felt the issue of contract was a matter of principle as far as I was concerned and, therefore, I was willing to do it on that basis."

JEFFERS: "Let me read you something, if I might. This appears at pages 2565, –66, and –67 of the trial transcript. It was during the voir dire examination of the jury. . . . It's a statement by Mr. Miller:

"'Now, one of the people whose name was mentioned to you as a person who has agreed to assist Pennzoil in this case and to appear as an expert witness, a man who has been identified to you as Mr. Tom Barrow . . . I recall Mr. Terrell speaking to you about him and telling you that he is not accepting a fee. We're glad to have that information, but I want to give you some additional information about Mr. Barrow so that you can answer a few questions.'

"'As you all know, there's more that one way to receive compensation. It certainly could be in the form of money. It could be in the form of creating friendship and business opportunities. Compensation can take many forms . . . but understand that Mr. Barrow is a director of the Texas Commerce Bank and has been a director for many years. That's of interest to us, the lawyers who represent Texaco, because the Texas Commerce Bank is one of Baker & Botts's clients and we have been told, I think it's right, that the person who asked Mr. Barrow to appear as a witness in this case is a lawyer at Baker & Botts or was the managing partner of that firm, who also served on the Texas Commerce board.'

"Have you followed what I just read?"

BARROW: "Yes, I did."

JEFFERS: "Do you have any comment on it or do you perceive how you were deriving any compensation by virtue of the fact you serve on the Texas Commerce board, as does the gentleman who asked you to meet with us?"

BARROW: "It's hard for me to see any tie between the two. Baker & Botts is not Texas Commerce's principal attorney."

JEFFERS: "Are they your personal attorneys or have they ever been?"

BARROW: "I've never used Baker & Botts for anything."

JEFFERS: "And can you perceive, in this commentary of Mr. Miller's, any benefit you're deriving from this testimony?"

BARROW: "No, I cannot."

JEFFERS: "Mr. Barrow, tell us what personal relationship, if any, you have with Mr. Liedtke or Mr. Kerr from Pennzoil?"

BARROW: "I have met Mr. Kerr once, maybe twice in the past. I would hardly recognize him if he were in the courtroom.

"Mr. Liedtke and I have been acquaintances for quite a few years. We were at the University of Texas at the same time. After the creation of Pennzoil, I was also in Houston during a period of years, and our children went to the same school.

"I've seen him at business functions occasionally. I've never been in his home. He's never been in mine. He's a business acquaintance."

JEFFERS: "Does that pretty well say it?"

BARROW: "Pretty clearly."

JEFFERS: "Is your relationship with any one of the people associated with Pennzoil any closer than it is with people associated with Texaco?"

BARROW: "No. Quite the contrary."

JEFFERS: "You have friends at Texaco, long-time acquaintances?"

BARROW: "Mr. McKinley and I have been friends for quite a few years. We both lived in the New York area together. We belonged to the same hunting club, and we saw each other in that respect quite often. I've been on various hunting trips with him."

JEFFERS:	"Mr. Kinnear is in the courtroom. Have you known him previously?"
BARROW:	"We were both residents of Greenwich at the same time, and our children were also in the same school."
JUDGE FARRIS:	"Is that Old Greenwich or Greenwich Village?"
BARROW:	"Technically, Judge, neither. But Mr. Kinnear's children, [his] son, particularly, has been in my home and my daughter has been in his, unlike Mr. Liedtke."

At the very beginning, Jeffers had to deal with issues raised by Texaco that might lead the jurors to question the motives of the witness. The answers Barrow gave in response to these questions were easy and unforced, and set the tone for the remainder of his direct examination.

When he spoke of not recognizing Baine Kerr, he was seated thirty feet from him. Similarly, he spoke of Jim Kinnear, who was at the Texaco table where he sat throughout the trial. Kinnear smiled at the mention of his name. He would not be smiling as much by the time Barrow concluded what I considered powerful testimony.

Jeffers then offered in evidence Barrow's two-page personal resume. After admitting the exhibit, Judge Farris had a question about the schools listed.

JUDGE FARRIS:	"Mr. Barrow, what is the basic difference between Phillips Andover and Phillips Exeter?"
BARROW:	"Phillips Andover is located in Andover, Massachusetts, and Phillips Exeter is located in Exeter, New Hampshire.
	"Both are private schools and founded by members of the same family, the Phillips family of Boston."
JEFFERS:	"Is one any better than the other?"
BARROW:	"I believe I'm prejudiced in that. My daughter went to Exeter and I went to Andover."

Barrow's areas of expertise apparently extended beyond contract formation and damages. Jeffers initially had Barrow tell the jury how he had received the first Ph.D. in petroleum geology ever awarded by Stanford. From 1951 to 1972 he worked for Humble Oil, serving as president for the last three years. When Humble was consolidated within Exxon, he was a senior vice-president and director of Exxon. After retiring in 1978, he became chairman and CEO of Kennecott Copper. When Kennecott was acquired by Sohio in 1981, Barrow became vice-chairman of that concern.

Jeffers noted that he had served on the boards of Exxon, Kennecott, and Sohio. He led the witness through other board affiliations listed on Barrow's resume, Cameron Iron Works, American General Insurance, Stanford University, Baylor Medical School, the Texas Medical Center, . . . the list went on and on.

Jeffers asked Barrow if his experience serving on these various boards had involved him in the negotiation of transactions and contracts at the highest levels. Barrow answered in the affirmative.

JEFFERS: "Your Honor, at this time we would tender Mr. Barrow as an expert witness for Pennzoil on the issues of contract formation and damages."

KEETON: "On the issue if damages, depending on what he says, I think the man is certainly knowledgeable in some areas.
             "The issues of contract formation, I think, are those the jury is supposed to decide . . ."

Keeton and Jeffers wrangled over whether Barrow could testify about contract formation. Judge Farris admitted the witness as an expert on damages and said he would reserve his ruling on contract formation until he had heard the questions.

Jeffers had Barrow testify that he had reviewed the Memorandum of Agreement and the Copley notes, and then asked him how he interpreted the motion that preceded the 15-to-1 vote.

This brought Keeton back out of his chair, objecting that the question called for the witness to read the mind of Harold Williams, who had made the motion recorded in the notes. Jeffers started to respond, but the judge cut him off.

JUDGE FARRIS: "I have read and I have listened to this witness's vitae, and it shows a lengthy life of serving on various boards.
                  "I will allow this line of questioning based on his experience as having served on various boards. Not as a mind reader."

The description of the Getty board vote in the Copley notes was followed by this entry.

JEFFERS: "Now, if you will move to the last page, page 68. You see where it says, 'The meeting was reconvened at 6:55 P.M. Eastern Standard Time, January 3, 1984, following which

Mr. Siegel advised the directors that Pennzoil had indicated to him that it would accept the counterproposal presented by the board.'

"Have I read that correctly?"

BARROW: "Yes, you have."

JEFFERS: "And what significance, based on your experience, did you attach to that portion of the minutes or notes?"

BARROW: "That the Getty board had made an offer and that Pennzoil had accepted it."

Jeffers had Barrow comment on the fact that Harold Williams, signator to the Memorandum of Agreement, had made the motion leading to the 15-to-1 vote. He then handed him the January 4 Getty press release, reflecting the result of that 15-to-1 vote.

JEFFERS: "Confining yourself to those documents and based on your experience, what did you conclude that the board of directors of the Getty Oil Company had done at the conclusion of its meeting on January 3?"

KEETON: "Again, Your Honor, we would object to the witness expressing a conclusion on this matter, since he is clearly taking the role of a juror. . . . He's not testifying in any expert manner. He's only giving a conclusion that he reaches from the reading and considering limited evidence."

JUDGE FARRIS: "Your objection is overruled. I will allow this line of questioning and the answers thereto just as I will allow your cross-examination of the same witness."

BARROW: "Based on the evidence that I have seen and only as an experienced executive who has served on boards, I would come to the conclusion that the Getty board had reached an agreement with Pennzoil."

The contract formation segment of Barrow's testimony also included the mechanics of how a board of directors makes a binding agreement. As an example, Jeffers asked him questions about the sale of Kennecott Copper to Sohio, which Barrow had negotiated.

Moving on to the subject of damages, Jeffers showed the witness some of the charts admitted during the testimony of Dr. Ron Lewis and Clifton Fridge. Barrow commented on the exhibits, adding an independent voice that bolstered the bona fides of witnesses who were, after all, Pennzoil executives.

Jeffers ran into a problem with two of the charts that had been distributed to the jury.

JEFFERS: "Your Honor, at this time, with the Court's indulgence, I'd like to ask the jurors if they could find their copies of Plaintiff's Exhibits 401 and 403 . . . the title which should have been on 401 is on 403 and the title that should have been on 403 is on 401.

           "The title of 401 should be 'Pennzoil's Worldwide Finding and Development Costs—Years 1980 to 1984.' Your Honor, Mr. Terrell takes full responsibility for this."

TERRELL: "I do not, Your Honor. Mr. Jeffers made me do it."

JUDGE FARRIS: "Why don't you take joint responsibility?"

TERRELL: "I'll take joint responsibility, Your Honor."

JUDGE FARRIS: "I'm waiting for Mr. Jamail there."

JAMAIL: "I don't even know what it is."

I include this little comedy to illustrate the extent to which Joe Jamail had delegated the nuts and bolts of the case on damages. Jamail wasn't kidding; he *didn't* know what this particular exhibit was. His expertise was not in the area of evaluating oil and gas reserves.

The testimony of Lewis, Fridge, and Barrow was necessarily technical; Texaco wasn't going to allow Pennzoil to stack up a bewildering pile of evidence to confuse the jury into awarding billions of dollars. Instead, we were subjected to a series of extremely detailed presentations on how these reserves are quantified and valued. Converting a billion cubic feet of natural gas to a barrel-of-oil equivalent (BCF to BOE) can be a mind-boggling exercise the first time you approach it.

I won't try to pretend it was interesting; it was deadly dull. But Lewis and Barrow were experienced in explaining these intricacies to those not schooled in the field.

After living practically all my life in Houston, I learned more about the oil business in these few weeks of damage testimony than in all the years my wife had been employed in the industry. I would recognize various terms that were used from past explanations of projects she had worked on, but in general, the experience confirmed the wisdom of my decision to pursue a career in another arena.

The fact that there is far from unanimous consent within the industry on the Pennzoil damage claim cannot be solely attributed to antipathy for its case. There are real differences of opinion, even among experts, about how these reserves are valued. The SEC reporting requirements provide only a very broad outline of the true reserve picture; the data is all subject to interpretation.

Jeffers questioned Barrow on the replacement model.

JEFFERS: "Based on your experience, could you comment on the difficulties of Pennzoil or any company in finding, in this day and time and in the future that faces, this billion barrels of proved reserves over time?"

BARROW: "In the United States, I would say it would be very difficult to find a billion barrels. I say that in light of the fact that I'm one of the few that ever have, so I know how tough it is. I don't think it can be done very likely in the future."

JEFFERS: "So they'd have to be found worldwide?"

BARROW: "I would say it would have to be worldwide."

JEFFERS: "And as you search for oil in the so-called frontier areas, the deep waters, do the costs escalate?"

BARROW: "Very rapidly."

What Barrow was saying here is that even the $7.53 billion in the replacement model could be inadequate to restore the billion-plus barrels.

Jeffers had Barrow describe the rich reserves of Getty Oil, as documented in the SEC filings. Based on his analysis of these reserves, Barrow had figured two alternatives to the replacement model of damages.

The first was based on acquisition of proved reserves from other companies, accompanied by a chart showing the cost per barrel of recent oil company mergers, including Marathon, Cities Service, and Superior.

The second model was based on a discounted cash flow analysis of three-sevenths of the Getty reserves. This calculation is not particularly hard to understand, but the explanation takes at least twenty minutes and I choose to omit it here. You should thank me; if you want the details, drop me a card and I'll run it down for you.

Jeffers asked Barrow a series of questions to sum up his expert opinion from the three different damage models he had considered.

JEFFERS: "What can you conclude for us about the range of damages here?"

BARROW: "Well, I think in doing this kind of exercise, there are different ways you can look at it, and I am more comfortable if I look at it on several different approaches, and then look at how much of a range do I have between the different answers I get . . . I think the 7.5 number is a reasonable answer, presently within the range that's correct.

"I think the 6.6 is in the correct kind of range. I have done other studies and maybe ranged from a low of 5 to a high of 10. . . . I think the 6 ½ to 7 ½ or 8, somewhere in that range is a reasonable answer, considering the extremely complex problem that we got to present."

The figure that Pennzoil had established through its in-house witnesses had been sniffed, tested, and finally blessed by Thomas Barrow: $7.53 billion was a reasonable measure of Pennzoil's damages.

I pulled Barrow's resume out of my exhibit folder to give it another look. Was this guy too good to be true? His credentials were extensive; his presentation was authoritative but at the same time dispassionate. Listed among his affiliations was trustee of the Woods Hole Oceanographic Institute. As luck would have it, that week a research vessel from Woods Hole had located the wreck of the *Titanic*.

They say the truth has a certain ring to it, and Barrow's testimony was all bells.

JEFFERS: "Do you feel you appreciate what three-sevenths in Getty Oil Company would have been worth to Pennzoil?"

BARROW: "I would put it . . . it would have made it a completely different kind of oil company."

JEFFERS: "We pass the witness."

After a brief recess, Richard Keeton rose to cross-examine Barrow. He quickly asked him a series of questions to establish that the extra $2.50 a share that Liman claimed he had painfully extracted from Liedtke to make the Pennzoil deal would have added only 8 cents to the cost of each barrel of the Getty reserves.

Keeton asked Barrow if he had any idea how difficult it had been to get Pennzoil to raise the price from $110 to $112.50. Barrow said he did not have enough information to make that determination, but Keeton had made his point anyway.

He then asked Barrow about the values he had assigned to Getty's nonoil and gas assets like coal, copper, and gold. When he mentioned oil shale, it brought this response:

JUDGE FARRIS: "Mr. Keeton, before you go any further, it has occurred to me that while the jury, I'm sure, knows what gold is and oil and copper, and so forth, they might not

	know what oil shale is. Would you get that before the jury?"
KEETON:	"Certainly, Mr. Barrow, would you tell the ladies and gentlemen of the jury what oil shale is?"
BARROW:	"These particular deposits or bodies of clay-type material [are] in a restricted area of Colorado. They carry varying amounts of a material known as carogen, from which you derive an essentially petroleum-like substance.
	"This is a very large energy resource that the government had a large synthetic fuels program trying to start the development of . . .
	"Judge, I don't know if I'm helpful or I'm confusing them."
JUDGE FARRIS:	"I'm sure that if they're confused, that one of them will raise his or her hand and let you know.
	"Go ahead, Mr. Keeton."

Occasionally, the Judge would assume the mantle of Professor Farris, conducting a symposium on the largest corporate contretemps in history for the edification of himself and his sixteen students.

Keeton questioned Barrow about other Getty assets, including real estate, farm land, and vineyards.

BARROW:	"I am not an expert on vineyards."
JUDGE FARRIS:	"Is there one in the house?"
MILLER:	"I think Mr. Kronzer knows quite a little bit about them."

Kronzer looked barely amused by Miller's crack; Jamail predictably shook his head in disgust.

Keeton asked Barrow a series of questions that made the assumption that even though the Trust and the Museum had presented the signed Memorandum of Agreement to the board and voted for it as *directors*, they had in fact reserved their approval of it as *shareholders*. Barrow demurred.

Texaco liked this argument, however, and pressed it with a number of Getty witnesses. It never really caught fire. If the Trust and Museum did not want to participate in the Memorandum of Agreement, why did they sign it?

Total the time spent on this one point with a variety of witnesses and it adds up to many hours; of such issues are six-week trials transformed into nineteen-week marathons.

Keeton attacked Barrow's calculations on the replacement model of damages, saying he had not taken into account the relative value of the reserves. Barrow conceded some ground on this point.

KEETON: "You would have to know more facts, wouldn't you?"

BARROW: "I would have to have more facts."

KEETON: "Where the crude's located?"

BARROW: "That would be an important fact."

KEETON: "Type of crude?"

BARROW: "That would be important."

KEETON: "Production costs?"

BARROW: "That would be important."

KEETON: "Development costs?"

BARROW: "Yes."

KEETON: "How long it was going to take you to actually bring the crude to market, whether it was off shore or whether it was on shore, with a pipeline nearby?"

BARROW: "Yes."

KEETON: "Whether it was oil or gas?"

BARROW: "Yes."

KEETON: "All those things, you would need to know.?"

BARROW: "You would need to know those things to determine the value of a barrel of oil."

KEETON: "But yet, in this chart, it appears to say that Pennzoil lost seven and a half billion dollars and it includes barrels that are foreign, it includes a bunch of barrels that are heavy oil of lower value, and no attempt has been made to show this jury what the comparison with the quality of reserves are between the Getty reserves that Pennzoil claims it would have had and the reserves that Pennzoil actually found with that money.

"You have not done that, have you?"

BARROW: "I am not sure whether that's an argument or a question, Counselor."

JEFFERS: "I object."

KEETON: "Well, that's probably a good point, Mr. Barrow. Have you done that?"

JEFFERS: "I object. Has he done what? I don't understand the question."

KEETON: "Have you, Mr. Barrow, attempted to, in some way, equate the Getty barrels to the Pennzoil barrels found?"

BARROW: "No, and I didn't represent that in that document."

KEETON: "That's fine. Mr. Barrow, let me turn your attention to an area that we were discussing yesterday . . . "

The remainder of Keeton's cross-examination focused on Barrow's testimony about the negotiations that led to the Pennzoil-Getty deal. No further questions were raised concerning Barrow's evidence on damages.

Jeffers spent some time on redirect having Barrow repeat the essential points of his testimony on both contract formation and damages. I'm not sure whether Jeffers felt Keeton had eroded the strength of the testimony or just thought Barrow's high points were so advantageous to Pennzoil that they were worth repeating.

This extended Barrow's testimony to a third day, when Keeton rose to cross-examine him yet again.

KEETON: "Mr. Barrow, as a businessman, you said that once there's a meeting of the minds on all essential terms, then the parties have reached an agreement. Do you recall that?"

BARROW: "Words to that effect."

KEETON: "In point of fact, there are two requirements for a binding contract to come into being. Don't you know that as a businessman?"

BARROW: "Mr. Keeton, what little I know on the subject is what your father taught me. I was an observer in his class. I'm not a lawyer; I never took a law class."

KEETON: "Well, maybe you didn't audit enough of the class."

BARROW: "I may not have."

KEETON: "Maybe my father will be in court and we can find out from him."

Obviously, this was not a major point, but Barrow's response to Keeton summed up a major Texaco problem with this witness. With his background, what motive would he have for manipulating the jury? Or for allowing himself to be manipulated in front of the jury, which really amounts to the same thing?

When he indicated he learned about contracts more than three decades before from Keeton's father, who Judge Farris had pointed out was the former dean of the University of Texas Law School, Barrow just added another star to the firmament that was the depth and breadth of his experience.

His testimony concluded shortly thereafter, with the legitimacy he had bestowed on the Pennzoil damage claim hardly dented. Texaco's damage experts would have their work cut out for them.

"The real antagonists": Lead attorneys Joseph D. Jamail, left, for Pennzoil and Richard B. "Dick" Miller, for Texaco battled mightily throughout the trial. (Photo by Steve Brady)

Baker & Botts attorneys Irv Terrell, left, and John Jeffers
played major roles in Pennzoil's courtroom victory. (Photo
by Virtle F. Bennett.)

Pennzoil President Baine
Kerr, who Jamail said was
"mild-mannered but tough
inside, like a piece of iron."

Judge Anthony J.P. Farris of the 151st District Court, a self-described "sixty-four and one-half year old ex-Marine." (Photo by Steve Brady.)

Key members of the Texaco brain trust, Vice-Chairman James Kinnear, left, and President Alfred DeCrane. (Copyright © 1985 Janice Rubin.)

Pennzoil Chairman, J. Hugh Liedtke, who Texaco attorney Miller called "one of the smartest men in America and one of the shrewdest and toughest." (Copyright © 1985 Janice Rubin.)

Merger maestro Martin
Lipton, a key Texaco witness
whose testimony cinched the
case—for Pennzoil.
(Copyright © 1987 Steven
Mark Needham.)

Former Getty Director,
Laurence Tisch, would later
help CBS avert a takeover—
and ended up running the
network himself. (Copyright
© 1966 Time Inc. All rights
reserved. Reprinted by
permission from Time.)

Gordon Getty as the jury saw him—on TV.

Getty's attorney , Moses Lasky, was visible during parts of the video deposition.

Gordon displayed a full complement of gestures and expressions.

Pennzoil witness Arthur Liman, right, would gain national fame as Senate counsel in the Iran-Contra hearings, where his memorable clash with Lt. Col. Oliver North electrified television viewers. (Photo by Virtle F. Bennett.)

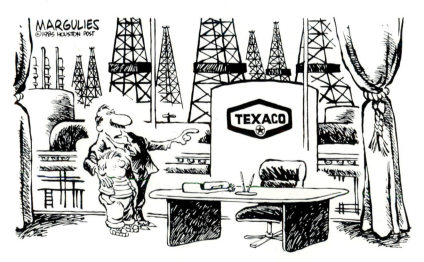

" SOMEDAY, SON ... ALL THIS WILL BE PENNZOIL'S ... "

(Copyright © 1985 Jimmy Margulies, Houston Post.)

21

During the trial, the jury literally never knew what would happen next. This was partly due to the ongoing intrigue between the opposing sides about which witnesses would be called and when. As a result, we were totally in the dark about what was coming when we walked into the courtroom.

On one afternoon in late August, John Jeffers rose amid the by-now familiar array of monitors and cables.

JEFFERS: "We will proceed with the video deposition of Mr. Gordon P. Getty."

At last, we would get a look at the man whose restless ambitions had set this whole drama in motion.

On the screen, we could see Gordon seated at the head of a large conference table, neatly attired in a coat and tie. He had a full head of curly brown hair and looked younger than his stated age of 50. Baine Kerr had commented that Gordon was an unusually tall man (he's 6'5"), but there was no way to determine this from the video.

Gordon's attorney was at his side; Moses Lasky, who had also represented J. Paul Getty for many years, was then nearly 80 years old. Judge Farris, himself approaching 65, once referred to him as "old man Lasky."

Jeffers had handled the deposition for Pennzoil, with backup from Jamail and Terrell. Miller and Keeton were present for Texaco.

JEFFERS: "Mr. Getty, for the record, would you give the reporter your full name?"

GETTY: "Gordon Peter Getty."

JEFFERS: "What is your residence address, Mr. Getty?"

GETTY: "It's San Francisco. Should I disclose the address? Is that
           a good idea? I will if I must, of course."

LASKY: "Not if you don't want to."

GETTY: "It's a home in San Francisco."

JEFFERS: "Is that the only home you maintain?"

GETTY: "It's the only home I maintain, but it would be less than
           fully responsive of me not to add that I have now acquired
           an apartment in New York and have built a home in Wheat-
           land, California."

MILLER: "That's more houses than you have, isn't it, John?"

Jeffers ignored Miller's barb and asked Gordon when he became a
trustee of the Sarah C. Getty Trust (as a result of the death of his father).

GETTY: "That would be June 6th, London time."

This answer was typical of Gordon's testimony, which included extremely precise details about things of no particular import as well as
vague to nonexistent recollections of significant events.

This was understandable, considering anything Gordon said in this
deposition could be used by some member of his litigious family in
another proceeding. Lasky was suitably protective of his client, since
as principal attorneys for the Trust, his firm would have to defend
against any lawsuits that might arise.

It wasn't exactly like marching to Russia, but Jeffers had to wrangle
with Lasky throughout the deposition. Most of this was edited out of
the version played for the jury.

Jeffers did manage to ask Gordon questions confirming the Standstill Agreement, the back-door incident at the November 11 board
meeting, and the Consent. Gordon described learning about the Penn-
oil tender offer and receiving a telephone call from Liedtke.

JEFFERS: "What can you recall of that conversation?"

GETTY: "That his intentions were cooperative and we should meet
           soon."

JEFFERS: "Did you express approval of the idea of meeting with him
           soon?"

GETTY: "Absolutely."

In the wake of the tender offer, Gordon recalled that Museum President Harold Williams had come to his home in San Francisco.

JEFFERS: "What do you recall about the conversation?"

GETTY: "I recall a few things. Again, memory can be tricky. We re-
            member some things that may not be the most important,
            but for some reason we remember them.

            "I do recall him making the point that at $100 a share or
            higher, the Museum was definitely a seller and not a buyer,
            which was a rhyme.

            "Put me among the circle of poets for that. That's the
            lowest circle in hell, by the way."

In the courtroom, Judge Farris responded to Gordon's quip. His unmistakable voice rose out of the darkness, saying, "I heard that."

Gordon had also talked to Martin Siegel, then vacationing in the Car-
ribbean.

JEFFERS: "Can you recall your conversation with him while he was
            still in the Virgin Islands?"

GETTY: "Yes, I can recall a conversation or two with him, yes, sir."

JEFFERS: "Can you recall what the two of you said?"

GETTY: "Well, once again, we remember odd things, but I remem-
            ber he thought that the Pennzoil tender offer was a very
            helpful and promising development. He didn't see it as a
            threat; he saw it as an opportunity."

At Getty Oil, the response to the Pennzoil tender offer was not as
benign. Management proposed a variation of the self-tender for a portion of the public shares that had been talked about for the last half of
1983.

JEFFERS: "Do you recall that there was some effort on the 30th and
            the 31st to see if it were possible to structure an agreement
            between Getty Oil Company and the Trust and the Museum
            which would, in effect, be an alternative to the Pennzoil
            tender offer?"

GETTY: "Something like that, yes, sir. . . . There was a proposal un-
            der which the company would buy some of its *own* shares.
            The Trust would not sell to the company.

            "The result would be to prevent Pennzoil from attaining
            control, whereupon Goldman, Sachs would sell the com-
            pany for the best terms it could arrange within 90 days."

JEFFERS: "Is it correct that you did not agree to such a proposal on
            the 30th or the 31st?"

GETTY: "That is correct. Let me clarify that the proposal was in the process of development. Nothing was ever clarified to the point where one would agree or not."

What Gordon seemed to be saying here is that he never endorsed such a proposal, though it clearly had been made. In fact, witnesses for the company and the Museum would later testify they thought that Gordon had consented to the self-tender. That impression may have come from Siegel or Cohler, or perhaps it was just wishful thinking.

Jeffers then steered the questioning to the January 1 meeting with Liedtke at the Pierre Hotel.

JEFFERS: "Was the meeting a friendly one?"

GETTY: "Certainly."

JEFFERS: "What can you recall Mr. Liedtke saying?"

GETTY: "There was a period in which he spoke of past ties between himself and my father and the Getty companies and Mr. [Jack] Roth . . ."

JEFFERS: "Who was Mr. Roth?"

GETTY: "Mr. Roth was about the number two man [during] my brother George's time. I forget his title."

JEFFERS: "Mr. Liedtke had known him?"

GETTY: "Yes, he'd been close to Mr. Roth, apparently."

LASKY: "That's what Liedtke said."

GETTY: "Yes, that's right. Then he came around to the Pennzoil tender offer and repeated what he had told me on the phone, namely, that his tender offer was progressing very well. It would probably be oversubscribed, that no one thought anyone would ever come in over a hundred dollars and he wondered if he should have been so bold as to go as high as a hundred dollars . . . now, maybe I'm putting words in his mouth there. That was the gist of what he said, that a hundred dollars was a handsome figure, and I'm not quoting him, just the gist of—"

JEFFERS: "I understand."

Gordon described how the meeting had progressed to discussions of a partnership between Pennzoil and the Trust. Siegel had proposed the four-to-three ratio of ownership, which would be accomplished by Pennzoil increasing its tender offer to 24 million shares and the company buying back 24 million shares of the outstanding shares. These proposals had been first discussed the day before in a preliminary meeting between Siegel and Arthur Liman.

GETTY: "And, also, Marty Siegel said that the price would have to go up, that a hundred dollars just wouldn't meet the competition."

JEFFERS: "I gather that this was something you and Mr. Siegel had talked about before the meeting?"

GETTY: "Yes, sir."

JEFFERS: "And he made that suggestion with your full endorsement?"

GETTY: "I made it."

JEFFERS: "Correct. That was your idea?"

GETTY: "It was my idea, sir, as far as I know. I'm not saying it wasn't his, also. It might have been we all thought of it independently, but as far as I know, it was my idea."

JEFFERS: "What do you recall Mr. Liedtke's response to have been?"

GETTY: "Well, he said he took a very dim view—that's not a quotation—of any price over a hundred. A hundred was being heartily subscribed, very well subscribed, so no need to go higher.

"Marty Siegel said, 'Well, it's going to have to be a hundred ten.' When the response to that was negative from the Pennzoil side, Marty Siegel said, 'Well, I know what you feel about a hundred ten, but let's just keep using it for a figure, for a number.' And from that point forward, a hundred ten was the number used.

"I'm not certain I've done full justice to your question. Let's assume that I have, if you don't ask more."

As we had learned, the discussion on January 1 had led to the Memorandum of Agreement. Jeffers next showed Gordon the two copies of this document bearing his signature. Lasky asked if he was saying the two documents were identical; Jeffers replied that the second version was somewhat revised. The jury had both versions that Gordon had signed; the differences between the two were minor and did not materially affect the terms of the agreement.

Lasky and Gordon stopped to conduct a line-by-line comparison. In the courtroom, Jeffers rose and told Judge Farris that counsel for Texaco had agreed to fast forward the tape past this silent comparison. Bob Brown asked if he could have a word with Jeffers.

JEFFERS: "Mr. Brown requests that we reflect for the record
                that Mr. Lasky and Mr. Getty compared those two doc-
                uments for four minutes."

JUDGE FARRIS: "All right. Mr. Lasky is going to be tired, particularly
                at his age."

JEFFERS: "He didn't show any sign of it, Your Honor."

JUDGE FARRIS: "I notice that when Mr. Getty wants to think, he puts
                his glasses on and when he wants to read, he takes
                them off."

In fact, Gordon did have a full complement of gestures and expressions. As Judge Farris had noted, Gordon would sit back in his chair
and listen to the questions. When he was shown an exhibit, he would
remove his glasses and study it closely. When the next question was
asked, he would put his glasses back on, as if they increased his hearing or brain function. When handed a document, he would sometimes
hold it up to the camera like some strange TV pitchman hawking prod-
ucts of a dubious nature.

In general, his testimony was a welcomed change from the almost
uniformly somber parade of lawyers and bankers we had heard thus
far. This was enhanced by the sheer novelty of seeing the central figure
in the case whom we had heard so much about since the first day of
the trial.

The unedited version of this deposition also reveals the inevitable
clashes between Jamail and Miller, who would trade barbs and insults
over some objection or comment.

During these exchanges, Gordon would sit back with a big grin on
his face. His head would pivot back and forth; left, right, left, right, as
if he were a front-row spectator at Wimbledon. Lasky was also amused,
but appeared puzzled at the vicious but familiar way they went at each
other. At one point, he told Gordon, "I'm not sure if [Jamail and Miller]
are old friends or old enemies." Almost all this sideshow was excluded
from the portion offered to the jury.

One such outburst occurred when Jeffers asked Gordon what he
thought the effect was when he signed the Consent. Gordon responded
that he would prefer to let the document speak for itself, prompting
this response:

JEFFERS: "We're trying to make a verbal record here that may be
                tried before a court and jury, and I'd think we'd like your
                own—"

MILLER: "I think that it's certain that it will be."

LASKY: "What did you say?"

MILLER: "He said that it *might* be tried. I can assure him that it *will* be."

JEFFERS: "I didn't mean to express any less certainty on that point than you."

LASKY: "I'm off to the side of that discussion."

MILLER: "Well now, you people have said—"

JAMAIL: "It's not talk. You can count on it, Dick. Now keep quiet. Hush and let the man finish his questions."

When Lasky and Gordon finished their comparing the two versions of the Memorandum of Agreement, Gordon offered this answer to the question of whether he had in fact signed these documents that bore his signature.

GETTY: "Well, sir, both those signatures do appear, and again, the blur of the events of those days . . . . Do I have a vivid recollection of the signing of documents? I can't say I do."

JEFFERS: "But you don't have any doubt that you did, do you?"

GETTY: "I can't think of any reason to doubt it, no, sir."

After a few more questions about what differences there were between the two documents, there was another flare-up, which was also edited from the tape we saw:

MILLER: "What is the reluctance to state what the difference is? Why is there this reluctance?"

JAMAIL: "You can ask him all that when it gets to be your turn."

MILLER: "Well, I think it ought to be in the record so it's clear."

JAMAIL: "You can ask him anything you want to."

MILLER: "This business of hiding the ball on the witness—"

JAMAIL: "There ain't nobody hiding the ball. Your paranoia is such that it just becomes ridiculous. Now let the man ask his question."

TERRELL: "Dick, he took all the time in the world to read them, which is fine. It's his right, but I don't think anybody's hiding the ball on him."

Jamail then offered Gordon an explanation of Miller's outburst.

JAMAIL: "He's got such an extreme case of paranoia. First, he claims that somebody's got a hit list."

Earlier, when Jeffers was asking Gordon about individual directors, Miller had in fact warned him that Pennzoil had a hit list of persons who it claimed had done wrong in the Getty deal.

MILLER: "I didn't say I was at the top of the hit list, Joe. I'm probably where I usually am, bringing up the rear."

JAMAIL: "That slander is merely typical Miller paranoia. I can't think of anybody that would think you were important enough to be on any hit list, if there *were* a hit list. Let's get on with it."

MILLER: "I'm sure you're right."

JAMAIL: "I am."

Back on the record, Jeffers continued the questioning, receiving a mixture of direct and elliptical answers from the witness.

JEFFERS: "Mr. Getty, taking you back to the meeting at the Pierre January 1, where Mr. Liedtke and Mr. Kerr came to see you—

GETTY: "Yes, and Mr. Glanville. And I'm not sure there wasn't also a man from Baker Botts. I'm just not sure whether there was or not."

JEFFERS: "I believe you said that Mr. Siegel did most of the talking for you or something like that?"

GETTY: "Yes."

JEFFERS: "Do you recall in fact whether you said anything other than pleasantries?"

GETTY: "I do recall saying something at the time when Mr. Liedtke stated his intentions about what the management should be . . . that I would be chairman, that he would be president and chief executive officer and Mr. Kerr would be chairman of the executive committee. And although it seems to me I didn't clearly assent, I think I said something to the effect that his position had been noted."

During the deposition of Gordon Getty, the trial passed the six-week mark. It had long since become apparent to the jury that the initial six- to eight-week estimate was far off the mark.

Pennzoil's case had been very detailed and protracted, in part by Texaco's vigorous cross-examination of Baine Kerr and Arthur Liman. The video depositions themselves were a special form of torture, surpassed in their tedium only by the reading of nonvideo depositions.

We had come to recognize the reporters for the two Houston newspapers, the only visible signs of outside interest for most of the trial. On mornings when they would leave as soon as the proceeding started, I knew we were in for a very long day indeed. On those occasions, the attorneys would take thick volumes of transcribed depositions and read the questions and answers aloud, often for hours at a time.

Among the jurors, the weeks had started to take their toll. Lillie Futch had taken ill, running a high fever for several days. Her doctor told Judge Farris she wasn't contagious and could sit around at the courthouse as well as she could at home, so her convalescence was spent in the jury box.

Shirley Wall had some problems with phlebitis in her leg, but not enough have her excused from the trial. She just moved down to one of the seats that had been placed next to the jury box to accommodate the four additional jurors. There she could sit with her foot propped up.

Although I had taken a seat in the jury box itself, I was keenly aware of my status as an alternate juror, albeit the first alternate as the thirteenth juror selected. The measures being taken to retain afflicted jurors dramatically reduced the possibility that any of the first twelve would be excused, giving me a seat in the deliberations.

Against this background, Ola Guy took sick during the playing of Gordon's deposition. She was obviously miserable, and reported that fact to the Judge. He told her to produce a note from her doctor to provide a professional opinion of the state of her illness. The start of the trial would be delayed 45 minutes the next morning to accommodate her doctor visit. I was glad to be able to sleep a little longer before making the trek to the courthouse the following day.

The next morning, Judge Farris discussed the matter with the attorneys before the jury was brought in.

JUDGE FARRIS: "That juror's name is Mrs. Guy. She is Juror No. 4, but she sits in the box over there with the alternates, [the] black lady over there."

JAMAIL: "Yes, I know who you are talking about, Judge. You are asking me how I feel about it?"

JUDGE FARRIS: "Both sides."

JAMAIL: "I think we should wait for the doctor. I think she's a qualified juror, and if the doctor says she can't serve, she can't. If the doctor says she should, she should."

MILLER: "We don't have any objections one way or the other. If the Court excuses her, that's agreeable with us. If she's qualified, we're happy to have her.

	It's a matter we prefer Your Honor to decide."
JAMAIL:	"I don't."
JUDGE FARRIS:	"Well, the issue is joined."
JAMAIL:	"I didn't mean that the way it sounded to you, Judge, but I think she's a qualified juror."
MILLER:	"If he's going to act like that, I'll just take the other side of it then."
JAMAIL:	"I'm saying that if she's healthy enough to serve, she should; if she's not, she shouldn't."
MILLER:	"You're making a mistake on her."
JAMAIL:	"I don't think so."
JUDGE FARRIS:	"Hopefully, she will be back at 10:00. Now, we come to the next item on the same subject. If she's not back at 10:00, how long should we wait?"
MILLER:	"Say about 10:01."
JAMAIL:	"He really likes this juror, Judge."
JUDGE FARRIS:	"That's his favorite."
JAMAIL:	"Judge, I think it's this: If she's not back by then, you should call or have the clerk call her doctor to see when she will be back."
CLERK:	"Ms. Guy is here."
JAMAIL:	"That ought to answer it. Tough luck, Mr. Miller."
MILLER:	"Nice try."

Outside of the trial, Pennzoil attorneys contended that Texaco hadn't wanted any blacks on the jury. Texaco's lawyers denied that, and said Pennzoil didn't want jurors with much education. If there was any truth in these allegations, neither side had been successful. Both groups were amply represented on the panel.

Ironically, Jamail was wrong about Ola. During deliberations, she made it clear that she didn't care much at all for Pennzoil. Left to her own devices, I feel reasonably certain she would have delivered a much different verdict from the one that ultimately resulted.

The jury was brought back in, and the second day of the video deposition of Gordon Getty began.

As his deposition progressed, it became clear that Gordon's memory on some points was just not there. For example, Jeffers questioned him about the $5 "stub" for ERC that was part of the motion approved by the 15-to-1 vote.

JEFFERS: "Do you recall speaking against that stub in any way?"

GETTY: "I don't presently recall having said anything about the stub at the board of directors meeting. I'm not swearing I didn't, I just don't presently recall."

JEFFERS: "Do you recall that the Trust was for it?"

LASKY: "The Trust is he."

JEFFERS: "Mr. Lasky's point is well taken. When I refer to the Trust, I'm referring to you, Gordon Getty, the trustee."

GETTY: "The Trust is me. Thank you, sir. That's true."

On the question of his memory, in fairness to Gordon, it must be noted that this deposition was taken nearly five months after the board meeting; there was little doubt that the Six Day War had involved a chaotic tangle of events for all involved, especially Gordon.

The questions turned to that celebratory champagne toast.

JEFFERS: "I believe it's correct that you went home from the board meeting after it adjourned on January 3 and drank some champagne?"

GETTY: "I believe so."

JEFFERS: "Do you recall what you were celebrating by drinking champagne?"

GETTY: "Yes, sir. Seemed to me there had been a lot of progress in my ambition to raise the value of the Trust."

JEFFERS: "Can you be more specific about what you were celebrating?"

GETTY: "I thought the 14-to-1 vote was progress; that would be specific. I wish I could tell you more clearly what the vote was for, but I can't. But I nonetheless thought it was progress."

Even with this equivocation, Gordon had confirmed that he had celebrated the climactic 15-to-1 vote with champagne. It was incorrectly characterized as a 14-to-1 vote here because of some confusion in an earlier question. In fact, director David Mitchell had arrived at the board meeting late on January 3, in time for that vote.

The day after this toast, the Getty press release would announce the details of the Getty-Pennzoil agreement, including Gordon's elevation to chairman and majority owner. Was *this* what he was celebrating? I think the question answers itself.

The champagne was another piece in the evidentiary mosaic being assembled by Pennzoil. Like the handshakes, the champagne was not the turning point of the case. But these things did start to add up, especially when considered in light of the documentary evidence.

Another point Jeffers raised was a visit Gordon made to Larry Tisch after the Getty-Pennzoil deal was announced about the bill from Kidder, Peabody for the services of Martin Siegel. The fees would total more than $15 million.

JEFFERS: "Was the purpose of your going to see Mr. Tisch to ask his view of the appropriateness of the Kidder, Peabody fee to the Trust?"

GETTY: "Something like that, yes, sir."

JEFFERS: "Do you recall what [Tisch] said?"

GETTY: "Again, snippets of information that might not be terribly useful. I do recall he said that Marty Siegel's a very bright guy, but he is no Albert Einstein."

LASKY: "Relatively speaking."

The relevance of this point is not that Siegel's fee was high; it was the fact that Siegel's calculation of the fee on January 4 was based on the presumption that the deal had already been completed, as was claimed by Pennzoil in this lawsuit.

Inevitably, Gordon was questioned about the subject of the Texaco contingent's arrival at the Pierre on the night of January 5.

Not long after, Museum lawyer Martin Lipton had shown up. There was conflicting testimony on how he had been summoned, but the events of that evening left no doubt that he had been there.

JEFFERS: "Would it be fair to say that when Texaco showed up at this meeting . . . [and] Mr. Lipton for the Museum, and you found out that the Museum was going to sell to Texaco, that you felt that you had no choice but to do so?"

GETTY: "I think that's a fair statement."

Was this scenario the strategy of the DeCrane notes come to life? Bruce Wasserstein, who had been the source of many of the strategic suggestions recorded in the notes, was present in the suite at the Pierre.

JEFFERS: "During this meeting on January 5 in your rooms, did Mr. Wasserstein, Texaco's investment banker, tell you that speed was very important, that it was important that you reach an agreement with Texaco as soon as possible?"

GETTY: "Yes, I believe he said words to that effect."

For good measure, Jeffers had Gordon affirm the Texaco squeeze play yet again.

JEFFERS: "Weren't you going to be a locked-out minority if you didn't sell to Texaco at that point?"

GETTY: "Yes, that was the result. If I hadn't sold to Texaco, the Trust would be a locked-out minority. I think that's correct."

From Gordon, Jeffers had obtained direct confirmation, of sorts, of Pennzoil's theory of the case.

A round of questions from Dick Miller followed.

Miller's questions, like those of Jeffers, covered a wide range of subjects. He started out with some simple background.

MILLER: "Had you ever heard of the Pennzoil Company before this tender offer?"

GETTY: "Yes, sir, I had."

MILLER: "In what connection?"

GETTY: "Well, I am afraid that I visualized it as a company that sold lube oil with Arnold Palmer on television."

MILLER: "You think Arnold Palmer might be chief executive?

GETTY: "No, I was not under that illusion."

During the portion of the video containing Miller's questions, Gordon's clothes suddenly changed color. Jeffers told Judge Farris that the remainder of the deposition had been recorded the day after what we had seen to that point.

Miller had Gordon describe the four directors he had nominated for the Getty board in the wake of the Consent.

GETTY: "Three of the names I did propose were highly respected, famous entrepreneurs, and [those] were Mr. Buffet, Mr. Tisch, Mr. Taubman.

"Then I added Mr. Graham Allison, a man of whom I had the highest opinion, who was the dean of the School of Government at Harvard, and he happens to have a keen interest in business affairs, financial affairs, even though he's not a practitioner."

MILLER: "Not an entrepreneur in the ordinary sense?"

GETTY: "No, he's not."

MILLER: "I suppose if he buys a new car, that's a big deal for him."

GETTY: "It is for me, too."

Miller had a tricky path to navigate in his questioning of Gordon Getty. He had simultaneously to elicit testimony that bolstered the Texaco contention that Pennzoil had taken advantage of a naive, inexperienced businessman while at the same time keeping any such aspersions from being cast in Texaco's direction. In addition, he had to avoid questions that might result in Gordon blurting out some damaging admission.

After leading him again through the Memorandum of Agreement and the "Dear Hugh" letter (to no great effect), Miller's questions turned to Gordon's January 5 meeting with Texaco.

MILLER: "In the discussions that night, I'm speaking of the night of the 5th, did you talk with Mr. McKinley?"

GETTY: "I would say that his remarks to the group of us were addressed principally to me, if that's an answer. And that my remarks to the group of them were addressed principally to him."

MILLER: "I think I understand what you are saying. Were you all in your sitting room at your suite?"

GETTY: "Yes."

MILLER: "And just so that we'll have some understanding, how large a room was that?"

GETTY: "About the size of this one, I would think, sir. If that doesn't help the reporter, I suppose this one is about 15 by 25."

MILLER: "All right, something like that, for sure. That looks like a very close match.
            "And how many people were in the room when you and Mr. McKinley were having your discussions? That would be you, Mr. Woodhouse, Mr. Siegel, Mr. Tisch, correct?"

GETTY: "Uh-huh."

MILLER: "Mr. Lipton?"

GETTY: "Yes."

MILLER: "Mr. McKinley?"

GETTY: "Yes."

MILLER: "Mr.—"

GETTY: "Kinnear."

MILLER: "Kinnear. Mr. Weitzel?"

GETTY: "Yes, [and] Mr. Wasserstein . . . that may run the list. I'm not certain."

MILLER: "Well, I didn't count them. That's eight or ten people? Would it be accurate to say the room was fairly crowded?"

GETTY: "I wouldn't think so . . . I wouldn't describe it as crowded intuitively."

MILLER: "Were people both standing and sitting?"

GETTY: "Not because they had to stand."

MILLER: "No, I understand that, but I mean in groups of that kind, some people prefer to stand."

GETTY: "Well, I think most of them were sitting most of the time."

Gordon seemed to react as strongly to the faint implication that he was an inadequate host as he did to anything else he was questioned about.

Miller cast his net cautiously in a few other directions, but didn't catch anything worth keeping. He then tied up one last loose end.

MILLER: "You mentioned in the course of your answers to Mr. Jeffers's questions something about a conversation with Mr. Liman on the subject of Mr. Medberry.
            "What did Mr. Liman say?"

GETTY: "I hate to say it. The lady is present."

The chivalrous Gordon didn't want to offend the dignity of the court reporter seated right behind him. Miller interceded to obtain her consent for Gordon to speak the unspeakable.

MILLER: "This is nothing other than a search for what happened . . . can you grit your teeth?
            "She's sure she can handle it."

GETTY: "Shall I say what he called him?"

MILLER: "Yes."

GETTY: "He called him a duplicitous prick."

LASKY: "That was mild enough."

MILLER: "That's not with Mr. Medberry present?"

GETTY: "No."

Following that last response, the screens went dark.

JEFFERS: "Your Honor, that concludes the video deposition of Gor-
          don Getty."

With Gordon's testimony, Jeffers had managed to obtain some corroboration for a significant part of Pennzoil's theory of the case.

Gordon's descriptions of the negotiations with Liedtke had generally
confirmed the testimony of Baine Kerr on that same subject. In the
same way, his recollection of the January 2–3 board meeting had no
serious conflicts with Arthur Liman's testimony.

The admissions about the Trust being forced to sell to Texaco
seemed to confirm the grand scheme of things outlined in Jamail's voir
dire presentation. Other circumstantial pieces of evidence like the
champagne toast had not been totally disavowed by Gordon, leaving
Pennzoil enough room to apply the interpretation they had advanced.

For Miller, these admissions had been just enough to prevent him
from effecting the stalemate that resulted from most of the video depositions. Again, Pennzoil's presentation of its case was not exactly setting the courtroom on fire. Still, the tide seemed to be shifting in its
favor, however slightly.

This realization was tempered by the certain knowledge that the
trial had a long way to go.

After a break for the technicians to remove the video equipment, we
lined up to come back into the courtroom and hear the next live witness. In the hallway, I groaned and told Bailiff Carl Shaw, half-jokingly,
that I didn't think I could stand another accountant.

Carl had a deadly serious look on his face as he told me in a low
voice, "Hang on. We're about to go to warp speed."

As we filed into the courtroom, I immediately understood what he
meant. Standing next to the plaintiff's table was J. Hugh Liedtke.

22

Standing next to Liedtke at the Pennzoil table was Joe Jamail. After we took our places in the jury box and sat down, Jamail addressed the court in a loud, clear voice.

JAMAIL: "Your Honor, at this time I call Mr. Hugh Liedtke as a witness for Pennzoil.

"Mr. Liedtke has already been sworn; he was sworn the first day, Your Honor."

JUDGE FARRIS: "Very well."

Even more than Gordon Getty, the name of Hugh Liedtke had been invoked as both a blessing and a curse since the trial began. Unlike Gordon, however, we had seen Liedtke for a week during voir dire. Now, we would hear him speak; the questions would come from Joe Jamail, in his first leading role since voir dire. The spectator section had filled up, and there was an air of expectation. I glanced at Miller; like the entire jury, his gaze was riveted on Liedtke.

JAMAIL: "I want to ask you some questions so that the jury can get some of your background and be better situated to analyze your testimony, judge you, and judge your credibility.

"So I am going to talk to you for a while about your background, where you went to school.

"It may appear it has nothing to do with this case, but it gives the jury a better insight into you and what you are about and gives them a better opportunity to judge your credibility and analyze your testimony.

"So tell us, where were you born?"

LIEDTKE: "Tulsa, Oklahoma."

JAMAIL: "And the year?"

LIEDTKE: "1922."

JAMAIL: "Where did you go to school?"

LIEDTKE: "Well, I went to grade school and then I went to a place called Cascia Hall for four years. It was an Augustinian school, a Catholic school run by a priest out of Philadelphia.

"I happen to be Presbyterian, but about 60%, I guess, of us were Protestant and 40% Catholic.

"Then I went to Amherst, Massachusetts, for four years. I graduated from there and went to Harvard Business School in a compressed course.

"I graduated from there; I was in the Navy for three years, three and a half."

JAMAIL: "Did you serve with the United States Navy in the Pacific during World War II?"

LIEDTKE: "Yes, on an aircraft carrier."

JAMAIL: "What did you do then?"

LIEDTKE: "Then I went to Texas University Law School. . . . In my second year of law school, my wife and I were married."

JAMAIL: "And her name?"

LIEDTKE: "Betty Lynn Derrickson was her maiden name."

JAMAIL: "And is she in this courtroom?"

LIEDTKE: "Yes."

JAMAIL: "You have children?"

LIEDTKE: "Five."

JAMAIL: "Would you tell us their names and ages, please?"

LIEDTKE: "I wish you wouldn't ask me that in front of my wife."

JAMAIL: "Well, do you know their names?"

LIEDTKE: "Yeah, I know their names . . ."

JUDGE FARRIS: "Let me confess, *mea culpa*, I have a hard time remembering our wedding anniversary date."

With these standard preliminary questions, Jamail was putting a very human face on the redoubtable chairman of Pennzoil. Liedtke's deep, rumbling voice wasn't what you would call animated, but we would soon discover that it aptly transmitted a full range of emotions.

JAMAIL:	"You and I have known each other a long time, haven't we?"
LIEDTKE:	"That's true."
JAMAIL:	"And we are friends?"
LIEDTKE:	"That's right."

The rapport between Jamail and Liedtke was itself a subject of curiosity: the high-powered corporate empire builder and his friend, the high-powered attorney who had risen from the ranks of what many in the profession consider ambulance chasers to become the undisputed King of Torts.

Despite my initial wonderment at the retention of a personal injury lawyer to press this litigation, by now there little doubt that Hugh Liedtke's vaunted instincts had served him well once again. Jamail's sense of courtroom artistry had cast this case as a morality play from the first day of the trial. Now, through the testimony of Liedtke, he would transform his contention into a living, breathing entity.

When discussing the Getty-Pennzoil deal during his voir dire presentation, Jamail had made the point that "it wasn't a building that made those promises, it was *people*." The embodiment of the people on the Pennzoil side was now seated on the witness stand.

To get the ball rolling, Jamail had Liedtke recount how he got started in the oil business as a young lawyer in Midland, Texas, in the late 1940s. Together with his brother, Bill, he had handled the buying of leases for an investor from his hometown of Tulsa.

They had done so well with this program that soon they were fronting for a group of investors, ultimately forming their own company.

JAMAIL:	"Now, tell me, what's the name of the little company you say you formed when you and your brother set out?"
LIEDTKE:	"Well, eventually, we formed, with two other men, a company called Zapata Petroleum Corporation."
JAMAIL:	"And who are the other two men that formed that company?"
LIEDTKE:	"One of them was George Bush and one of them is John Overbee."
JAMAIL:	"Would that be the George Bush that is now the Vice President?"
LIEDTKE:	"Yes."
JAMAIL:	"And for how long did you operate under the name of Zapata and what did it do?"

LIEDTKE: "Well, Zapata started out buying—its purpose was to buy
             producing properties.

             "Actually, we bought one that was about, as I recall,
             6,000 acres that had six wells on it, widely dispersed, and
             we spent all our money on one deal, which is not very
             smart.

             "But, that one turned out to be lucky and I think we
             drilled somthing like 127 wells on it with no dry holes."

The jury, indeed the entire courtroom, was engrossed by the tale
Liedtke was weaving. In the Texas of my youth, stories of oil riches
were more likely to revolve around some farmer or family with a
wretched piece of land that turned out to have some potential for production than any of the Jett Rink fantasies concocted by Edna Ferber.

Liedtke's 127 wells without a dry hole represented the ultimate version of that story. The corollary to such a tale concerns the guy down
the road with a similarly situated piece of land that contained no oil
at all. It wasn't enough to be good; you had to be lucky. Apparently,
Liedtke was both.

Jamail had Liedtke continue with the story of his early business exploits after that first major discovery.

        LIEDTKE: "We were in debt then, of course, and so we raised
                     some more money and started buying additional prop-
                     erties.

                     "We formed an onshore drilling company called Za-
                     pata Drilling Company in, 1957 or 1958.

                     "At a later date, we formed Zapata Offshore Com-
                     pany which is here in Houston, and at still a later date,
                     George Bush came to Houston and took the drilling
                     company part and we owned 43% of that, we sold off
                     to his uncle's firm and some of his associates."

         JAMAIL: "Did Mr. George Bush continue to be a partner with
                     you after that date or not?"

        LIEDTKE: "No. . . . When he went into politics, he was nobody's
                     partner in the oil business. He sold out, totally."

         JAMAIL: "Okay. I am not sure you meant that the way it came
                     out."

 JUDGE FARRIS: "Do you want the court reporter to read it back?"

         JAMAIL: "No, I am going to leave it alone, Judge. I don't want
                     it."

This last exchange provoked much laughter in the courtroom, but the points about George Bush were well taken. After making a lot of money in business with Liedtke, Bush had come to Houston to enter politics. He had served as a congressman from a west side district, one of the few Republican officeholders at that time. In fact, Bush is still generally regarded as a moderate in his adopted hometown. On balance, the Liedtke-Bush connection struck a generally positive note in the jury box.

In the early 1960s, Liedtke would make the connection that would be his entry into the big time—the South Penn Oil Company.

JAMAIL: "Before we get to the South Penn Oil Company, which I think had a direct connection that we want to connect up with the Getty transaction ... did you know anybody at that time connected with the Getty Oil Company?"

LIEDTKE: "Oh, yes."

JAMAIL: "Would you tell us who they were?"

LIEDTKE: "Well, I knew J. Paul Getty. I knew George Getty, his oldest son, and I knew a man named Jack Roth who was the senior vice-president; he was the operating head of the company.

JAMAIL: "Tell us, if you will, how you happened to know J. Paul Getty."

LIEDTKE: "Well, I saw him, I guess, first in Tulsa when my wife and I were—before we were married, when some neighbors had a party and he was there and I saw him there.

         "Her father knew him. And then I saw him still later in London, outside of London where he lived."

JAMAIL: "How did you become involved in South Penn?"

LIEDTKE: "Well, I had been studying just various companies and I would come across South Penn. It looked like kind of a dead situation. It had a number of oil properties. It had a good name for making motor oil under the name of Pennzoil, which was a subsidiary. But it hadn't been going anywhere. It was paying out, making $1.90 and paying $2.00.

         "And so then I noticed that Paul Getty owned 9.9%, just under 10% of it."

Liedtke described how he became convinced that South Penn had potential for greatly increased earnings with a more aggressive management. He went to Jack Roth and George Getty and asked them to consider letting him have a crack at running it. This message was relayed across the Atlantic to J. Paul Getty.

LIEDTKE: " . . . And the word came back, 'Tell him to buy as much stock as I have got and I will put him in.'"

JAMAIL: "Could you do that?"

LIEDTKE: "No. I said, 'Jack, I can't raise that kind of money. I don't have it and I can't raise it.' And he told Mr. Getty that. Mr. Getty said, 'Well, you tell him to buy till I say stop, and then I will think about it.'

"So we did. We put together a little group in a company called Choctaw, which was just a private group of people, and bought some stock and told him that we had gone as far as we could go.

"So then he said, 'Well, all right. Put them in.'"

This abbreviated story of how Hugh Liedtke became chief operating officer left out a few small details; for example, the fact that there was an existing management in place at South Penn that was hardly thrilled when he showed up in Pittsburgh to take the reins.

But Liedtke had again proven prophetic. He turned South Penn around, more than doubling the stock price within two years. At that point, J. Paul Getty "allowed" Liedtke to buy him out; his confidence in the young oilman from Tulsa had not been misplaced. This was the birth of the modern corporation which Liedtke would rename Pennzoil. Liedtke moved the headquarters to Houston in the mid-1960s and set off on a program of expansion.

With a minimum number of questions, Jamail allowed his client to tell his story.

LIEDTKE: "One of the principal things that I have been interested in over the years is trying to cause Pennzoil to grow. Different people like different things, and for me, and for a long time, my brother and Mr. Kerr, our interest has been in growth and not in liquidation."

Through acquisitions and discoveries, Pennzoil had indeed grown and taken its place in the *Fortune* 500. As an oil company, however, it still lagged far behind the industry giants, Exxon, Shell, Mobil, and Texaco.

In 1983, the rumblings that were shaking the foundation of the Getty Oil Company were detected at Pennzoil's Houston headquarters. Liedtke ordered his lieutenants to gather as much information about Getty Oil as was available, mostly from public filings and news accounts.

From this information, the financial people at Pennzoil were instructed to perform some rudimentary fiscal analysis of Getty Oil, what has been called a pro forma or a "what if" calculation. One of the limited group assigned to that effort was Clifton Fridge.

Recently, the Pennzoil accountant had occupied the seat where his chairman now sat. Fridge had been subjected to heated questioning by Dick Miller about one such pro forma.

JAMAIL: "Did you ask for one that called for $120 a share of stock?"

LIEDTKE: "No."

JAMAIL: "Did you get mad when they showed you one calling for $120?"

LIEDTKE: "Yes."

JAMAIL: "Why?"

LIEDTKE: "Well, look. If you were going to go to an auction, let's say you and your wife are going to go to one of these auctions where they sell furniture and she likes some kind of a table or a dining room table. Well, you don't call up the auctioneer and say, 'Hey, I might bid two hundred bucks on that thing.' You don't tell him anything. You don't want him to know what you are thinking about.

"We didn't want to take any chances at all that somehow a rumor would get started that we were prepared to go a lot higher . . . that would be like shooting yourself in the foot.

"Quite frankly, Mr. Jamail, you can do these pro formas pretty well on the back of an envelope and to have that type of thing floating around. . . . We've got all sorts of people in and out of our building, and there are a whole gang of people in the investment business who make their living in what they call arbitrage and they're arbitrageurs, they call them, and they make their money by getting little hints as to what people are going to do. . . . They are reputed, at any rate, to go through your trash or every other darn thing looking for hints."

Risk arbitrage was not an unknown quantity in 1985 by any means, but the revelations of the Ivan Boesky affair that would widely disseminate details about this curious trade were still a year away.

Liedtke's fear of the "arbs" notwithstanding, the $120 pro forma was still a stone in his shoe. Jamail asked him if he concluded from that pro forma that it was still a workable transaction at that price.

LIEDTKE: "Yeah, it was doable at a hundred and twenty. It was doable

at a higher figure than that, but I didn't need to be told
that, I didn't think."

JAMAIL: "And since you had not requested that, you became some-
             what irritated, I take it, about it?"

LIEDTKE: "I think sometimes people get too zealous and they can—"

JAMAIL: "As a matter of fact, it looks like I'm going to irritate you
             all over again."

LIEDTKE: "Well, I don't think they intended to be. They are very
             bright guys. They're very helpful."

JAMAIL: "Well, let me ask you this: Did you tell them you didn't
             want that floating around?"

LIEDTKE: "Sure."

JAMAIL: "Okay. And you've told us the reasons why?"

LIEDTKE: "Yes."

The jury would never have bought Hugh Liedtke as a kindly old
shopkeeper, but the full-bodied portrait presented in this testimony
had a depth and complexity that might have seemed risky in another
type of case.

Here, however, it was incumbent upon Pennzoil to demonstrate just
what damage had been inflicted when it lost the Getty deal. Taken in
that context, this story of the building of the company provided a context for measuring what that loss of Getty had meant.

Did Liedtke have a few rough edges? Of course he did. He was blunt
and direct and definitely understood the exercise of power. In that
sense, he reminded me of Lyndon Johnson, a man whose unique qualities were obscured during so much of his presidency. Like LBJ, Liedtke
made things happen.

Miller's insinuations to previous witnesses that Liedtke was some
kind of a corporate bullyboy were addressed by Jamail in due course.
He asked Liedtke a series of questions that brought a step-by-step account of Pennzoil's actions into the record. These events had been described by Kerr and Liman to varying degrees, but, as told by Liedtke,
the story lived again, with some emphasis on the ethical foundations
of these actions that Miller found lacking.

JAMAIL: "Explain, if you will, for us—we've heard it before and we
             don't want a detailed or scientific explanation of the tender
             offer that Pennzoil made in this regard. What was it?"

LIEDTKE: "Well, I'm not a very good one to explain tender offers to
             you.

"A tender offer, quite frankly, any time you buy any-thing, it's a tender offer. . . . You tender. You make an offer, a proposal. And when somebody says, okay, I'll sell you my hundred, you've got an agreement.

"Now, usually they don't call, you know, offers to buy a hundred shares or even a thousand a tender, but that's what it is technically."

The Pennzoil tender offer had resulted in that meeting between Liman and Siegel at Glanville's office in the last days of December. Liedtke described his reaction when the results of the meeting were relayed to him.

LIEDTKE: "Mr. Glanville said that they were interested in having a structured meeting.

"And I said, 'Well, what on earth is that?' And he said, 'Well, they want to present to you an outline of what you think you ought to talk about, what you want to talk to Mr. Getty about and then they'll go up and review it and take up their review with Mr. Getty.

"And just to be very blunt about it, I said, 'Like hell they will. I'm going to deal directly with Mr. Getty and I'm not going to deal with these third parties. I don't want them telling Mr. Getty what I think or what I'm offering. I want to sit right across the table from him and tell him myself, so there won't be any question about who said what.'"

JAMAIL: "Did you have any objection to Mr. Getty having his law-yers and financial advisors with him when you talked with him?"

LIEDTKE: "None whatsoever. And, in fact, he did."

Arthur Liman had testified that Liedtke did not deal with agents, a fact that was thunderously reaffirmed by this testimony.

Jamail had tried to introduce into evidence a *New York Times* article written in the wake of the Consent of December 5. Miller had vehe-mently objected, forcing Jamail to postpone introduction of the exhibit until he could argue the point outside the presence of the jury.

The next morning, Jamail tried again.

JAMAIL: "Mr. Liedtke, yesterday I asked you a question about some of the things that you had read . . . that led you to believe that there could be an arrangement of an agreement between Pennzoil and the Getty Oil Com-pany. Do you recall that?"

LIEDTKE: "Yes."

JAMAIL: "And one of the things you told the jury and Judge Farris was that you read a *Wall Street Journal* article dated December 11, 1983. Do you recall that?"

LIEDTKE: "Yeah I recall it . . . It was *The New York Times* of December 11 . . .''

JAMAIL: "Is this the article that helped you formulate your motivation and intention to make a study or continue a study acquiring some of the Getty assets?"

LIEDTKE: "Yes."

JAMAIL: "Your Honor, I introduce Plaintiff's Exhibit 436 at this time."

MILLER: "We wish to enter an objection on the grounds that the document is hearsay."

JUDGE FARRIS: "Mr. Jamail."

JAMAIL: "Your Honor, I thought that yesterday you ruled that he could testify about it. . . . It is not hearsay. It goes to Mr. Liedtke's state of mind.

"Texaco has made an issue as to the motivation of Pennzoil. . . . It is relevant. It is not hearsay. It is a compilation of historical facts and data that Mr. Liedtke and Pennzoil used in making up its mind to acquire part of Getty."

JUDGE FARRIS: "Anything else, Mr. Miller?"

MILLER: "No, sir."

JUDGE FARRIS: "Four thirty-six will be admitted."

JAMAIL: "Now, Mr. Liedtke, I want you to look at 436. We are not going to take the time to read it. The jury has it. They can read it when they desire to read it.

"But, I want you to look at page 1, paragraph 2. Do you recall reading that?"

LIEDTKE: "Yes."

JAMAIL: "I want to have you read outloud paragraph 2."

LIEDTKE: "'These riches, however, have not brought peace to the world of Getty. Quite the opposite. The beneficiaries of J. Paul Getty's worldly endeavors, the company, the museum, and associated heirs are embroiled in a Byzantine, confusing, and still unfolding battle for Getty Oil. Many principals remain silent, but inter-

> views with their lawyers and financial advisors on both coasts and a study of Court documents that are quickly piling up reveal a high-stakes and high-tension board room drama.'"

Nearly two months into the case, Pennzoil managed to get into the hands of the jury this lengthy newspaper article recounting the whole Getty saga that spilled the blood in the water. Under the thin guise of showing Liedtke's "state of mind" at the time of the tender offer, the jury was given what seems to be sweeping confirmation of much of the earlier testimony, from *The New York Times,* no less.

Jamail had Liedtke read the paragraph detailing Gordon's effort at the London meetings to persuade the Museum to join with the Trust to control the company. The aim was to deflect Texaco's contention that the "Dear Hugh" letter was an unconscionable—and unprecedented—attack on the Getty board.

JAMAIL: "Did you have any hand in suggesting in October of 1983 that Mr. Getty attempt to oust the current Getty board of directors?"

LIEDTKE: "No."

Having Liedtke read that paragraph to answer that question was somewhat ridiculous, but Texaco continued to point to the "Dear Hugh" letter as a mark of shame. While it was hardly Pennzoil's finest hour, it was certainly understandable in the context of the prolonged struggle between Gordon and the other Getty directors.

Jamail had Liedtke sum up these answers on his "research" by describing the conclusion he reached:

LIEDTKE: " . . . it was obvious something was going to happen. . . . You could see this conflict between the three different parties; you could see the breakdown in morale of the company."

All of this was supposed to make Pennzoil's tender offer for 20 percent of the Getty stock seem almost inevitable, I guess. Some of the arguments were obviously more persuasive than others, but they all found their brief place in the sun during this marathon trial.

Another possible Texaco defense that strained credibility concerned paragraph 5 of the Memorandum of Agreement, which simply stated that the agreement must comply with "applicable regulatory requirements."

Apparently their idea was that the deal wasn't binding because these regulatory requirements had not yet been fulfilled (a federally mandated waiting period was required before the stock purchase could be finalized).

Jamail sought to nip this idea in the bud, again by having Liedtke ridicule it. Of course, he had to deliver these quasi-legal opinions as "a businessman."

JAMAIL: "Based on your experiences in the business world and your education in that field, can you tell us, please, whether or not the Memorandum of Agreement . . . as you understood it at the time you signed it, was it necessary that 5 be complied with before you yourself intended to be bound by this agreement?"

LIEDTKE: "I felt that we were bound. By we, I meant Pennzoil Company and I think that the Getty Trust—"

MILLER: "That's all the witness was asked. We object to anything further."

LIEDTKE: "Well, now, let me finish the answer to the question, please."

JUDGE FARRIS: "Just a minute, sir."

JAMAIL: "That ended the question for me, Judge. I'll ask him another question."

JUDGE FARRIS: "All right."

JAMAIL: "Now, based on your experience, had you seen language such as expressed in 5 in instruments such as this?"

LIEDTKE: "Yes, Mr. Jamail. Let me give you an example. You all around here are all Marines. They used to assign you occasionally to serve on a summary court martial board, I got that every now and then. They'd come in and they'd charge some poor guy with something, and the first thing they'd always say is, 'Is the complainant in due form and technically correct?'"

"Now, that's what that kind of thing means to me. The guy was going to be charged all right, but you've got to have the thing in accordance with the rules and regulations, in that case, of the Marine Corps."

Jamail and Liedtke were on a long march through the Pennzoil case, and through their prism, everything became clear.

Later, the clock was turned back to 1934 as the original circumstances of the creation of the Sarah C. Getty Trust were recalled. The fifty-year-old document was in evidence; every member of the jury had a copy.

LIEDTKE: "It is an agreement between J. Paul Getty and his mother, Sarah C. Getty. . . . Mr. Getty's mother is reputed to have felt that [he] was one of the last of the big-time livers. He had gone through two or three wives by this time. He had five, I think, total and she felt that to conserve the estate of her husband, George Getty, who had spent some fifty years building up the Getty Oil Company, who had died in 1930, this [trust instrument] was filed in 1934—"

JAMAIL: "Is it a public document?"

LIEDTKE: "Oh, yes."

JAMAIL: "And anybody can get it if they want to get it?"

LIEDTKE: "Yes, you can get a hold of it very easily.
          "Anyhow, they entered into an agreement [that] gave him complete latitude [with] the corpus or the assets of the Trust. But she made one provision . . . the only reason they could sell was to avoid a substantial loss to the Trust. That is a financial loss."

JAMAIL: "Don't sell unless faced with dire economic loss?"

LIEDTKE: "That's right."

JAMAIL: "Well, was the Pennzoil proposal one that would have required the Trust to sell any of its stock?"

LIEDTKE: "No. It was designed to *not* cause that. The Trust would have to sell no stock."

JAMAIL: "Under the Texaco proposal, what happened to the Trust stock?"

LIEDTKE: "In my view, the Trust was forced to sell and frightened into selling."

JAMAIL: "Well, was the Trust stock sold?"

LIEDTKE: "Yes, it was."

JAMAIL: "And if it were sold in compliance with this Trust, in your opinion as a businessman, based on your knowledge and your training in your chosen profession, if the terms of the Trust were being conformed with, was the Getty Trust stock sold to Texaco because it was faced with a dire economic loss?"

LIEDTKE: "I think that's what the Trustee obviously believed. Otherwise, he would be violating, one, the wishes of his deceased father, and number two, be violating the law because this Trust has the force and effect of law, I think."

The DeCrane and Lynch notes would also reflect specific knowledge of this provision of this Trust instrument, fretting that "there seems to be no way to get Gordon on base first" if he complied with the terms of the Trust. Jamail had Liedtke cement that provision into the case with this exchange. The intricacies of the Trust are relevant since, as trustee, Gordon was constrained by them.

This harmonizing between the chairman and his friend the lawyer was comparable to two jazz musicians exchanging leads. Jamail would blast a handful of staccato notes from his trumpet, to be answered by a towering riff from Liedtke's baritone sax.

Nowhere was this more apparent than when Jamail had Liedtke recall what he had learned about how Texaco got Gordon to sell that night at the Pierre Hotel. After getting the Museum lined up first, they went to see Gordon with the friendliest of intentions. When that didn't work out . . .

LIEDTKE: " . . . up comes Mr. Lipton and he tells him. Obviously, he wasn't up there just to pass the time with good old Gordo in the middle of the night.

"Having made a deal with Mr. Kinnear and with Texaco to sell his stock, [Lipton] goes up there and tells him, 'Gordon, if you don't sell we are going to put the pants on you.'"

JAMAIL: "Leave you with paper?"

LIEDTKE: "Leave you with paper and leave you in the big trouble . . .

"Now, ask yourself what other reason could Mr. Lipton have been up there for? What other reason did they dispatch Mr. Kinnear first to get Mr. Lipton lined up for? It's perfectly obvious what they were doing and they *did* scare Gordon Getty."

JAMAIL: "Well, let me ask you something. The testimony in this case from Mr. Getty's lips, and I made a note of it, was that Mr. Lipton truly was there, that Mr. Siegal was there, that his other people were there, that after conferring with Mr. Lipton and the Texaco people, that he was forced to sell his shares.

"Would that be what you are now telling us. . . . Would he then be able to be a majority holder or he would be relegated to a minority?"

Of course, a question like that isn't really calling for an answer at all. Through the direct examination of Liedtke, Jamail made his strongest affirmative case of the trial. In a very real demonstration of the teamwork that marked Pennzoil's presentation, Jamail had Liedtke's words resonate off the earlier testimony elicited by Jeffers (from Kerr and Gordon) and Terrell (from Liman, Boisi, and DeCrane).

What Texaco actually did was spelled out in explicit terms, again and again.

> JAMAIL: "Can you think of any other reason for your loss of opportunity and agreement with the Gettys, other than Texaco's involvement and inducement?"

> LIEDTKE: "No. Had they not interfered in the way that they did, and guaranteed everybody what they thought was more money, and guaranteed that if they were sued for damages by us that they'd pick up the tab on everybody, I think that is what caused it and I think that they forced Gordon Getty to sell.
>
> "You know, Gordon Getty, the Trust, paid a billion one hundred million dollars in taxes it would not have paid under our deal."

Not only was Gordon forced to sell, Liedtke charged, but the Trust was socked with a hefty tax bill that even the higher price could not offset.

As for the Texaco claim that it had been asked to bid by Getty Oil's investment banker, Jamail had Liedtke remind him of the Getty bylaws spelled out in the Consent that required the approval of fourteen of sixteen directors for such an action.

> JAMAIL: "Do you see *anywhere* in those Getty minutes compiled and edited by Mr. Copley ... that the Getty board authorized Mr. Boisi or any investment banker or *anybody* to go out and solicit bids to sell Getty Oil Company?"

> LIEDTKE: "No. In fact, there's [a] denial in there by Mr. Petersen, that that's not what he was doing even though that's what he *was* doing."

> JAMAIL: "But is there any authority that you see anywhere in those minutes?"

> LIEDTKE: "No, sir, not a scintilla. Not a bit of it."

> JAMAIL: "Would a company like Pennzoil have any interest or incentive to enter into an agreement such as it's entered into

with the Getty interests, as it was approved by the board of directors on all its essential terms, if the only meaning of such agreement was that the other parties could use it as the starting point to go shop it and trade it?"

LIEDTKE: "Of course not. At some point, the trading stopped. And when you have a deal, when you've agreed to something, which we had done. We had met the requirement; we'd done what they asked us to do, the shopping stopped. You have a deal, that's the end of it."

Self-serving testimony? Absolutely. But in light of the evidence Pennzoil had paraded past the jury for almost two months, powerful testimony, and for one reason more than any other: *it made sense*. Of course, the fact that it immediately followed the video deposition of Gordon Getty was no accident, and this testimony tended to portray Gordon's plight in dramatic terms.

The Jamail-Liedtke duet was no mere dog-and-pony show, although those elements were involved. Rather, it was an intricate, sophisticated tapestry of fact, opinion, and conjecture that rested on a broad evidentiary base. In addition, it was riveting theater after too many long days of concentrated tedium.

Liedtke had even managed to growl at his own attorney more than a couple of times.

JAMAIL: "And I have discussed this matter with you on many occasions, is that not correct?"

LIEDTKE: "That's true."

JAMAIL: "Could anybody tell you what to say up there?"

LIEDTKE: "Mr. Jamail, I'm supposed to tell the thing as best I know it and nobody is going to tell me what to say or put words in my mouth, if I can avoid it."

JAMAIL: "Well, I'm getting pretty convinced of that."

To conclude his direct testimony, Jamail handed Liedtke several of the exhibits introduced during the testimony of Dr. Lewis, and then asked him a series of questions that allowed him to review Pennzoil's actual damage claims. Liedtke also testified about the $48 billion Texaco saved over what it would have cost to find the oil it bought from Getty.

Finally, Jamail had Liedtke confirm the motive Pennzoil had attributed to Texaco.

JAMAIL: "Mr. Liedtke, as an executive faced with that production record that Texaco has, can you tell the jury whether or not in your opinion that higher cost than average to find and develop oil and the revisions downward of eight hundred million barrels, whether or not in your opinion that would motivate Texaco management to grant indemnities and special warranties and enter into such an arrangement as they did with Getty?"

LIEDTKE: "I think so, Mr. Jamail. In fact, I would be positive in my own mind. That's a desperate situation."

JAMAIL: "Pass the witness, your Honor."

Dick Miller rose to begin the cross-examination of Hugh Liedtke, quite possibly the turning point of the trial. He had been vigilant throughout the direct examination by Jamail, but had been able to do little to stop their smooth exchange. He had bided his time, knowing that his turn would come. He confronted the Pennzoil chairman.

MILLER: "Could you tell us what documents that you reviewed in preparation for your evidence?"

LIEDTKE: "I reviewed my own testimony. I, a long time ago, looked at what Mr. McKinley had to say and what Mr. Getty had to say.

"I have read the sanitized minutes of Mr. Copley. I have read the board minutes of the Pennzoil meeting of the 19th and also the 5th. I have read the Texaco board minutes of the 5th.

"I have had available, Mr. Miller, all the documents and I may have looked at others from time to time. But I don't recall any great emphasis on them."

MILLER: "I believe you remarked during your examination by Mr. Jamail about those records and I think you referred to them then as 'sanitized' minutes, did you not?"

LIEDTKE: "Yes, that's right."

MILLER: "All right. I am going to ask you some questions about where you got your information that the minutes were, as you say, 'sanitized.' Now, you mean by 'sanitized,' I suppose, cleaned up."

LIEDTKE: "That's right."

MILLER: "And you say 'sanitized' in the sense that the minutes have been distorted?"

LIEDTKE: "That's right."

MILLER: "And probably matters have been removed from them?"

LIEDTKE: "That's correct."

MILLER: "Or possibly matters put in them that did not occur?"

LIEDTKE: "That's correct."

MILLER: "And who do you charge has done that?"

LIEDTKE: "My feeling, Mr. Miller, from what I can tell, is that after visiting Texaco on the 7th and 8th [of January], why, Mr. Copley went down and met with Dechert, Price.

              "In any event, they then destroyed the original minutes. That's what the records show. And when people destroy stuff that is in evidence, that's in a case where there is a lawsuit pending and they know it's evidence, I think that's a very strong case that they sure don't want them around for some reason.

              "That's not legal, as I understand it. Maybe it is. The court can instruct on that, of course, and I am sure it will."

MILLER: "Has Pennzoil ever caused any investigation of Mr. Copley to take place?"

LIEDTKE: "No, of course not."

MILLER: "Is there anything that Pennzoil has discovered about Mr. Copley that you think he would do a criminal act like that? I mean, has he got a criminal record to your knowledge?"

LIEDTKE: "Well, you asked me. I will answer that question.

              "On the 4th, Mr. Kerr and I discussed with Mr. Getty [about] certain Getty officials who we might want to keep in the picture, if we possibly could. . . . When we got to Mr. Copley, Mr. Cohler suggested that we keep him on for a while and then get rid of him because you couldn't trust him.

              "Now, that was told to me. So that might answer your question, Mr. Miller."

MILLER: "And the person who told you this is this lawyer from San Francisco?"

LIEDTKE: "[He] was Mr. Getty's attorney . . ."

MILLER: "All right, sir. So do you know anything other than that gossip about Mr. Copley that would cause you to think he would destroy evidence, anything?"

LIEDTKE: "Nothing other than his actions, which I think speak much louder than any words I might hear about him."

MILLER: "All right, sir. Now that constitutes all of the evidence that Pennzoil has got against Mr. Copley?"

LIEDTKE: "That's it."

MILLER: "All right, sir, and as I understand what you said, that Mr. Copley then took the notes to the regular counsel of Getty Oil Company, the Dechert, Price firm, and they got in on it then. Is that correct?"

LIEDTKE: "I think it happened at that time."

MILLER: "Do you accuse any particular lawyer there of having done this?"

LIEDTKE: "I'm not accusing anyone, except saying that the notes were destroyed. I don't know who flushed them down the toilet. Maybe they were put in a shredder, but they got rid of them.'

MILLER: "As a matter of fact, when we took your deposition, Mr. Jamail said they set fire to them, didn't he?"

LIEDTKE: "Well, maybe they did . . . they *were* pretty hot."

Miller couldn't win with this approach. Liedtke was never going to concede an inch on this point, nor could he afford to. The Copley notes were an important piece of evidence, and they contained something to offer comfort to each side. Pennzoil's advantage with their attack on the destruction of the original notes was that it prevented Texaco from simply offering the defense that the words "binding contract" do not appear in the notes.

Still later, Miller went to ask Liedtke a question about some event described in the Copley notes. When he went to hand the notes to the witness, Miller couldn't locate them.

"Why do we have so much trouble finding this exhibit?" Miller wondered aloud.

"Those notes do have a way of disappearing," Liedtke growled with a perfect deadpan.

After the jury had been dismissed that afternoon, Miller complained to Judge Farris that the family and friends of Liedtke had packed the courtroom and were seated adjacent to the jury, leading the laughs in response to Liedtke's frequent wisecracks.

This provoked a response from Jamail:

JAMAIL: "Is Mr. Miller upset that they've taken his seat over there by the jury?"

MILLER:	"I thought that was mine."
JUDGE FARRIS:	"Mr. Jamail."
JAMAIL:	"Of course, I haven't heard a cheering section except the people from Texaco, those chickenshits sitting back there from Texaco. Get that on the record as I said it."
JUDGE FARRIS:	"Do you know how to spell it, Miss Miles?"
COURT REPORTER:	"Yes, sir."

The war of nerves continued unabated for the rest of the trial. Jamail's reference to the Texaco cheering section would have made sense to the jury, if we had heard it.

During the long weeks of video depositions and technical witnesses, the courtroon was frequently bereft of any spectators not connected with the trial.

The Texaco contingent, which included several company lawyers and observers who did not openly participate in the trial, would take seats spread through the small spectator section. When they would respond to a Miller witticism, the effect would be a basso chorus of laughter.

If their intention was to appear as a spontaneous group of unrelated chucklers, they never quite made it. Every morning as we sat on the benches out in the hall awaiting our summons to file into the jury box, we would see this group get off the elevators with Kinnear and Miller and his partners.

On the other hand, if it was a sore subject with Jamail, maybe they accomplished their goal after all.

When the cross-examination resumed, Miller went right back after Liedtke. He took the same *New York Times* article he had fought to keep out of evidence and asked some questions designed to rebut Liedtke's unflattering description of Getty chairman Sidney Petersen. Liedtke's description of his own past links with J. Paul Getty had also rankled Miller.

MILLER:	"And the last sentence of the article is a quotation from Mr. Petersen that says during the last five or six years of Mr. Getty's life he was fairly close to him and that he was a product of that environment."
LIEDTKE:	"That's Mr. Petersen's view as he expressed it in that article."

MILLER: "And when the remark was not being made to try to prove something in court?"

LIEDTKE: "I don't imagine it was."

MILLER: "Yes."

LIEDTKE: "Probably made to bolster his ego."

Liedtke had maintained that Getty Oil's outside ventures were not prudent business policy for an oil company. He attributed those ventures to Petersen.

Miller pointed out that, in fact, those acquisitions had been made under Petersen's predecessor Harold Berg, whom Liedtke knew and had cited as a good oilman.

LIEDTKE: "Well, at the same time, it was after Mr. Berg had brought in the financial man, Petersen, who didn't know anything about oil and gas. He was a financial man and that's the type of thing that he would urge Mr. Berg to buy, in my opinion. And I would attribute that acquisition to him."

MILLER: "Well, Mr. Petersen had been with the company twenty-five years. He certainly ought to know something about the oil business."

LIEDTKE: "Well, I suppose he knew how to post profits, whether to put it on the debit side or credit side of the ledger. I suspect he knew that."

Liedtke didn't give ground gracefully, if at all, on this point.

LIEDTKE: "... I think the public record of oil companies getting out of their field is pretty well known, and I think the facts there speak for themselves as to the wisdom of that as the long-term policy."

Another theme Miller had sounded early and often was the matter of protection for the Getty employees. Texaco's acquisition of Getty Oil had meant a large reduction in the Getty Oil work force. The merger agreement had spelled out termination benefits and other limited protection measures for the Getty employees.

The Texaco argument was since the drafts of the Pennzoil-Getty merger document didn't contain such provisions, the important issue of employee protection was not addressed.

Liedtke responded that under the Pennzoil plan, none of the Getty employees would be terminated—with the exception of top management, who were already protected. He contrasted this with the fate of the Getty rank and file under Texaco.

Nevertheless, Miller continued to press the issue of protection for the Getty employees:

MILLER: "That would be an important and essential element in any negotiation, would it not?"

LIEDTKE: "Yeah, and protected more than just termination pay.
           "You know, termination pay is just fine; it's perhaps generous. A termination plan is just great, but you know, you can die by lethal injection. That's fairly painless. Or you can be electrocuted. That may hurt a little bit more.
           "In either case, you are dead if you get fired."

Another of Miller's objectives was to respond to the Pennzoil characterization of the missing handwritten originals of the Copley notes as "destruction of evidence."

This response took the form of a broad-based assault on the credibility of many *Pennzoil* documents and witnesses. Miller claimed the board minutes of Pennzoil's December 19 board meeting were deceptive because they didn't mention any pro formas. He confronted Liedtke with the document:

MILLER: "Well, in addition to that, if a witness gets on the witness stand, and this is the only piece of paper you have, he's free to say, 'I don't know. I don't remember.'"

LIEDTKE: "Well, I must say that's pretty farfetched. These minutes are legally sufficient minutes. They were designed this way for the purpose of maintaining confidentiality.
           "They were not written with the idea of in any way impeding your cross-examination of me or any other Pennzoil witness. That was the purpose of it and I think you know that, Mr. Miller."

During the cross-examination of Fridge, Miller had started to advance the idea that Pennzoil was the real destroyer of documents in this case. The reference was to those $120 pro formas that had been shown to Liedtke before that board meeting. Miller charged that Liedtke had concealed the information in his earlier deposition.

LIEDTKE: "I have told you about the pro formas repeatedly, Mr. Miller. The only thing I told you [was] that I didn't recall their being presented in writing, but I have told you repeatedly that the board was aware of them."

MILLER: "You've told me that in this case, but not when your deposition was taken, Mr. Liedtke. Now, you know better than that."

LIEDTKE: "Well, Mr. Miller, I can't . . .

JAMAIL: "If he is going to continue this argument without asking questions, he can point out specific instances where he may have asked him about pro formas and show it this way. But this is totally unfair to this witness to sit here and argue with him, and I object on the basis that it's argumentive.

"If he's got a question, I ask the Court to ask him to ask the question."

MILLER: "Then, I think it's fair comment to say that the minutes were drafted in an artful way by Mr. Perry Barber, who's not listed as a witness in this case, to conceal the fact of the existence of the pro formas."

JAMAIL: "He's listed as a witness. That is another Miller inaccuracy! Mr. Miller is trying to insinuate to this jury that Mr. Liedtke held something back on him. Mr. Miller never once asked him about a pro forma and I asked him to point out in that deposition where he ever asked Mr. Liedtke about a pro forma.

"What I'm objecting to is his argumentative posturing before this jury trying to say that Mr. Liedtke misled him. Mr. Liedtke never misled him."

MILLER: "Now, Mr. Jamail is quite right. I didn't ask you any questions because I didn't know about them. You people have destroyed every single copy. Isn't that true?"

JAMAIL: "Your Honor, we're getting into this argument. He wants to make another argument. May I respond?"

JUDGE FARRIS: "Yes."

JAMAIL: "If Mr. Miller has a question, he should ask this witness a question. If he wants to make jury argument at this point, prolong this trial needlessly, and that's exactly what he's been up to for three weeks, I object to the argumentative nature of this, and if he's going to argue, I ask the Court to lift this ban on the witness and let the witness respond in argument."

Miller and Jamail battled mightily throughout the trial, but the intensity was increased while Liedtke was on the stand.

Another facet to the pattern of deception Miller was alleging concerned the Purpose clause (paragraph 11) of the tender offer, which he had already called misleading and false. In addition, he was now charging Liedtke had concealed his own role in the preparation of that clause in his deposition.

MILLER: "Well, I asked you two questions about paragraph 11 on page 301. You told me that your reading of it the first time was in my presence.

"I asked you again the following day . . . two more questions about the authorship and contents of paragraph 11, and you never, not one time, told me that you had drafted on that paragraph, did you?"

LIEDTKE: "Mr. Miller, again, I repeat that you are talking in terms of this document, did I see the final "Purpose of the Offer," paragraph 11? I testified I did not see that document, but you never asked me if I had worked on that and it didn't occur to me to say, 'Well, Mr. Miller I worked on the purpose clause of it.'

"If you'd asked me, I would have been happy to tell you. It was no big secret. I don't see anything wrong with it."

MILLER: "But you did not identify yourself as one of the authors, did you?"

LIEDTKE: "You didn't ask."

MILLER: "I'm not supposed to ask your questions for you. The Judge has so instructed."

LIEDTKE: "I'm sorry, Mr. Miller. I wasn't paying enough attention to your golden words."

Liedtke was more than capable of counterpunching Miller's jabs, although Jamail generally preferred to do it for him. At one point well into the cross-examination, Miller returned from a break with his fly unzipped. As he walked back and forth in front of the jury, some of the female jurors were driven to distraction. A note was hastily scrawled and passed to the bailiff. Carl passed it on to the Texaco table, where Keeton summoned Miller over to pass the word. Miller turned to the judge.

MILLER: "I'm told I need to make some repairs. May I be excused momentarily?"

JUDGE FARRIS: "You can make repairs any time, Mr. Miller."

After stepping out into the hall for a minute, Miller returned with a grin on his face.

MILLER: "Well, I thought somebody would have spoken a little bit louder."

JUDGE FARRIS: "You know, one day one of the people in the audience suggested that I ought to adjust my wig. I need one, but I don't have one."

The high drama of the trial was frequently broken by low comedy of this sort.

I would later read that Jamail said Miller had staged the unzipped fly incident, and that it was a rip-off of a tactic he (Jamail) had pioneered. It did succeed in distracting the jury for a few minutes, which may have been to Miller's advantage.

Little else in the cross-examination of Liedtke could be construed that way, however. It wasn't so much that Liedtke was cutting Miller to ribbons; for the most part, he wasn't. It was that almost every effort of Miller's to cast Liedtke as the heavy inevitably backfired.

One such incident occurred late one afternoon as Miller tried to get Liedtke to acknowledge that the negotiations with Getty Oil were tainted by the fact that the Pennzoil tender offer was still in place, a situation Miller had said was described as a "gun at the head" in takeover parlance.

Miller had tried the same line of questioning on Baine Kerr, who wanted no part of it. After an initial denial, Kerr had finally said he may have heard the term before.

When the question was put to Liedtke, he said he had never heard of it. Miller unwisely tried to impeach Liedtke with the less than ringing affirmation of Kerr. This brought Jamail to his feet.

JAMAIL: "Your Honor, I make this objection with a statement that these are Mr. Miller's terms that he attempted to get Mr. Kerr to agree to, but Mr. Kerr said, 'I may have' after Mr. Miller's questioning.

"These are Mr. Miller's terms, and if the question is clear in that regard, I have no objection to Mr. Liedtke answering it the way it's put."

JUDGE FARRIS: "Mr. Miller."

MILLER: "Would you say that it's an advantage or a disadvantage to have a gun at somebody's head while you are negotiating an agreement?"

LIEDTKE: "Well . . ."

MILLER: "That's really not too hard, is it?"

LIEDTKE: "No, it really isn't, but I must comment . . ."

JUDGE FARRIS: "Is that one of those 'say what' questions?"

MILLER: "State what, yes."

LIEDTKE: "Or 'stop beating my wife.' Quite obviously, Mr. Miller, I don't believe that a tender offer is holding a gun at someone's head, and I have testified this morning I never heard that phrase before.
            "I don't believe it's true, and I don't believe it's used frequently."

MILLER: "I wondered about that."

LIEDTKE: "Well . . . I have testified, I'm under oath that I have never heard that before you brought it up today."

MILLER: "Well, it's clear, isn't it, Mr. Kerr has heard of it?"

LIEDTKE: "Well, after you badgered and badgered him, he said he *may* have heard of it."

MILLER: "Badgered and badgered him? I asked him two questions about it!"

LIEDTKE: "Mr. Miller, I don't want to argue with you."

MILLER: "Would you like to look at it?"

LIEDTKE: "I don't need to look at it. It's been presented to the jury. They can make up their own minds about your tactics."

MILLER: "I'm sorry, I didn't hear you."

LIEDTKE: "I said the jury can make up its own mind about your tactics."

MILLER: "Mr. Liedtke, you have me at a disadvantage, sir. I cannot answer you."

JAMAIL: "Your Honor, I object to any response to that, because the jury *will* have to make its mind up about Mr. Miller's tactics."

MILLER: "That's agreeable."

JUDGE FARRIS: "I can understand you not liking that last remark from this witness, Mr. Miller, but the fact remains that the jury has heard all of this discord, yours, Mr. Jamail's, reading of a Q and A of Mr. Kerr and the witness.

"Now we have reached the end of the day, so we will stop on that bittersweet note."

This wasn't the end, of course. Miller would cross-examine Liedtke for something like seven days in all. The Pennzoil chairman was obviously no shrinking violet, but he grew weary of Miller's repeated provocations.

The intensity of this testimony was having a wearing effect on the jury as well. The no-holds-barred excitement of the early days of the Liedtke testimony had given way to something of a siege mentality.

The pressure was building on every side. The first cracks appeared in the jury box, of all places.

The long lunches and numerous interruptions in the trial resulted in the jurors spending many more hours together outside the courtroom. Like many central business districts, downtown Houston can be an alienating environment. Outside the workplace, there is little reason for the average Houstonian to come downtown. There are virtually no residences there, nothing you could really call a neighborhood feeling.

Harris County is so large that jury duty is one of the few occasions for many people to come downtown. Having worked there for ten years, I didn't have that problem. I was comfortable downtown and knew how to take advantage of its hidden amenities. The sense of isolation in the midst of chaos, however, caused problems among those who had no other reason in the world to spend countless weeks in the immediate vicinity of the courthouse complex.

A social order seemed to emerge among a group of jurors who spent a lot of this idle time together. With my daily trips to my office and the library, I remained outside of this circle.

Diana Steinman, the shy 29-year-old temporary office worker from Long Island, had been unhappy at the way the trial had progressed. When the original six- to eight-week estimate had passed with still no end in sight, it exacerbated the unhappiness she had been feeling.

I would frequently catch a ride with Diana in the afternoons; her route home went past my street. Although we couldn't discuss the case, we did talk some about the people involved, what we thought of Judge Farris, the lawyers, and so on.

She was one of the group that was, in effect, stranded together when the trial was not in session. Sometimes on the way home, she would express anxiety about the way some of the other women on the jury would ask her questions during lunch. It seemed innocent enough to me, but Diana said she felt she was being tricked into making personal revelations.

There was more to it than that, however. As a temporary office worker, she had been accustomed to sporadic employment. She was not working when the trial commenced. When selected as a juror, however, it was certain she was going to be out of work for a long time. Unlike many members of the jury, including myself, she wasn't drawing a salary while she watched this story unfold.

After seeing how Lillie, Ola, and Shirley had not been excused from the jury despite physical hardships, it was obvious that Judge Farris wasn't going to discharge anybody for any reason short of a major crisis.

Near the end of Miller's cross-examination of Liedtke, we showed up one morning and were greeted with the announcement that "Juror No. 7 was unable to continue." Judge Farris added that she was under strict instructions not to discuss the case with anyone.

That evening, I called Diana at home to thank her for the rides she had given me. When she answered the phone, she told me that she couldn't talk to me and she had to call the court to tell them I had called her. I said, "I hope you're all right, and by all means call the court."

Before the trial began the next morning, Carl came and told me Judge Farris wanted to see me. I knew exactly what he was going to say. Seated behind his desk, he boomed, "Mr. Shannon, I thought I made it quite clear yesterday that no one was to contact the juror that was excused."

"No, Your Honor. You said she was prohibited from discussing the case," I explained. "I called her to express my thanks for giving me a ride home many times during these past two months. I explicitly complied with your instructions, as I have since the trial began. The case was never mentioned."

Judge Farris and his little lecture could not diminish my spirits this day, however. I was certain that getting off the jury would soothe whatever wounds had been inflicted to Diana's psyche.

The announcement yesterday had taken me completely by surprise. Despite Judge Farris's shell game with the jurors and alternates, I had known from the day they picked the jury that as the thirteenth juror selected I was the first alternate. The tension of the Liedtke testimony was now doubly enhanced by the knowledge that I had just made the final twelve. When it came down to the end, I would have a say in the matter after all.

The cross-examination of Hugh Liedtke ended with several attempts by both Miller and Jamail to negate the effectiveness of the other. The heavy advantage here went to Jamail, since he had the witness on his side. Liedtke had been a strong advocate for Pennzoil's case, which was nearing the end.

Immediately after Liedtke stepped down, Jamail was on his feet: "I call Mr. McKinley as an adverse party." Texaco's chairman would be the final witness for Pennzoil. As befit Jamail's reputation for style and daring, he would conclude his case with "dueling chairmen."

23

John McKinley had gone to work for Texaco in 1941, became president in 1971, and now reigned as chairman. Like Liedtke, McKinley had been present during the jury selection. As has been noted, Jamail took advantage of this appearance to have him subpoenaed as a Pennzoil witness.

Strategically, this move had two advantages. First, it ensured that the Texaco chairman would indeed testify. Otherwise, Jamail thought Miller would find an excuse not to put him on the stand. Second, McKinley would face sharp questioning from Jamail before he was led through his story by his own lawyers.

Earlier, I commented that in many ways Hugh Liedtke reminded me of a Lyndon Johnson–type character. I am now obliged to point out that McKinley bore an uncanny physical resemblance to LBJ.

Jamail set out to establish the lay of the corporate landscape with McKinley.

JAMAIL: "As chief executive officer of Texaco, Mr. McKinley, would you be the one who would be in charge of company acquisitions?"

McKINLEY: "Yes, in the sense that the chief executive is the ultimate responsible employee-officer in that connection."

JAMAIL: "And you would be the one, in this instance, that would be in charge of the Getty acquisition, would you not?"

McKINLEY: "Yes."

JAMAIL: "Is it Texaco's policy, sir, the interfere with other people's contracts?"

McKINLEY: "No."

JAMAIL: "All right, sir. We'll come back to that.
"And you would stop that, and it's your duty as chief

executive officer to do so, if there is evidence that leads you to believe that Texaco is interfering with someone who has a prior agreement with the same party that Texaco is negotiating with? Is that correct or not?"

MCKINLEY: "I think the answer is yes, I would stop it. The question or statement got so long that I might not have understood all the variations of it."

JAMAIL: "All right, sir. Let me ask you something. Does agreement in principle mean anything to you?"

MCKINLEY: "Yes, it means agreement in concept, an agreement to negotiate."

JAMAIL: "It's an agreement, a meeting of the minds of the parties to negotiate and work out whatever is left to be worked out?"

MCKINLEY: "The meeting of the minds part I don't . . . I can't endorse that because I don't know the legal meaning and so forth of that. You asked me what I thought it was."

JAMAIL: "Sure."

MCKINLEY: "I told you, but I didn't know the legal meaning of an agreement in principle. I'm not a lawyer. But to me it means an agreement in concept and an agreement to go forward with negotiations."

JAMAIL: "In good faith?"

MCKINLEY: "I don't know that there is in the agreement in principle the concept of good faith. I don't know that. But to me personally it means fairness."

JAMAIL: "What is your definition, if you will just give it to us, of good-faith dealings?"

MCKINLEY: "It's being fair in your business."

JAMAIL: "And that's it? Could you elaborate?"

MCKINLEY: "No, I don't know anything else to say about it."

JAMAIL: "Let me ask you: Is being deceitful, in your opinion, a definition of good faith?"

MCKINLEY: "Not in the way I would behave."

McKinley's words seemed harmless enough, but Jamail would manage to attach an air of menace to the soft-spoken engineer who ran this powerful corporation.

After asking a few rounds of these preliminary questions, Jamail focused on the way Texaco became involved in its bid for Getty. At every step of the way, he elicited responses detailing McKinley's intimate involvement.

> JAMAIL: "Did the words good faith ever come up by any of these bankers you talked to that told you Getty was free to deal?"
>
> McKINLEY: "I don't recall that specific word, but they were pointing out that they were legally open and were seeking bids."
>
> JAMAIL: "Well, I'll come back to that in a minute, but let's stay on what. . ."
>
> McKINLEY: "They approached me."
>
> JAMAIL: "Mr. McKinley, you've said that several times. But are you taking the position that if someone approaches you, an educated man that's reached the top of the ladder in your business, in your company, that if someone approaches you with a suggestion or a proposition that you should do something wrong, are you saying that you're just going to go ahead and do it because he approached you first?"
>
> McKINLEY: "I'm certainly not saying that, and I didn't do that at all."
>
> JAMAIL: "Well, you keep saying that. . ."

This is how Jamail kept McKinley on the defensive. For the jury, the first exposure to the top man at Texaco came with Joe Jamail as our tour guide.

He repeatedly framed questions that made McKinley uncomfortable. He asked him to characterize the effect of the Texaco bid:

> JAMAIL: "What Texaco finally did ended up resulting in a defeat of whatever this agreement was that Pennzoil had with Getty?"
>
> McKINLEY: "You like the word defeat. I don't have any problem with it, but it was not a purpose of ours and it wasn't my word. We were out to make a bid."
>
> JAMAIL: "Well, let's zero in on Pennzoil.
>   "Texaco successfully prevented Pennzoil from acquiring Getty . . . is that not correct?"
>
> McKINLEY: "The completed definitive merger of Texaco and Getty certainly would not permit the anticipated concept of that agreement in principle to take place. Both things couldn't be done at the same time."

JAMAIL:	"Texaco's success, very simply, defeated whatever this arrangement that Pennzoil had with Getty."
MCKINLEY:	"I don't have any problem with that, if you'd like to phrase it that way."
JAMAIL:	"Is your answer yes it did?"
MCKINLEY:	"Well, you keep using the word defeat. We weren't out to defeat Pennzoil and we were out—"
JUDGE FARRIS:	"Mr. McKinley, if you understand the question and it is an obvious yes or no question, answer it yes or no, and if you care to explain it, explain it. But first answer yes or no."
MCKINLEY:	"All right, sir."
JUDGE FARRIS:	"Ask it again, Mr. Jamail."
JAMAIL:	"Texaco's success, and because of Texaco's conduct in this matter, the result is Pennzoil's agreement or arrangement of whatever you want to call it for yourself was defeated? That was the result of Texaco's conduct, wasn't it?"
MCKINLEY:	"Yes, but as I pointed out to you, Mr. Jamail, Texaco wasn't interested in defeating Pennzoil or any other potential bidder. . . . The answer to your question was, I suppose, yes."

In light of the extensive testimony on the case the jury had already heard, McKinley's answers had an effect quite the opposite from that intended. His flat Alabama drawl displayed none of the emotion that we knew had marked this conflict from the very beginning.

Clearly, McKinley wasn't going to break down and confess on the witness stand, nor did Jamail expect anything of the sort. He *had* gotten McKinley to acknowledge his awareness of the agreement between Pennzoil and the Getty entities announced on January 4.

Before the first hour of his testimony was over, McKinley had also been compelled to say that Texaco's move had "defeated" the Pennzoil deal. Although he reiterated his firm belief that Pennzoil had no binding agreement with the Trust, the Museum or Getty Oil Company, McKinley virtually admitted knowledge and interference. If the jury found there *was* a binding agreement, this became very dangerous testimony indeed for Texaco—and it came out of the mouth of its own chairman.

In an earlier deposition, Jamail had been strident and combative with McKinley but ultimately cut the deposition short. If the Texaco chairman was anticipating a repeat performance, he was disappointed. Jamail's questions ranged from pointed to obscure, but he didn't really take hammer and tongs to the witness. In the jury box, we realized that for Jamail this practically amounted to being on his best behavior.

At the same time, the contrast between the recently departed Hugh Liedtke and John McKinley was sharp. Liedtke had rarely equivocated on matters large or small. McKinley's testimony, on the other hand, was handicapped by his repeated insistence on including the basic tenets of Texaco's defense in so many answers ("they assured us they were free to receive an offer," "we told them we only engaged in friendly takeovers," etc.). This repetition of his defense was frustrating to the jurors.

Jamail asked McKinley about his conversations with Sidney Petersen when Texaco was making plans for their move on Getty. McKinley predictably said Petersen told him Getty Oil was free to receive an offer.

JAMAIL: "Well, did he describe for you what he [said] in an interview in *Fortune* magazine . . . about what he thought at the time, sir?"

McKINLEY: "No, he didn't describe that to me on the 4th and 5th when I talked to him; and I would be glad to look at the *Fortune* magazine if there's something there."

JAMAIL: "Would you read it, please."

McKINLEY: "Certainly. It says, 'Petersen: We thought there was a better deal out there, but it was a bird-in-the-handish situation. We approved the deal, but we didn't favor it!'"

JAMAIL: "Does that tell you that at the time Mr. Petersen believed he had approved the deal with Pennzoil and that's the deal he is referring to?"

McKINLEY: "Well, I don't know what it's referring to."

JAMAIL: "But you never were aware that Mr. Petersen . . . had already participated with the Getty board in approving the deal?"

McKINLEY: "Again, we're using the word 'Pennzoil Deal' and that that was the approval that was being referred to, if any, which was the agreement in principle."

JAMAIL: "But it doesn't say agreement in principle in that *Fortune* interview, does it, Mr. McKinley?"

McKINLEY: "This is a *Fortune* article. It says that here."

JAMAIL: "Well, he says something else, doesn't he, Mr. McKinley? 'We thought there was a better deal out there, but it was a bird-in-the-handish situation.' Do you know what that means?"

MCKINLEY: "Yes, I think so. I think he means that he had a possibility of negotiating a definitive merger agreement."

JAMAIL: "What does bird-in-the-hand mean to you?"

MCKINLEY: "It's just what I described to you that he thought . . . that he had a possibility in these negotiations [of] arriving at such an agreement."

JAMAIL: "Does not bird-in-the-hand mean to you, sir, something that I have captured? 'I got it.' Isn't that what it means?"

MCKINLEY: "Well, he might have thought that he could negotiate it. I can't answer all the things that he was thinking of."

JAMAIL: "Well, I'm asking what you think bird-in-the-hand means."

MCKINLEY: "Well, it could mean a variety of things to different people, and it's not an expression that I can verify what someone else exactly thought he had . . ."

JAMAIL: "I was asking what you mean by it."

MCKINLEY: "Well, I don't recall using the expression, but I think bird-in-the-hand means, to a lot of people, that's something I think I can get."

JAMAIL: "Or that I already have?"

MCKINLEY: "Well, that was what you added, sir."

It seemed unwise for McKinley to be so evasive about the meaning of "bird in the hand."

JAMAIL: "Well, sir, my question: Do you know of anything that would have prevented the completion of this if Texaco had not made an offer at that time?"

MCKINLEY: "Yes, I know of possible things, but I don't have any specific thing that I could relate to because I was not privy to those discussions."

JAMAIL: "Let me rephrase it. I'm not looking for possible things, I'm looking for actual knowledge, Mr. McKinley.

"And the truth is, is it not, sir, if Texaco had not stepped in and made its bid at the time, you know of no reason why this definitive merger agreement would have not have been completed, do you?"

MCKINLEY: "I don't know of any specific reason why it might not have been completed."

JAMAIL: "Have you ever heard the expresion before, 'We've got to stop the train' or 'Stop the train'?"

MCKINLEY: "I have heard that in the charge to the jury and I'm not certain but I think that a set of notes were shown [to] me on one of the depositions and I was asked if they were my notes and I said, 'No.'"

JAMAIL: "Do you agree or not agree that 'Stop the train' means . . . for Texaco to do whatever they could to keep the definitive merger agreement from being signed?"

MCKINLEY: "No, it does not mean that."

JAMAIL: "You weren't trying to stop the signing?"

MCKINLEY: "No, we were not trying to stop any signing. We were considering making an offer, a competitive offer."

JAMAIL: "And nobody at Texaco was trying to stop the signing. Is that your testimony?"

MCKINLEY: "No one that I know of was attempting to stop the signing. They were attempting to make a competitive bid."

Jamail asked McKinley to turn to page 8 of the Lynch notes.

JAMAIL: "What's the first thing you see in Mr. Lynch's notes?"

MCKINLEY: "The words 'STOP SIGNING.'"

JAMAIL: "Are you telling this jury you did nothing to stop the signing of this Pennzoil-Getty definitive merger agreement?"

MCKINLEY: "I did not tell them we did nothing. I said we investigated and made a competitive bid."

JAMAIL: "And that's all. You never actively asked anyone not to sign this agreement?"

MCKINLEY: "Number one, it's an agreement in principle. I should answer you, no, I did not ask anyone at any time to interfere with the signing of any agreement."

JAMAIL: "You call it an agreement in principle. The jury will decide whether it's binding.

             "Is it that you're saying you did nothing to encourage or say to anyone of the principals, 'Stop the signing' or 'Don't sign the definitive merger agreement'?"

MCKINLEY: "I did not say to any of the principals to stop the signing or words of that type. I don't mean to be repetitive, but that simply wasn't what I did.

             "I investigated and proceeded, as we've discussed before, to make an offer, a competitive offer."

Again, Jamail had managed to use a reluctant McKinley to solidify Pennzoil's claims of Texaco's knowledge and interference.

There was a sworn deposition that seemed to contradict Texaco's benign account of McKinley's visit to Gordon's suite at the Pierre on the night of January 5. While trying to confront McKinley with this deposition, which was from Bruce Wasserstein of First Boston (who was at the Pierre that night), Jamail found an apparent tender spot that provoked yet another clash with Miller.

JAMAIL: "I would like to show you Mr. Wasserstein's sworn testimony at page 125, [the] question at line 8:

"'At the meeting with Gordon Getty, did you or any representative of Texaco at any time say to him that if he were not prepared to make a deal with you, that you would put him in a minority position at Getty?

"And the answer was: 'That he would be at a *disadvantaged* position, not minority position. That he would be disadvantaged.'

"... Are those the words you used or did you tell him how he would be disadvantaged? Have I read that right?"

MILLER: "Your Honor, may it please the Court? In the interest of ..."

JAMAIL: "I haven't finished, Your Honor. I'm going to read all of this."

MILLER: "Well, may it please the Court, I object to him stopping there because he's taking that question and separating it out from the page that follows and not giving the witness the opportunity to see the context in which this was said.

"I think he ought to go ahead and read all of it now and then he can ask whatever he wants to ask."

JAMAIL: "I plan to read this, Your Honor, in sections, all of it. But I don't think he can prescribe how I examine this witness."

In fact, under the rules, counsel may insist on the inclusion of a more complete offering from such a deposition. Miller begins to read it out loud, interrupting Jamail.

MILLER: "I think it's under the Rules of Optional Completeness. It's the witness who says. 'Mr. McKinley is not a threatening man ...'"

JAMAIL: "I'm going to read all of this, Judge."

MILLER: "'... He was very cordial, but I think the point that was made is, it was stated the other way. It was to Gordon's advantage.'"

JAMAIL: "Your Honor, excuse me, but ... this is a typical 'Tricky Dick' trick and I think it's time that he understand the rules. He cannot teach me how to cross-examine this witness."

MILLER: "Well, somebody has to."

JAMAIL: "He's never been able to before. So, I'm saying Your Honor, I should be allowed to ask this witness if he recalls those statements being made. I have it all underlined."

JUDGE FARRIS: "Mr. Jamail, are you quite through with your sidebar remarks ..."

The judge had finally stepped in, this time choosing to place the blame on Jamail, who had come up with the more colorful outburst.

The issue was not dropped, however. Jamail continued his effort to confront McKinley with Wasserstein's deposition, which seemed damaging to Texaco. If Wasserstein had indeed told Gordon he would be "disadvantaged" if he didn't sell to Texaco, that put a very different face on that January 5 meeting. The fact that DeCrane had established Wasserstein as the source of many of the hardball tactics recorded in his notes underscored this testimony.

For his part, McKinley responded to this line of questioning with a game of one-upsmanship with Jamail, a game he was to lose.

JAMAIL: "Mr. McKinley, this top part that Mr. Miller was so insistent on reading, do I not have it underlined?"

McKINLEY: "It's not underlined. You marked over it in yellow."

JAMAIL: "And is this the same yellow?"

McKINLEY: "No, sir, it's not. You got certain red marks on the first page."

JAMAIL: "Well, it's the same yellow, isn't it? Let's just show it to the jury."

McKINLEY: "I apologize. I'm not trying to be argumentative, but you ask these things very quickly."

JUDGE FARRIS: "After eleven weeks, we're all beginning to see red marks."

JAMAIL: "The highlights on this page are the same types of highlights that Mr. Miller just read to you, aren't they?"

McKINLEY: "They are both highlights, but there are more highlights on the first page."

JAMAIL:	"I'm going to come to that. You seem to be very observant."
McKINLEY:	"Well, I can see red and yellow."
JAMAIL:	"Sure. So can I."

Jamail read the next segment of the deposition aloud, as McKinley followed the text.

JAMAIL:	"Have I read that right?"
McKINLEY:	"You read that right and that's highlighted. Do you want me to tell you about the highlights?"
JAMAIL:	"Mr. McKinley, I want you to do anything you want to do."
McKINLEY:	"The highlights are in blue on this page."
JAMAIL:	"Right."
JUDGE FARRIS:	"Had I known we were going to have such a good time this afternoon, I would not have planned to stop at 3:30."
JAMAIL:	"Mr. McKinley, you are very astute, and had you used some of that astuteness to inquire of some of the outstanding writings that would have evidenced the agreement between Pennzoil and Getty, perhaps you wouldn't be here today."
MILLER:	"Is this a question or what?"
JUDGE FARRIS:	"I don't know."
JAMAIL:	"It's going to be one, Judge."
JUDGE FARRIS:	"Mr. Miller, it's the first half of the question."
MILLER:	"Maybe so. I'd like to have him ask a question and not pray over the witness."
JAMAIL:	"Mr. McKinley, didn't you tell us just before the break that unless all three of the Getty parties agreed, you wouldn't do any of it?"
McKINLEY:	"That's correct. We would offer for all of it."
JAMAIL:	"Well, you were not interested in the majority. You wanted all of it."
McKINLEY:	"We wanted all of it."
JAMAIL:	"And you have represented to this jury, among other things, that it was not Texaco's intention to in any way put Gordon Getty in a bind, so to speak?"

McKINLEY: "That's correct. Yes."

JAMAIL: "And you also tell us you wouldn't buy from anybody if Gordon Getty did not want to sell?"

McKINLEY: "Yes, that was if he had not indicated his desire to sell we would not have made the other offers."

JAMAIL: "Were these resolutions shown on Plaintiff's Exhibit 76 adopted pursuant to your recommendation?"

McKINLEY: "Generally, yes."

The resolution passed by the Texaco board that authorized management to acquire Getty listed the three Getty entities in the same order the First Boston strategy had indicated. Jamail moved to emphasize this by asking McKinley what the resolution intended.

JAMAIL: "The [next] resolution, sir, can you tell us generally what you were trying to tell the board to do without any great detail, just in general terms?"

McKINLEY: "Well, then, I think the best way I can do that is by what the paragraph says.

"It says, 'Resolved: That the acquisition be effected by means of private purchases from the J. Paul Getty Trust, from Gordon P. Getty, the trustee of Sara Getty, and from Getty as authorized, buy unissued shares and from other parties holding shares.'"

Gordon had said that after the Museum shares were sold to Texaco, he was forced to sell the Trust's shares. Texaco contended it was pure coincidence that its board resolution listed the J. Paul Getty Trust, which was the Museum, first. It also denied making a deal with Lipton for the Museum shares before confronting Gordon.

Jamail returned again to tie the Texaco board minutes around McKinley's neck, with "questions" like this:

JAMAIL: "Mr. McKinley, it has been admitted by your lawyers that you recommended and approved the resolutions contained in these minutes that you're looking at."

MILLER: 'I didn't admit it or state it as a thought. To put these scare words in here like I 'admitted' it, is just a bunch of baloney."

JUDGE FARRIS: "Mr. Jamail, ask the question aside from the baloney. Mr. Miller is not in it at all."

JAMAIL:	"Judge, we will need a muzzle."
MILLER:	"I'll get one if it will speed things along."
JUDGE FARRIS:	"Are you gentlemen ready for me to fill out the form for the first time?"
MILLER:	"No sir, I'll sure behave if I can get some cooperation."

The form the judge referred to was a contempt citation. The jury had heard him on several occasions say that all he had to do was fill in the name, amount of the fine, and number of days jail time. Miller and Jamail had somewhat escalated their war of words since Liedtke had taken the stand over two weeks ago. McKinley may have had a touch of battle fatigue himself. The testimony had been grueling.

Jamail handed him the Siegel affidavit, filed late in the afternoon of January 5.

JAMAIL:	"This telexed affidavit to the Los Angeles Court was sent prior to the time of your meeting with Mr. Siegel in Mr. Getty's suite, correct?"
McKINLEY:	"You told me that, and I accept that as being correct."
JAMAIL:	". . . Would you read along with me, sir? Would you just read Paragraphs 4, 5, and 6 aloud?"
McKINLEY:	"'On January 2nd and 3rd, 1984, I attended a special meeting of the board of directors of . . . Getty Oil. . . . During the special directors' meeting, the Getty Oil board of directors approved a corporate reorganization transaction [the transaction] among Getty Oil, Pennzoil, the J. Paul Getty Museum, and the Trustee as a result of which the Trustee will own approximately 57% and Pennzoil approximately 43% in Getty Oil. Pennzoil, the Museum, and the trustee have all agreed in principle to the transaction approved by the Getty board of directors.'"

The jury was by now on intimate terms with Marty Siegel's sworn statement that was sent to California. Having McKinley read it to us was, I guess, Jamail's idea of another brick in the wall.

One thing about having him read a document aloud was that it made it harder for McKinley to say "they assured us they were free to receive an offer and we only engage in friendly takeovers" while he was reading.

The subject of Lipton's role at Texaco's meeting with Gordon on January 5 came up. Jamail asked McKinley how Lipton came to be at a meeting between the Trust and Texaco.

MCKINLEY: "I don't know the purpose. I assume that he was invited there by Mr. Getty."

JAMAIL: "You don't know that do you?"

MCKINLEY: "I don't know that."

JAMAIL: "Well, could Mr. Lipton assist Texaco in persuading Mr. Getty that he would be disadvantaged if he didn't sell his Museum trust stock?"

MCKINLEY: "I can't testify what Mr. Lipton may have done, but he didn't do that on my behalf."

JAMAIL: "Mr. Lipton, if he had made an agreement with Texaco, that would have put a different posture on Gordon Getty's transaction with Pennzoil, wouldn't it?"

MCKINLEY: "The answer is yes, but I didn't make any proposal to Mr. Lipton that would have allowed him to do that."

Throughout the trial, Texaco clung to the position that it made no deal with Lipton before Gordon consented to sell his shares. At times, more effort was expended to prove this point than perhaps any other. I didn't buy it then; I don't buy it now. The deal for the Museum shares was no doubt contingent on Gordon selling out, but the facts overwhelmingly point to the prior existence of a Texaco-Museum agreement.

Even if there was some sort of a wink or nod with Lipton that Texaco felt allowed the firm to put a veneer of truth on this fiction, the fact remains: Gordon sold out because he was convinced that the Museum had gone with Texaco, persuaded by Wasserstein and Lipton weaving their scenarios of dire consequences for the Trust.

Before Jamail passed McKinley to the comfortable arms of his own attorney, he went to the heart of McKinley's oft-repeated contention that the Getty parties were free to deal. If that were so, why the indemnities and warranties that specifically named Pennzoil?

JAMAIL: ". . . Texaco, as you say, accepted their word up until this time. And when they were required to put it in writing, Mr. McKinley, again these people, the Museum and the Trustee, would not put in writing that there was no claim against their selling their shares of stock, would they?"

MCKINLEY: "I cannot answer you that. They did not. They negotiated for it as an important item to them and have it in the contract, the indemnity, shown there, and we investigated it, as I have indicated, and we proceeded and we think that was entirely proper."

> JAMAIL: "Your Honor, I have no more questions of Mr. McKinley and I pass the witness."

The jury's prolonged exposure to McKinley before Dick Miller ever got to ask him a single question had placed the defense lawyer at a disadvantage. Miller spent some time trying to repair the damage, but mainly tried to get the jury to look on McKinley as a man, as opposed to what one wag would later call him, the Godzilla of Big Oil.

He joshed with McKinley, whose sense of humor seemed minimal. Not that Jamail had given him much to laugh about.

> MILLER: "Mr. McKinley, I want to come back to this perhaps a little bit later on. Do you mind answering a couple of questions about this?"
>
> McKINLEY: "Not at all."
>
> MILLER: "Not a thing in the world you can do about it?"
>
> JUDGE FARRIS: "I heard that."

McKinley's stare indicated that he had missed Miller's little attempt at fun. Earlier, when Jamail asked him when he first seriously considered Getty Oil, a puzzled McKinley had replied "I'm always serious." That was certainly believable. Gamely, Miller tried again.

> MILLER: "I believe Texaco owns a building down here on Fannin. When was that building built?"
>
> McKINLEY: "I think it was in 1914."
>
> MILLER: "Isn't it about time for a new building?"
>
> McKINLEY: "We have one out at Bellaire. That's one of our headquarters."
>
> MILLER: "I didn't mean to get on you about it."

This routine about the building had fallen flat also. John McKinley examined each question with a scientific detachment worthy of the engineer he was, but revealed little about himself as a person.

> MILLER: "How about yourself, Mr. McKinley, your progress in the company. How did a guy from Alabama ever get to be the head of the Texas Company?"
>
> McKINLEY: "On a serious answer, I don't suppose anybody really knows how these things happen."

The witness stand is not a psychiatrist's couch nor is a trial a prolonged group therapy session (usually). But when conflicting stories
meet, giving the jury a glimpse at the people behind the conflict can
shed light on the case.

This trial was not a personality contest by any means, but the
guarded nature of many of the Texaco witnesses made it somehow easier to accept evidence depicting them as participants in a naked power
play.

"Evidence" is the key word in that last sentence, not "personality."
The jurors are charged with judging the credibility of the witnesses
and deciding what weight to give their testimony. We weren't looking
for jokes or folksy anecdotes, just an indication that when a witness
testifies to a given fact, we understand what that fact is.

McKinley's reluctance to acknowledge the definition of expressions
like "bird in the hand" or the word "alliance" did not devastate his
testimony, but it definitely blunted the sharp edge of verisimilitude
Texaco needed to derive from his appearance. Perhaps this is why Jamail had theorized that Miller would never put him on the stand.

Miller had McKinley review some of the publicly documented figures on oil reserves that had come in during the testimony of Lewis
and Barrow. McKinley the engineer was in his element here, but he
spoke of "estimated" reserves, slightly different terminology from that
employed thus far. This soon caused confusion in the jury box. Rick
Lawler asked a question to try to understand the jumble of terms.

JUROR LAWLER: "Judge, when he says 'estimated,' is that the reserves
                 we've been talking about all through the case?"

JUDGE FARRIS: "Mr. Miller?"

MILLER: "I think so. I think reserves by their nature are al-
                 ways estimated, Your Honor. If I'm not wrong about
                 that."

JUDGE FARRIS: "Ask him."

MILLER: "Is that correct, Mr. McKinley? When we're speaking
                 of reserves, are we always speaking of estimated re-
                 serves?"

McKINLEY: "Yes."

MILLER: "There isn't any other kind, is there?"

McKINLEY: "I believe the correct answer is no, if you are talking
                 about oil reserves . . . "

JUROR LAWLER: "Your Honor, the phrase that's been used is 'proven reserves.' Are these estimated proven reserves? See, I'm confused.''

MILLER: "I think that's an excellent point.''

JUDGE FARRIS: "That's all right. I think most of us are confused, so you are not alone. Mr. Miller, would you clear that up, please?''

After having McKinley reconcile the terminology, Miller moved on to another area. The subject of Texaco's operations in ARAMCO prompted a question from the bench.

JUDGE FARRIS: "What about recent announcements about Saudi Arabia [and other] members of the OPEC nations as to the number of barrels that they will be producing in the immediate future?''

MILLER: "Your Honor, I think you are probably referring to recent newspaper articles pertaining to Saudi Arabia's efforts to increase the sales that it makes of crude oil in the world.''

JUDGE FARRIS: "Well, I don't remember where I read it. As the fellow judges say, 'I don't read the business section,' but maybe it wasn't in the business section.''

MCKINLEY: "Yes, sir. Their level of sales have been below their agreed-upon quota in OPEC. Their agreed-upon quota under present conditions is approximately four million barrels a day. And they have been in recent months producing and shipping less than that.''

JUDGE FARRIS: "Then the newspaper articles were incorrect?''

MCKINLEY: "I didn't see the articles in which they said they were trying to sell less.''

JUDGE FARRIS: "Well, since it's not my job, I didn't cut them out.''

MILLER: "Texaco is not a part of OPEC, is it?''

MCKINLEY: "No, sir.''

MILLER: "Thank God for that.''

JAMAIL: "Well, Your Honor, I object to that. He has just proven that Texaco is a partner to the leading oil producer in OPEC.''

JUDGE FARRIS: "That's what confused me.''

MILLER: "Well, I am glad you brought that up because I don't want anybody claiming that we are members of OPEC.

> "I guess that would be about as quick a way to lose a case as you could figure."

Jamail couldn't sit in silence while Miller tried to distance Texaco from what Jamail would later call "that old Saudi connection that Texaco has." Miller managed to fend off this attack, which was so far out of the realm of the case that it was funny to everybody except Joe Jamail (and, quite possibly, John McKinley).

The issue of the Getty employees invariably came up. Jamail had asked McKinley about reports that Getty employees were disgruntled at how they had fared under Texaco. At this point, Miller had McKinley describe the way Texaco now maintained the existing benefits, giving all Getty employees equal treatment.

The concept threw me, because earlier testimony had focused on the severance packages of the top executives. Emboldened by Rick's earlier questions, I raised my hand when the fog grew too thick.

MCKINLEY: ". . . Texaco purchased from Metropolitan Life irrevocable annuities for all Getty employees, present and retired, for all of their past service in Getty, whether or not they had been with Getty long enough to be vested under the Getty program."

JUROR SHANNON: "Excuse me, Your Honor. I am listening to all these questions; is the testimony that the benefits provided to the terminated employees were basically equivalent to those that top management got—what we have called golden parachutes? Or is that question out of line?"

MILLER: "May I ask that question, Judge? I think that's a good point and I think I ought to ask some questions about that.

"These benefits that were provided to the top executives were contracts that were made in, I think, mid-1983; is that correct?"

MCKINLEY: "Yes, I believe that's correct."

JUROR SHANNON: "I don't want to conduct your examination . . ."

JUDGE FARRIS: "That's all right. You will."

JUROR SHANNON: ". . . but I thought we have had the board minutes read that the Compensation Committee went beyond those contracts that predated the Getty deal. I am confused. I am trying to understand what you are saying."

MILLER:	"I had forgotten that, if that's the case."
JUDGE FARRIS:	"We sure don't want the jurors confused. They have got to decide the facts."
MILLER:	"I'm trying to hold my own here. Here are the minutes or notes of Mr. Copley."
JUROR SHANNON:	"It's towards the end."
MILLER:	"The way I read this, Mr. McKinley, and I don't know . . . I think this was subject to action by the committee that was appointed . . .''
JUROR SHANNON:	"Is that the people you said that had the preexisting employment contracts?"
MILLER:	"Yes, sir . . . There was mention made in an article . . . 'Some Getty executives feel that protected top executives forgot the rest of the company's employees and wrote insufficient safeguards for them into the merger agreement.'
	"Did I read that all right?"
McKINLEY:	"Yes."
MILLER:	"Well, the jury can decide if the top executives finked out on everybody else or not. They now know what the top executives did for the employees. They know what the top executives got. So I guess they can make up their minds about that."
JUROR SHANNON:	"Excuse me, Your Honor. We know what they got? I haven't heard that. That was my point of confusion.
	"Mr. Miller, I am not trying to get cantankerous with you."
MILLER:	"You couldn't hurt my feelings. I think you said a total of $12 million for those nine people, did you not, Mr. McKinley?"
McKINLEY:	"Yes, I said approximately."
JUROR SHANNON:	"It was a mention of the contracts that confused me, because you or somebody mentioned that the contracts were in effect a year, two, three years before.
	"I didn't understand that the $12 million you cited included the contract and the modification."
McKINLEY:	"I believe that is correct."
JUROR SHANNON:	"Thank you, sir."

That ended the questioning of John McKinley from the jury box, but I harbored misgivings about the appearance of it. Had it seemed like the jury was trying to cross-examine McKinley? I resolved to seek guidance from Judge Farris.

When court recessed for the day, I told Carl I needed to speak to the judge for a minute if possible. He returned and motioned me into the hallway. Instead of going to the judge's chambers, we turned into the courtroom. Judge Farris was still on the bench; Jamail, Miller, the court reporter, and all the other attorneys looked expectantly at me.

JUROR SHANNON: "Your Honor, I would like to apologize for speaking up. I wasn't trying to start anything."

JUDGE FARRIS: "You don't have to apologize. Let me point out that I am no shrinking violet. If I think you have overdone it, I will tell you."

JUROR SHANNON: "I figured you would . . . I just did not understand. I was paying attention."

JUDGE FARRIS: "You have a right to understand, and I am glad you were paying attention."

JUROR SHANNON: "Thank you, Your Honor. I apologize."

MILLER: "No sweat. Glad you did it."

JUDGE FARRIS: "Is that a legal term, Mr. Miller?"

MILLER: "That's well known in my circle."

This public exorcism wasn't exactly what I had in mind, but there you have it. If I seem a trifle submissive in this scene, an unexpected trip before the Bar of Justice can have that effect.

Judge Farris's assurances to the contrary, he soon announced that there would be no more questions from the jury. He said this was the first time he had ever permitted it, and it would be the last.

The testimony of John McKinley would conlcude the next day, his seventh day on the stand.

To the last, Miller gamely tried to depict McKinley as something of an average guy who just happened to be head of one of the most powerful corporations in the world. It was a valiant effort, but necessarily doomed to fail. Finally satisfied that he could do no more, Miller stopped.

MILLER: "We don't have any more questions. One other point, just so you don't get divorced: Is your wife from Texas?"

McKINLEY: "Yes."

MILLER: "Okay. Thank you."

JUDGE FARRIS: "Mine is from San Angelo."

Miller had questioned McKinley at some length about how the Getty employees had been dealt with in the wake of the Texaco takeover, trying to contrast the good treatment they received with the lack of any such provisions in the drafts of the Pennzoil-Getty final merger document.

He had raised many of the same points with Liedtke. Like Jamail, Miller wanted to show the contrast between the two chairmen, but McKinley had not been nearly as adept at playing along with his lawyer.

Jamail wasn't going to let what McKinley *had* said go unchallenged. He stood up to pose some additional questions to the Texaco chairman about this subject.

JAMAIL: "... You knew that there was overlapping of employees
             and some Getty people would have to be fired, termi-
             nated, or let go?"

McKINLEY: "Yes, it was my hope that their share would be some re-
             dundancy of personnel and that they could be termi-
             nated, yes."

JAMAIL: "That was your *hope*?"

If Jamail had written the script, he probably wouldn't have been bold enough to have McKinley speak those lines. But the words spontaneously poured out of the witness's mouth: he *hoped* some people could be fired.

McKINLEY: "It was my knowledge that there would be efficiency,
             synergisms in joining two organizations."

JAMAIL: "And you would let some of them go because that would
             save money?"

McKINLEY: "The purpose of productivity reduction is to ... make the
             company more competitive."

JAMAIL: "And you were hoping to do that by letting some of the
             Getty people go?"

McKINLEY: "Yes, with the proper termination payments."

JAMAIL: "Well, they're still fired when they're fired, aren't
             they? ... They're gone?"

McKINLEY: "They're not fired, sir. It's quite a different matter."

The distinction was lost on me. Jamail also tried to tie Texaco's motive in the Getty acquisition to the failure of an expensive exploration project called Mukluk.

JAMAIL: "Mr. McKinley, was the Mukluk misadventure or failure a part of the reasoning that you made up your mind to go after Getty?"

McKINLEY: "No, sir."

JAMAIL: "Didn't play any part of any kind?"

McKINLEY: "It played no part of any kind, no, because we had been interested in acquiring reserves by purchase much earlier than that, and we still are."

JAMAIL: "So Mukluk had nothing to do with it?"

The Mukluk "misadventure," as Jamail called it, had been frequently mentioned as a failure of catastrophic proportions for Texaco. What I had learned about the oil business in this trial, however, convinced me that the element of risk attached to exploration would prevent any company from betting the farm on a single prospect.

Nevertheless, Jamail managed to get McKinley to agree that Mukluk had played a "subconscious" role in his decision to go after Getty Oil. Again, it was not a major point, but it seemed consistent with Pennzoil's description of Texaco's motivation.

When the prolonged testimony of John McKinley finally ended, Jamail proclaimed, "Your Honor, at this time I rest Pennzoil's case."

By the time Pennzoil completed the presentation of its case, more than nine weeks had passed. The trio of Gordon Getty, Hugh Liedtke, and John McKinley had covered the vast terrain of this involved affair.

While Texaco's case had yet to begin officially, the past week had been spent with the jury listening to its chairman. Unfortunately, it didn't appear that McKinley had made a very persuasive argument for the defense. His testimony had been almost uniformly proper and correct, but Jamail had been able to elicit certain answers that fit Pennzoil's case.

The advantage was definitely to Pennzoil at this point, although many details remained murky. The motivations it had attached to many of the players on the Getty and Texaco side would no doubt come under fire during the presentation of Texaco's defense.

That happened to a certain extent, of course, but in the end, the Texaco defense continued to boil down to two words: "No contract."

PART VIII

CRITICAL DOCUMENTS 2

Evidence of Knowledge and Interference

THE DECRANE NOTES

THE LYNCH NOTES (PAGES 8 & 13)

THE TEXACO-MUSEUM
STOCK PURCHASE AGREEMENT
(REPRESENTATIONS AND WARRANTIES CLAUSE)

THE LIPTON AGENDA

Texaco president Alfred DeCrane's notes were made on January 3–4, 1984. In these four pages it is possible to trace the evolution of the Texaco bid for Getty Oil.

Getty   3 Jan

1. Salomon (Higgins) call Sunday 1 Jan—represent Museum

DeCrane's first recorded contact was with Museum investment banker Jay Higgins of Salomon Brothers.
The next call recorded is from John Weinberg of Goldman Sachs.

2. Weinberg call 3 Jan—puts Jeff Boisi (?) Head of M&A on phone—They represent Getty Co—the mgmt and "original Bd"—In a recessed Bd Mtg—the sense of the Bd is, if there was an attractive offer for the entire company would sell—The Museum is a definite seller—Gordon may be a seller or a buyer (seller at a price)—he doesn't want to be a minority party

   → Is Tx interested in principle?
   → Raise concerns re insurance, cable TV, etc; do want assets, have not really studied—Can't value $100 here and now
   → If Bd came to Goldman, concluded they wanted to sell—contact people—would Tx want to be on that list?—
   → Assume would mean data available (Yes, a data room)
   → Yes, we would like the chance to look at it

DeCrane records his answers as well as Boisi's questions. The notes reflect the exploratory nature of this call, which came during the recess in the two day Getty Board meeting.

3. JKMcK—Morgan Stanley (Wilson/Flog, Flom (?))

These initials are those of Texaco chairman John McKinley, still on vacation in Alabama, and appear to refer to a conversation about retaining the investment banking firm Morgan Stanley. "Flog/Flom(?)" refers to Joseph Flom of Skadden Arps, takeover lawyers on a par with Wachtell, Lipton. Skadden Arps, but not Morgan Stanley, was ultimately retained by Texaco in their bid for Getty Oil.

Getty

1. Salomon (Heggen) call Sunday 1 Jan - represent Museum - William interested in selling - owned fejets - administer Museum - group meeting to try to resolve, probably con't —

   I rec'd, bidding against $100 w/o data are unfriendly, management, anti-trust, energy etc —

   — Williams as trustee must protect self —

   Heggen said he would call Tues - just wanted to alert.

2. Weinberg call 3 Jan - puts Jeff Basir (?) Head of NoA on phone - they represent Getty Co - the Mgt are "original Bd" — In a rescued Bd Mtg: the sense of the Bd is, if there were an attractive offer for the entire company would sell — the Museum is a different seller - 'Gordon may be a seller on a hunger (seller at a price) - he doesn't want to be a minority party.

   → Is Tx interested in principle?

   → Rare concerns re insurance, cable TV, etc; do want assets, have not really studied - Con't value $100 here and now

   → If Bd came to Goldman, concluded they wanted to sell - contact people - would Tx want to be on that list? -

   → Assume would mean data available (Yes, a Data Room) -

   → Yes, we would like the chance to look at it.

3. JKMcK - Morgan Stanley (Wilson / Flog, Flor(?))

Pages two, three, and four of the Decrane Notes were made on January 4, 1984, the day before Texaco made their bid for Getty. Tom Petrie and Bruce Wasserstein of First Boston appear to be the major source of the information recorded in this section of the notes.

Tom Petrie   4 June

Bruce Wasserstein—(says Flom was contacted several minutes after F.B. contacted him)

→ Getty bylaws require 14 of 16 to act
→ Relates general concern of company people
→ Bd: Tisch was key Bd Member—was a key "public shareholder" (?) Tisch said would look for a higher deal—
→ Goldman—mad, embarrassed—claimed it wasn't the fair value
→ Museum/Salomon/Lipton = felt getting squeezed since Pennzoil with 20% would be able to take over the leverage role with Gordon
  They felt they needed to act = Lipton says there are 24 hours—
→ Gordon, talked about a "sale" at 125 = did refer to buyer and seller at a price—
→ ALERT—Isn't an ordinary merger—putting the Museum stock in a revocable escrow agreement which will be a lockout for anyone
    Only have an oral agreement
→ If they don't reach an agreement (1 yr ?) the assets are to be divided! so the option for PZL on the treasury shares they get more assets

---

Joe Perella—Jim Maher—First B.
Morris Karmer—Skadden Arps   4 Jan 11 PM
Jim Elliott—valuation team

Who are the key players:
    1) Shell no—
    2) Chevron—yes—
    3) Mobil—Say can beat them
        Massod worked hard
        Christmas week

Tim Petrie

4 June

Cross:

1 __Wasserstein__ — (says Flom was contacted several minutes
after F.B. contacted him)

→ Getty bylaws require 14 of 16 to act.

→ Relate general concern of Company people —

→ Bd. Tisch was key Bd Member — was a key public
shareholder (?) Tisch said would look for higher
deal —

→ Siedman — mad, embarrassed — claimed it was not
the fair value

→ Russell / Salmon / Lipton — felt Getty squeezed
since Pennzoil with 20% would be able to
take over the leverage sale with Gordon —
They felt they needed to act — Lipton say there are
5 hours —

→ Gordon, talked about a sale at 125 — and refer to
being not a seller at a price

→ Alert — Isn't an ordinary merger — putting the
Pennzoil stock in a reversible escrow agreement
which will be a lock out for anyone —
Only know an oral agreement

→ If they don't sign an agreement (1 yr?) the assets
are to be divested! So the option for GO on the
treasury shares they get more assets.

Jo Pirelli — Jim Shannon — Fred B.   )
Dennis Kramer — Shapiro Corp.   ) 4 Jan 11. PM
Jim Elliott — Value Tech   )

Who are the players?   ① Shell no —
                        ② Chevron — yes —
                        ③ Mobil — they can beat them
                          Pennzoil would have
                          Chevron would

Page Three begins with a continuation of the list of "key players" that began on the bottom half of Page Two.

4) Gulf—could have had Marathon but blew it at the Hoopman level—Don't feel that they see a fit

5) Phillips per Wasserstein

P. says need to contact early am that there is a proposal possible and should have that deal after noon board deal
  → Should also give notice to Gordon!
  → Ask about Company = Tom Petrie talked to Petersen who 2 weeks ago "was defeated".
(F.B.)
  → Option to PZ = won't get it till merger agreement is signed!— could be a way to take care of Liedtke—
  → Gordon tax position = (that a law firm represent Tx and Simpson Thatcher represent them
  → Question of the trust (Gordon) = It can't sell at a tender—only at a merger.

Valuation = didn't value Northern Tier, Chemicals, fertilizer and other materials—
    Loaded at reserves—leases, etc.
Value ERC at 1.0 to 1.4 billion

The rest of page three and the top half of page four contain an informed analysis of the assets of Getty Oil. Again, the source appears to be Texaco's investment bankers, First Boston. The notes clearly demonstrate Texaco's intimate knowledge of the agreement negotiated between Getty Oil and Pennzoil, as well as the terms of the Sarah C. Getty trust instrument, whose constraints restricted what actions Gordon could take.

CBrey – "Could have our Marathon but like it at
the Gorgeous level – Do't feel that
they are – for
E Phillips for Warrenton !

_____

P. says need to contact ~~corporation~~ early am Not Reserve or
proposal possible we should have that
deal after noon board deal !
→ Stanel also que nature to Gordon
Ask about Company – chan Othu talked
to Peterson who 2 week ago "was deposited".

Otten Ruttman (FB)
Glu Jeffrey (Beaver)   → Option to P2 – won't get it till merger agreement
is signed ! – caused he a way to take care
of Decided –

Tom Petrie
Paul Freedman   → Baram tax position – (not a law firm represent
Seymour Phillips   Tx and Seymour Phillips represent them) –

→ Question of the trust (Baram) = It can't sell at a
tender – only at a merger.

Valuation = Didn't value Southern Pacific, Chemical,
fertilizer and other minerals –
Looked at reserves – term, etc –
Value ERC at 1.0 to 1.4 billion
On their $140/share gross –
Ohlue terms listed
at 3.31.83 = Gross amount = $2.2 B
before 1.3 q LT debt at that time
(Include ERC at $1.350) =
→
Own 6% ~~insurance~~ and capped at $75
Tmo at 6% we capped at 3/4 of Bthe all value

The list on the top of page four is headlined "Saleable Items." Heading the list is the Employers Reinsurance Company (ERC), the same Getty asset that was a bone of contention in the Pennzoil-Getty deal. Pennz-had finally raised its offer by $2.50 a share for the value of ERC.

At the bottom of this last page of the DeCrane Notes is another entry that confirms the extent of Texaco's intelligence about the Sarah C. Getty Trust. The stars, arrows, and word "Imp" in the left margin indicate DeCrane felt (or was told) this information was "important":

Trust: 1930 draft ←
   —"Only a merger"—and to avoid a loss to the trust
   So might have to go the tender offer route to merge him out or create concern that he will take a loss—Problem is there seems to be
   no way to get Gordon on base first-

   Have standstill agreement—need to
       look at application—do you have to get
       all three groups!

Texaco's repeated protestations that it only engaged in "friendly" deals rang false in light of entries like this, and how they seemed to fit fit with the actions Texaco would later take. Wasserstein, who DeCrane credited with most of this information, accompanied the Texaco contingent to the Pierre on the night of January 5 and told Gordon he would be "disadvantaged" if he didn't sell his stock. In this way, Wasserstein himself helped implement the strategy he devised for Texaco.

Saleable items:

1. ERC — have contacted insurers already =
   English insurance Co. — any of top 3?
   have captive sales force —
   → Stock looked at it 3 years ago for E.G.?
   Feel that the IB is a value figure — our ????
   are looking at a write up of their values
   on Getty purchase.

2. Other minerals = $250 (+ $250 coal)

3. Refining / marketing — $530 M

4. Ships & pipelines = $350 M
   $75 DWT
   $5000/mile —

5. Hot Dog Chain NOL 215 M

6. Marketing 6½% NSV = 70 M

7. Fee lands 150 M $750/a

8. Entitlement business = $50 M
   offer for sale last year — no takers

9. Japanese monopoly $30 M
   (refining) —

Trust 1930 draft
   — only a merger" — and to avoid a loss to the trust
   So ought have to go the tender offer route
   to merge him out or create concern that he
   will take a loss — Problem is that seems to
   be no way to get Gordon on board first

Have standstill agreement — need to
look at application — do you have to get
all three groups?

Texaco deputy controller Patrick Lynch attended many of the strategy sessions in White Plains on January 4–5, 1984. Unlike the tight, precise notes of Alfred DeCrane, Lynch's notes are less formal and sprawled over fourteen pages. There are several interesting passages, some of which are reproduced here.

Stop signing

Buy Museum shares subject to Hart Scott
Pennzoil tender offer
tender off—40%—100% (in lieu of control with Getty)

Ideal
1) Museum
2) Trust
3) Balance

This section of the notes is interesting because of two points: The "Stop signing" clearly indicates Texaco's knowledge of the Pennzoil–Getty deal moving toward execution of the definitive merger agreement.

Under the heading "Ideal," the Getty entities are listed in precisely the same order in which the Texaco board authorized their management to go after them. At the trial, Texaco witnesses said this was "a coincidence."

Board meeting notes

Trust	25.
Pension	24
Bond	35

- Stop signing

- Buy museum shares subject to Nat'l Sec'y
  financial tender offer
  tender off - 80% - 100% (in lieu of contested Battle

  Need
  ① museum
  ② Trust
  ③ Balance

any all - common payment on withhold date not wait
until proration date

Trust Document
cannot sell except

- merger
- to avoid loss to trust

Part of Merger Agreement with Getty each extended

Ⓧ H should contact J Getty direct
   TK   " "   L Tisch

As First Boston's operatives outlined the strategy Texaco should pursue, Lynch dutifully recorded it.

Timing

      call Lipton 8:30
        Larry Tisch (Company & Gordon)

The key role of Lipton and Tisch were apparently major themes of the First Boston strategy, because both men's names appear repeatedly in the DeCrane Notes and the Lynch Notes.

The passage at the bottom of this page spells out the dilemma that would confront Gordon Getty if there was a tender offer in which he did not participate:

        If there is a tender offer and Gordon doesn't tender
          then he could wind up with paper

"Paper" presumably means some sort of note or security of uncertain value.

Troung
    Carl Lighton 8:30
            Larry Koch (Company + Gordon)
Competition
    Socal - contacted by Betty, (Peter) would like to buy it
    Shell -

Gordon - Buyer 115
        Seller 125/130
    "Cont miss" "at right price Gordon was seller"
    Museum / Trust

Base Case
    Gordon + Museum will not tender - 15 R will look up
    Tender Offer - go for more than half
        1st Bidder vs 2nd Bidder   vs 3rd
                        50 @ 125   50 @ 130   Cont
            maximize 2nd offer   must be materially higher

*   If there a tender offer and Gordon doesn't tender
            then he could work up offer

When you sell your house or car, you must normally represent to the buyer that it is yours to sell—free of liens or other encumbrances.

The Museum makes such a declaration in this clause (Number 5), but interestingly enough, "no representation is made with respect to the Standstill Agreement, the Consent, the Stockholders' Agreement or the Pennzoil Agreement."

The other agreements excepted in this clause did not contemplate any change in Museum ownership of their stock in the Getty Oil Company, unlike the Pennzoil Agreement. This exception was written into the Museum's stock exchange agreement with Texaco at the insistence of Museum lawyer Martin Lipton.

Amounts hereunder shall be appropriately adjusted to take
account of any stock split or stock dividend on the Common
Stock occurring after the date hereof.

5.   The Museum represents and warrants to the Purchaser that (a) the Museum is duly authorized to execute and
deliver this Agreement and this Agreement is a valid and binding agreement, enforceable against the Museum in accordance
with its terms; and neither the execution of this Agreement
nor the consummation by the Museum of the transaction contemplated hereby will constitute a violation of or default
under, or conflict with, any contract, commitment, agreement,
understanding, arrangement or restriction of any kind to which
the Museum is a party or by which the Museum is bound (except
that no representation is made with respect to the Standstill
Agreement, the Consent, the Stockholders' Agreement or the
Pennzoil Agreement); and (b) on the date hereof the Museum
has, and at the Museum Closing the Museum will have (without
exception) valid and marketable title to the Museum Shares
free and clear of all claims, liens, charges, encumbrances
and security interests, and the transfer of the Museum Shares
to the Purchaser will pass good and marketable title to the
Museum Shares, free and clear of all claims, liens, charges,
encumbrances and security interests.

-7-

THE LIPTON AGENDA

Museum attorney Martin Lipton prepared this agenda prior to his meeting with the Texaco contingent on January 5, 1984—one day after the Pennzoil-Getty deal was announced.

The items listed bear a striking similarity to some provisions of the Pennzoil deal.

Texaco vice-chairman James Kinnear and general counsel William Weitzel met with Lipton late on the afternoon of January 5, and were presented with this agenda.

Item 1 is labeled "Price." The $112.50 figure is the same price agreed to in the Pennzoil-Getty deal.

Item 3 talks about putting the Museum shares in escrow. Trial testimony by Museum lawyers showed that an escrow agreement with Pennzoil had been drafted.

Item #6 is labeled "Indemnity," and seeks full indemnification for the Museum against any claims by or through, GET (Getty Oil Company), PZL (Pennzoil) and S.C.G. Trust (the Sarah C. Getty Trust).

1. _Price._   What premium over $112.50. Will all
                  shareholders be treated equally. What
                  is form of deal.

2. _Lock-up._   Will the Museum shares be purchased now.
                  Will stock or crown jewel options be
                  requested.

3. _Hart-Scott._ Will Museum shares be purchased in escrow.
                  Will Museum get the interest on the
                  escrowed funds.

4. _Antitrust._   Will the purchase from the Museum be hell-
                  and-high-water so that Museum gets price
                  no matter what.

5. _Top-up._   Will the Museum get favored-nation and
                  high-bidder protection.

6. _Indemnity._   Will the Museum be fully indemnified
                  against any claims by or through, GET,
                  PZL and S.C.G. Trust.

PART IX

"THEY NEVER HAD A CONTRACT"

Texaco's Defense

24

"Texaco calls as their first witness Mr. Barton Winokur," announced J. C. Nickens, a somewhat younger partner of Dick Miller.

Nickens, who had made no previous appearances before the jury, handled the direct examination of Getty Oil's chief outside legal counsel. Winokur, who in recent years had taken the role previously occupied by Lansing Hays, had been intimately involved in the whole "Blood in the Water" saga, which as we have seen, spilled over into the Six Day War. Unlike Hays, however, he had no relationship with Gordon Getty at all. In fact, in league with Getty Oil chairman Sidney Petersen, Winokur had taken a sharply adversarial role in the tense relationship between the company and the Trust, more specifically the Trustee.

Like some of those called to the stand by Pennzoil, Winokur was an all-purpose witness, able to give testimony over a broad range of events. Nickens had obviously done his homework with this witness, because he led him through the story with great dispatch, pausing in relevant areas for Winokur to deliver the appropriate blows to Pennzoil's case.

Irv Terrell handled the cross-examination, and he repeatedly objected to the way Nickens would have Winokur present his views.

Again, as with so many of the witnesses who were lawyers, there were multiple objections to them delivering anything that sounded like a legal opinion.

As with yes-and-no answers, legal opinions could be a particular sore spot with Judge Farris. To get by in his courtroom, this simple maxim had to be observed: *Always* give a yes-or-no answer; *never* give a legal opinion.

Winokur would acknowledge the admonishment not to give a legal opinion, and then blithely deliver one. Terrell was in and out of his chair more often than a streetcar conductor.

Nickens asked Winokur questions about his reaction to the Pennzoil tender offer after the Memorandum of Agreement was presented at the beginning of the board meeting. Specifically, he showed him the so-called Purpose clause Miller had recently spent so much time grilling Liedtke about.

NICKENS: "In light of what you had just been told about this agreement between Pennzoil and the Trust, what was your reaction to the statement of Pennzoil's purpose in that document?"

WINOKUR: "Well, it didn't ring true."

NICKENS: "Why is that?"

WINOKUR: "Well, if somebody was interested in participating in a constructive way in restructuring a company, one would have thought they would have talked to the company about restructuring.

"What's more, I think if you want to be constructive, you don't hold a gun to somebody's head. I mean, if somebody came to your front door and had a paint brush in one hand and a gun in the other and said, 'I heard your house needs painting,' I mean, you might suspect he's there for something other than painting your house."

This launched a section of the testimony where Nickens had Winokur define a variety of colorful expressions of the takeover trade, including "sharks," "blood in the water," "in play," and "bearhug."

NICKENS: "What situation does the phrase 'bearhug' describe?"

WINOKUR: "It's when a target company is approached by a potential acquiring company that in effect pretends to make love, so to speak, hugs the acquirer.

"The term 'bearhug' is meant to imply, unlike a normal human being you can disengage from, the bear is a lot stronger and once he hugs you, you're stuck."

This testimony wasn't intended as an abstract seminar in takeover tactics. Winokur portrayed Pennzoil as "sharks," who when they smelled "blood in the water," realized Getty Oil was "in play" so they put a "gun to the head" and got them in a "bearhug."

With that preamble, Nickens attempted to have Winokur tell the jury what advice he gave to chairman Petersen and the Getty board at the January 2–3 board meeting and then describe the action taken by the board in response to his advice. This touched off a bitter clash that continued throughout the first day of the Texaco defense.

Terrell leapt to his feet after the first such question and got the ball rolling:

TERRELL: "Excuse me, Your Honor. My objection is that this man purports to speak for Getty Oil Company . . .

"I want to bring to the Court's attention the fact that Mr. Copley prepared these edited notes in this man's law office; and this man's partner was right there when Mr. Copley threw the original notes away; and further, this man's law firm filed a lawsuit against Pennzoil on January 6th, before those notes were destroyed.

"It would be a total miscarriage of justice to let this man now get up and talk about something that's not even in his own edited notes with Mr. Copley, and I don't think he's got the right to do that."

Terrell was fighting to prevent Texaco from taking advantage of the Copley notes by having Winokur recite incidents and details not recorded in the notes.

TERRELL: "He can't speak for the company . . . I object to it. We need to know who said what to whom and not just some overview, 'Well, the Getty Oil Company thought this and the Getty Oil Company thought that.'

"Getty Oil Company—and I don't mean to make a bad joke—but it doesn't think. It's a corporation. It acts through its board of directors, and this man is not on the board."

Nickens struggled to find acceptable questions that still allowed Winokur to give testimony that, on its face, was very damaging to Pennzoil.

As a practicing lawyer, however, Winokur apparently couldn't keep from inserting legalistic-sounding comments in his responses on a wide variety of topics. Again and again, this effort led to squabbling between the attorneys. After a lengthy bench conference failed to resolve the impasse, the jury was sent to lunch.

The transcript records the following sharp exchange, sparked by Jamail's response to Judge Farris's ruling that Nickens could pursue this angle:

JUDGE FARRIS: "I will allow you to question this witness, but not in such a manner that the effect of his answers is going to be a legal opinion."

JAMAIL: "That is the problem. After he asks the question, this

real volunteering witness gives his answer, obviously
exuding hostility ... I can look at him and see it."

MILLER: "Wait, that's so unfair ..."

JAMAIL: "I don't need to be interrupted by Mr. Miller at all ...
Your Honor, it's too late at this point. Mr. Miller can
shout all he wants about any kind of shouts he wants
to make. It's obvious [Winokur] has come here to do a
hatchet job.

"He has come here to impeach these Copley notes
by things that only he, in his mind, can now dredge up
to say, and nobody can challenge it. .... This 'volunteer' is going to give an answer that cannot be corrected. The jury will have it, and therein lies the evil
of allowing this line of questioning."

MILLER: "Well, I would like to see the case conducted fairly for
both sides, and when we bring a witness in here, the
first witness who has appeared in our case, and he has
to take this kind of abuse from a lawyer who knows
little or nothing about the case, I think I can say
fairly ...

"To attack him with no reason whatsoever ... they
now, realizing the weaknesses connected with their
case, they can't stand for us to put on our evidence."

Miller had come to the defense of Nickens by unloading the heavy
artillery on Jamail. The comment about him knowing "little or nothing" revealed Miller's actual view, at this point, of Jamail's role in the
case.

While he had admittedly taken a back seat for much of Pennzoil's
case, his masterful questioning of Liedtke and McKinley had confirmed that Joe Jamail was very much in command of Pennzoil's legal
effort. It's hard for me to see why there was any doubt about that
anyway.

Miller continued to press his attack.

MILLER: "They sit there wounded and bleeding and want to go shut
him up and not let him present the evidence that he has,
when it's all subject to cross-examination ...

"To make these kind of gratuitous attacks ... where the
witness is not in a position to respond, you know, it doesn't
exactly deserve a medal for bravery."

JAMAIL: "I merely point out the obvious fact to this Court, the evil of
allowing this man to ramble on in essay fashion, to get in

through some back-door means what really is a legal opinion under the disguise of something else. . . . Your Honor has
not allowed it henceforth and I am asking the Court not to
allow it now."

Farris upheld his ruling, clearing the way for Nickens to continue
this line of questioning. The outraged objections of Terrell and Jamail
convinced Miller that Texaco had finally touched a nerve with Winokur's testimony.

Winokur sat passively while the verbal combat raged all around
him. After lunch, his testimony continued for a couple of hours with
minimal interruptions as he basically gave a narrative account of the
board meeting as Nickens asked questions from the Copley notes.

His characterization of the 15-to-1 vote approving the "Pennzoil proposal" as something so vague and tentative left me wondering how anybody could have known what they were voting for. He said when Harold Williams tried to make the motion, as described in the notes, he
got so confused he had to let Lipton explain it.

A problem for Texaco was the Getty press release the next day that
announced the deal in definite terms that didn't require an explanation
from Lipton.

Nickens attempted to deal with that issue by having Winokur review
something called the *New York Stock Exchange Company Manual*. The
thrust of this testimony was to show that some sort of press release
had to be made to comply with disclosure requirements even if there
was no deal.

These questions called for what was essentially a legal opinion. Winokur skirted the issue for a while, but when one of his answers included the term "legal obligation," a waiting Terrell pounced.

TERRELL: "Excuse me, Your Honor, there we go. The witness has
            been cautioned a number of times about giving legal
            opinions. He just gave one. . . . Your Honor, I'm not
            trying to blame Mr. Nickens, but the witness knows
            better.
               "I would ask the Court to admonish him not to slip
            in legal opinions in his answer and I would ask that
            the answer be stricken . . ."

NICKENS: "Your Honor, I don't believe he was offering legal
            opinion, he was describing custom."

JUDGE FARRIS: "I heard him, Counsel. The entire answer is stricken.
            Ask the question again and caution him not to slip it
            in."

The questions about the press release centered on the drafting session at the Paul, Weiss offices. Nickens had Winokur review an earlier draft version that was rejected. This was important testimony because Winokur himself had approved the January 4 press release for Getty Oil, along with Copley.

When answering a question concerning why the language about the Museum shares was deleted from the draft, Winokur strayed over the line once again.

TERRELL: "Excuse me, Mr. Nickens. 'There was no question in my mind that the board would be totally opposed to it.' I don't know how many times I had to get up and ask this man not to speak for the board or to talk about what's in other people's minds.

"He knows what the objections are. He is a lawyer, yet he keeps doing it over and over again, and I would ask Mr. Winokur not to do it. He knows the rules."

MILLER: "May we approach the bench?"

JUDGE FARRIS: "I have got a better idea. We will break for the day."

If Terrell's objection sounded vaguely familiar, it was because it bore a striking similarity to many of the objections raised by Miller during Jeffers's examination of Baine Kerr.

With the jury sent home for the day, the bitter lunch-time clash was soon renewed.

NICKENS: "Mr. Terrell gets up and says, 'Well, he's trying to state what's in the minds of the board members.' That's not correct.

"He has explicitly stated that he was stating what was in his mind, our mind, about what the board would do."

TERRELL: "I didn't know we had more than one mind, Your Honor. . . . He is talking about what's in 'our minds.'

"It's a deliberate effort on behalf of Mr. Winokur to evade the Court's rulings, and I don't know whether this is the plan they worked out with the lawyers from Texaco on how to do this or not. . . . He is not a person who has some sort of position that entitles him to speak for anybody but Barton J. Winokur, lawyer for Getty Oil Company."

NICKENS: "Your Honor, Mr. Terrell makes his allegations very easily with very little evidence . . . that somehow or

another Texaco has been involved in some misconduct ... Mr. Terrell's accusations are simply un-
founded."

TERRELL: "Your Honor, there is no board member present at
Paul, Weiss ... yet he pretends to speak for the
board. ... It is totally misleading to the jury. It's an
improper speculation about the minds of sixteen
members of a board of directors who are not there."

JUDGE FARRIS: "Mr. Terrell, I agree with you on everything except the
fact that counsel for Texaco planned it this way. I
don't believe that."

TERRELL: "I don't know."

JUDGE FARRIS: "Well, I don't believe that they did. However, I think
that the witness does understand that when I have
ruled, I have ruled. I am striking the entire answer and
I don't want this to happen again.

"I want you to counsel with him overnight, so that
it will not happen again. Because if it happens again
tomorrow morning, I will unload on him in front of
the jury. And I don't want to do that."

This whole exchange had aroused the ire of Dick Miller. His response, however, managed to stir the rumbling volcano of Mount Farris
into a full eruption.

MILLER: "I want to say something for the record: These comments that Pennzoil's counsel makes concerning the
conduct of the witness are totally uncalled for. He suggested earlier that the witness deliberately 'slipped
something in.'

"That's the expression he used, and I think the jury
can handle his comment. But Your Honor said something at the same time concerning the witness 'slipping
something in' to suggest that the witness had deliberately violated Your Honor's instructions.

"The witness is not familiar with the rules here in
Texas and I know him personally. He is an honorable
man and I know he would not do that.

"The jury now believes, because of the comment the
Court made, that the witness has deliberately attempted to evade the Rules of Evidence and somehow
to say something unfair, improper, which puts the witness in a very bad light in front of the jury.

"And I have seen Mr. Terrell and Mr. Jeffers grinning at the jury over there and making these motions as they are wont to do. It puts the witness in a very impossible position.

"I know the witness has not done anything improper. He certainly did not intend to do anything improper. I suppose he is at a very considerable disadvantage because he talks a little funny, if I can say so, and I know he won't take umbrage at that, but for there to be a suggestion that the witness has slipped something in—"

JUDGE FARRIS: *"Mr. Miller, are you quite through lecturing me?"*

MILLER: "I didn't mean to be lecturing you."

JUDGE FARRIS: "That's what you have been doing and I will have no more of it. Understood, sir?

"You want to make a bill, make it right now."

MILLER: "I only want the Court to instruct—"

JUDGE FARRIS: "Do you want to make a bill, sir?"

MILLER: "No, I want the Court to instruct the jury to disregard the Court's comment and I shall request it."

JUDGE FARRIS: "I will not do it, and I will further say, counsel with this man tonight because I will unload tomorrow morning if it happens again . . . *I will not be lectured to.*"

MILLER: "Sir, I was not lecturing you and I must take exception to Your Honor's remark."

JUDGE FARRIS: "All right. Take exception."

MILLER: "I intended my remark to . . ."

JUDGE FARRIS: "We are recessed for today."

MILLER: ". . . be courteous and polite to Your Honor as always."

This outburst by Judge Farris was totally outside the presence of the jury, but we were preoccupied with other concerns. Since the beginning of the trial the court had been operating on a fixed schedule, with no testimony after noon on Friday.

During this time, jurors would routinely take care of personal business, schedule doctor's appointments, and so on. After the first day of Winokur's testimony, we were informed that we would indeed have to stay all this Friday afternoon in an attempt to permit the witness to complete his testimony and return home.

This caused no small measure of unhappiness among many of the jurors. When we were seated on the benches waiting our call back into the courtroom, we exchanged grievances. Two jurors in particular had scheduled medical appointments for themselves or their children for that Friday and expressed their unhappiness to the bailiff, our link to the judge.

"Look, Carl, we are in our third month down at the courthouse," I told him. "Friday afternoons are the only time we can arrange to take care of our own business. Now, with virtually no notice we have to cancel whatever arrangements we have made for the convenience of this witness. Tell Judge Farris it's just not fair."

Several other jurors chimed in, and the beleaguered bailiff went to deliver the message to Judge Farris. Carl Shaw suffered mightily at the hands of the judge and jury throughout the trial, but fortunately he had a sense of humor. His role as liaison was a tough one, and on more than one occasion the judge told us to lay off the bailiff.

This minor rebellion by the jurors was not done in a fit of pique, but Winokur was the wrong witness for whom to request a special accommodation. The utter contempt he had for the plaintiff seemed to spill over to include the whole trial as well. As for the jury, he studiously avoided ever looking in our direction. It was if we were scullery maids and field hands let in to the Great Manor to see the master's debating society. Winokur never deigned to even look at the jury.

Did such "trivia" undermine his testimony? Of course it did, but to my mind it was not trivia at all. Like his conduct at the November 11 board meeting, Winokur's demeanor in the courtroom was not on the point of the case. But both situations spoke volumes about his inclinations and attitudes to the jurors who would have to weigh the credibility of his testimony.

I remember wondering how the Texaco lawyers could have possibly thought this guy would make a good witness.

25

Judge Farris relented, and court was recessed at noon that Friday, with the proviso that he could order the trial to continue on a future Friday afternoon without any advance notice to the jury.

It was nearly a week before Winokur would return to complete his testimony. In the interim, Texaco presented its only video deposition of the case.

Patricia Vlahakis was the attorney from Wachtell, Lipton who had played a key role backing up Martin Lipton in representing the Museum during the Six Day War. Now pregnant, she had a doctor's prohibition against travel.

Unknown to the jury, during the first month of the trial, Texaco had pleaded to be allowed to interrupt Pennzoil's case to call her as a live witness before her pregnancy advanced. Jamail had refused to even consider it, saying it would disrupt his case. The sympathy factor of putting a pregnant woman on the stand in front of a jury panel that included ten women was no doubt something Jamail wanted to avoid.

Miller had argued that her appearance was vital to their case, adding, "I didn't make her pregnant."

"You would have if you could," responded Jamail, continuing the war of words.

The final result was this video deposition, which had been taped one weekend during the trial.

A thick, bound book of the exhibits used in the deposition was distributed to the jury for easy reference.

Pat Vlahakis was a pleasant-looking woman who appeared to be in her late twenties. She had been present at most of the January 2–3 board meeting and had represented the Museum at the Paul, Weiss offices where the press release was drafted.

As Lipton's assistant, Vlahakis had responsibilities that far outweighed her age and experience. It was very apparent to the jury, however, that she had more than held her own with the ranks of high-powered attorneys she had to deal with.

As she described that session at the Paul, Weiss offices, the chaos and confusion seemed to come to life. One revealing point she disclosed was how she had gone into a room and found Martin Siegel on the phone apparently giving details out to some reporter on the West Coast. She hollered at him to get off the phone and shut up. As she recalled the incident, there was a flash of the wrath that she had directed at Siegel.

Vlahakis delivered strong testimony for Texaco, stating flatly that the Museum was never bound to any agreement with Pennzoil. She also insisted that nothing in the press release contradicted this statement.

She testified that on behalf of the Museum, she had approved the press release, after consulting with Lipton by telephone.

John Jeffers had managed to elicit some testimony from Patricia Vlahakis that caused Texaco problems in other areas, however, and made the deposition something less than the stalemate that had resulted from most of the other video depositions.

On the subject of the indemnity Texaco had given the Museum, Vlahakis testified that it was an essential element of any agreement between the two parties. She went over the Lipton agenda for his meeting with Texaco on January 5 that included the indemnity and a guarantee of the $112.50 price provided in the Pennzoil deal.

Further, she testified that the indemnity obligated her and Lipton to assist Texaco in its defense of this lawsuit. In other words, she seemed to be saying her testimony on behalf of Texaco was given because she was obliged to do so. By logical extension, this could be said to apply to each of the Getty witnesses Texaco put on the stand.

Vlahakis also answered questions about the drafting sessions on the final merger document involving lawyers for Pennzoil, Getty Oil, and the Trust. She had represented the Museum in these meetings.

Under Jeffers's questioning, she recounted how she had been on her way out the door to one of these sessions when Lipton told her not to go. "I need you here," she quoted him as saying, because the Texaco contingent was coming to call.

On one point, however, Jeffers scored what seemed to be a telling blow. On page 4 of the Memorandum of Agreement, Lipton had written some language under the signature line of Harold Williams: "UNDER CONDITION THAT IF NOT APPROVED AT THE JAN. 2 BOARD MEETING REFERRED TO ABOVE, THE MUSEUM WILL NOT BE BOUND IN ANY WAY BY THIS PLAN AND WILL HAVE NO LIABILITY OR OBLIGATION TO ANYONE HEREUNDER."

The jury was very familiar with these handwritten interdelineations from previous testimony.

In addition, she identified a note written in the top margin of Page One of the Memorandum of Agreement as being to her from her boss, Lipton. This reinforced her intimate involvement with this document.

Jeffers's question about Lipton's writing under Williams's signature was clever but to the point. By writing that, she said Lipton wanted to be "doubly sure" the Museum was not bound.

VLAHAKIS: "We wanted it to be absolutely, positively clear."

JEFFERS: "No board approval, no one is bound?"

VLAHAKIS: "If this is rejected, no one is bound. That's correct."

Jeffers beautifully underplayed the next line to a wary Vlahakis.

JEFFERS: "And what's the converse of that?"

VLAHAKIS: "Sorry?"

JEFFERS: "What is the converse of that?"

VLAHAKIS: "If it's not rejected, then people are bound."

JEFFERS: "If it's approved, then people are bound?"

VLAHAKIS: "That's correct."

JEFFERS: "And that includes your client, the Museum?"

VLAHAKIS: "That's the converse of what it says, yeah."

She had hesitated only slightly, but realized she had no alternative than to give the answer she gave.

The net effect of Vlahakis's testimony was to focus the attention of the jury back where it had been since the beginning of the case. What did the Getty board approve by that 15-to-1 vote?

The Vlahakis deposition had been played over a period of three days and had dwelled at length on many of the points already outlined. By the time it concluded, Winokur had returned for the third day of his direct examination.

J. C. Nickens continued to ask questions designed to chip away at Pennzoil's claim of a binding agreement. The focus this time was the negotiating sessions with the Pennzoil attorneys on the final merger document on January 4 and 5. Nickens had Winokur describe several different points on which there was no agreement, bolstering Texaco's claim that the agreement wasn't binding because it didn't address all the essential terms.

Terrell kept his objections on a less personal basis for most of the afternoon, focusing principally on the issue of legal opinions from the witness. For his part, Judge Farris ruled with an even hand and the questioning proceeded accordingly.

Toward the end of the afternoon, Nickens entered the home stretch.

NICKENS: "Mr. Winokur, in your role as a negotiator and advisor to the company, did the Getty Oil Company have an agreement with Pennzoil on January 3rd?"

WINOKUR: "No."

NICKENS: "Did it have an agreement with Pennzoil on January 4th?"

WINOKUR: "Absolutely not."

NICKENS: "Did it, at any time, have an agreement with Pennzoil?"

WINOKUR: "No."

NICKENS: "How many mergers and acquisitions have you been involved with?"

WINOKUR: "Hundreds."

NICKENS: "Are you familiar with the customs and practice with regard to whether such transactions are reflected in writing?"

WINOKUR: "Yes."

NICKENS: "What is that custom and practice?"

WINOKUR: "They are always reflected in writing."

NICKENS: "We pass the witness, Your Honor."

To his credit, Nickens had come back and taken Winokur through a relatively matter-of-fact recitation of his evidence.

In the jury box, I had to set aside my dislike of the witness and consider the totality of his testimony. With any corroboration, it appeared there were indeed problems on the horizon for Pennzoil.

Irv Terrell rose to conduct the cross-examination for Pennzoil, and yet another transformation occurred. The Barton Winokur we had seen thus far was sweetness and light compared to the man who confronted Terrell.

That first afternoon, however, there wasn't much time left. After a preliminary round of questioning, Terrell launched into the legal equivalent of "Previews of Coming Attractions."

TERRELL: "Mr. Winokur, we're going to break in a moment, and
                  I want you to know what we're going to talk about
                  first thing in the morning so there won't be any sur-
                  prises.
                       "We're going to talk about the November 11, 1983
                  Getty Oil board meeting. You remember that, don't
                  you?"

WINOKUR: "I remember the part that I was at, yes."

TERRELL: "*Yes*! You were only there for part of it, weren't you?"

WINOKUR: "That's correct."

TERRELL: "You were only there for the part that Gordon Getty
                  was not there for. Isn't that right?"

WINOKUR: "When I was there, Gordon Getty was not there."

TERRELL: "And when you left, he was still not there, was he?"

WINOKUR: "That is correct."

TERRELL: "And in fact, you're the one that came into the
                  room . . . and talked to the board about bringing a suit
                  against Gordon Getty, to have a co-trustee appointed,
                  aren't you?"

NICKENS: "Your Honor, once again, I renew my objection . . . it's
                  immaterial and irrelevant to any issue that will be be-
                  fore this jury and I object."

TERRELL: "This man's credibility is before this jury. This man
                  came in and did something that I think this jury ought
                  to know about. There's been prior testimony about
                  this.
                       "This man's credibility, he's sworn here, he's the
                  lead witness for Texaco, they are vouching for his
                  credibility, and I think the jury needs to know the
                  facts. I think it's entirely relevant."

JUDGE FARRIS: "Are you going to go into this line of questioning now
                  or tomorrow?"

TERRELL: "I'll do it tomorrow, Your Honor."

With that, the jury was dismissed for the day. After we left, Judge
Farris had a few words for Terrell.

JUDGE FARRIS: "I don't want you or indeed any lawyer on either side
                  to tell the jury what a bad guy somebody is. Just go
                  ahead and question him.

"If they think you're a bad guy after they hear the answers, so be it. But don't again go ahead and start slicing on him."

TERRELL: "I won't, Your Honor."

On his first full day of cross-examination, Terrell did indeed take Winokur back through the November 11 board meeting. Although the jury had heard several accounts of what was generally called the back-door incident, Winokur was the first witness who had actually come through the back door to come to court and testify about it.

TERRELL: "My question is, Did you discuss the co-trustee suit with him out of this room?"

WINOKUR: "We discussed the likelihood that a co-trustee suit would be brought . . . and what role the company would play if such a suit was brought."

TERRELL: "Indeed, you recommended that the company intervene in that suit, didn't you?"

WINOKUR: "I don't know that I recommended the company to intervene. I discussed with the board the alternatives, how the company could react to that suit.
            "My own view was intervention was the best alternative."

Not only did he confirm the scenario as Terrell had outlined it, Winokur still defended his actions against Gordon.

Terrell launched a broader assault on the sweeping claims that Winokur had made on behalf of the Getty board, namely, that they had specifically intended not to be bound to the deal with Pennzoil.

There had been extensive battles between Nickens and Terrell when it appeared that Winokur was trying orally to modify and enhance the Copley notes to boost the dominant theory of the Texaco defense, "No contract."

At this point, much of the earlier Pennzoil outcry over the Copley notes became understandable. Since the notes didn't spell out the Pennzoil-Getty agreement in an absolute, definitive way, they couldn't be left to open interpretation. If the typewritten notes were indeed a literal translation with no additions or subtractions in content, Copley had done Pennzoil a big favor when he destroyed the original notes.

For his part, Terrell hammered away at inconsistencies between Winokur's testimony and the Copley notes.

TERRELL: "You have given some very important testimony in this case now, Mr. Winokur, about what was agreed and was not agreed at this board meeting and what I want to know is what do you remember individual board members saying as expressions of outrage in response to the 'Dear Hugh' letter in the board meeting, in front of Mr. Copley, the board's secretary, who was sitting there taking notes?"

WINOKUR: "Well, I cannot remember exactly which statements were made in the board meeting when I was there and which were made to me immediately after the session."

TERRELL: "In fact, if we look at the Copley notes at page 57 where this 'Dear Hugh' letter was read, there is not one thing said by anybody in the nature of outrage about the 'Dear Hugh' letter; isn't that right?"

NICKENS: "Your Honor, I don't understand the question. Is he asking the witness to say what was said or what is reflected in the notes?"

JUDGE FARRIS: "I think the latter. What is it, Mr. Terrell?"

TERRELL: "The latter, Your Honor."

JUDGE FARRIS: "He's just asking what's written in the notes, as opposed to what was said."

TERRELL: "Your Honor, may I address this?"

JUDGE FARRIS: "Yes, go ahead."

TERRELL: "He asked him on direct and he gave a very different answer than he did when I asked him on cross. He said he just can't remember."

JUDGE FARRIS: "If the notes were taken during the board meeting, wouldn't that reflect it?"

NICKENS: "Your Honor, I think it's been clear that Mr. Copley, a single person, sitting in the board meeting with thirty-something other people took down as best he could what was asked. It's very clear that it was impossible for him to have taken down everything that was said."

JUDGE FARRIS: "Are you saying that he took things down selectively?"

NICKENS: "I'm not saying that, Your Honor. He just had to take down what he was able to get down."

JUDGE FARRIS:	"I'll allow the question. I want to hear the answer."
TERRELL:	"What's the answer, Mr. Winokur?"
WINOKUR:	"What was your question?"
TERRELL:	"My question is, do you find any expression of outrage by one single, solitary soul in that board meeting about that 'Dear Hugh' letter upon the reading of page 57 of the Copley notes?"
WINOKUR:	"I don't see it on 57. But, let me tell you as I read this, the reading that you're referring to on 57 was a reading that in fact occurred when I was in the board meeting. I thought you were talking about the first time this was disclosed to the board, which is when I was not in the board meeting."
TERRELL:	"You mean you heard something said when you were not in the board meeting?"
WINOKUR:	"No. I thought you were talking about the first time that the directors had been informed of this letter, after which there was a break and I had a discussion with the directors."

Texaco's first witness would later return to the stand for another two days of cross-examination. His combative stance during cross tended to dilute the substance of his testimony.

Winokur made Terrell work for every answer, and went to often ridiculous lengths to prolong the questioning. After three months in the jury box, we were definitely not amused. Terrell's questions tended to be long and sometimes obtuse, offering Winokur even more opportunities for exchanges like this:

TERRELL:	"Mr. Winokur, arbitrageurs keep their eyes and ears open for information in a merger transaction or [when] negotiations in a merger transaction are underway, aren't they? They look for the best price?"
WINOKUR:	"Are you asking me if they keep their eyes and ears open or do they look for the best price?"
TERRELL:	"First, do they keep their eyes and ears open?"
WINOKUR:	"They do."
TERRELL:	"Do they look for the best price?"
WINOKUR:	"Always."

Making Terrell ask direct questions was one thing, but giving answers like the following was really bordering on the ridiculous:

TERRELL: "Mr. Winokur, before our afternoon break I was asking you to put yourself in the shoes of Pennzoil buying oil at $3.40 a barrel.

"And let me ask you, sir . . . you're a lawyer. We all know that. Do you have any experience as a businessman in buying and selling of oil?"

WINOKUR: "You mean, in large amounts of oil?"

TERRELL: "As a businessman, do you have any experience in buying and selling of any oil, in barrels of oil, large or small?"

WINOKUR: "I'm sorry. I just was trying to distinguish between the amounts that I buy for my car and things like that."

TERRELL: "Okay. You've got me there."

JUDGE FARRIS: "I'm not supposed to comment on the weight of the evidence nor on the evidence itself. Of course, the judges in the ugly yellow building do it all the time.

"But I will say that was pretty good."

TERRELL: "Leaving aside the purchases that you and I make and all of us make in gas stations and wherever one buys oil for your car or gas for your car, you understand when I'm talking about crude oil, don't you, Mr. Winokur?"

WINOKUR: "I've been in some oil and gas investment programs, too."

Winokur would have to return yet again to complete his testimony. After the next day, the trial would recess for a week to permit Judge Farris to fulfill a long-standing commitment to attend a judicial conference in another city.

Around this time, it seemed like the trial was starting to unravel as the third month came to an end. The case had obviously taken a heavy physical toll on Judge Farris; his previous heart problems were a topic of discussion among the jurors. There had been references to the judge's condition since voir dire. On one occasion, Jamail had reminded the judge he needed to take his medicine. Another time, the start of the trial had been postponed half an hour because the judge had a doctor's appointment. On that occasion, Judge Farris returned to the bench and announced, "Members of the Jury, I have been to my cardiologist this morning and, courthouse rumors to the contrary, he informed me that I *do* have a heart."

Against this background, I can never forget the day, well into the third month of the trial, when I first saw Judge Farris outside of the courtroom. As he got off the elevator and made his way to his chambers in street clothes (a business suit), he seemed smaller somehow, a lesser figure than the authoritarian who frequently dominated the courtroom.

His decline was rapid. On the bench, he sometimes seemed tired and listless. To be sure, much of that tedious testimony would have fatigued a younger man. It had surely fatigued me. The breaks in the trial got longer, as the bailiff would inform us the judge was taking a nap during lunch or resting during the afternoon break.

I could sympathize. Early in the fourth month of the trial, I found myself in the grip of some bug that produced waves of nausea. Having come that far (and become one of the final twelve), I resolved to stick it out. At least seven of the sixteen jurors and alternates would suffer some tinge of illness during our protracted tour of duty. In the war zone, life went on.

When the trial resumed after what had turned out to be a nine-day break, Winokur was still on the stand. What followed was two days of testimony, almost entirely cross-examination by Terrell, that reminded me of the worst stonewalling of the Watergate hearings a decade before.

No semantical point was too small for Winokur to argue over, no commonly accepted definition of a term acceptable. It was a prime example of the type of excruciating word play that gives lawyers a bad name with so many people.

In the jury box, the longer the Winokur testimony continued, the more it seemed like fingernails scratching on a blackboard. For Pennzoil, this achieved the desired effect—to obscure that portion of Winokur's testimony that was damaging to its case in a downpour of verbiage. Ironically, it was the combative style of Winokur that indirectly helped Terrell achieve this end for the plaintiff.

26

After Winokur left the stand, Texaco called Geoffrey Boisi of Goldman, Sachs. Getty Oil's investment banker's video deposition had been played more than two months before, during Pennzoil's case.

Dick Miller handled the direct examination for Texaco, and it was obvious he thought Boisi was a strong witness for his client. Practically the first defense advanced by Miller had been, "We didn't crash this party, we were invited." Boisi, the young man seated on the witness stand, was the one who had issued the invitation.

By the time Boisi testified, the jury had heard a detailed account of how he had shopped the Pennzoil-Getty deal all over town. DeCrane's notes record a conversation with Boisi, who had been presented to Texaco by Goldman, Sachs chief John Weinberg.

Miller noted the presence of the Goldman, Sach's attorney representing Boisi who was seated in the courtroom. I remembered from his video deposition that an attorney had been seated by his side throughout that questioning also.

Boisi's quest for a price that would be fair to the public stockholders was the motive he had staked out to justify all his actions. There must have been concern within Goldman, Sachs that Boisi's actions could somehow have negative repercussions for the firm, because Boisi was extremely guarded in his answers, even to questions from Miller. This made him to appear to be walking a tightrope, and like Winokur, he sometimes took refuge in semantics.

That the 15-to-1 vote was for price only, Boisi readily confirmed, but the price of $112.50 was one on which he was not prepared to give a fairness opinion. Besides establishing the bona fides for Texaco's invitation, Miller sought in his brief examination to enhance the bully image he had been painting of Pennzoil from the beginning of the trial.

As you have probably figured out by now, this meant the "Dear Hugh" letter above all else. Gordon Getty's promise to seek to have the

370

Getty board ousted if its members blocked the Pennzoil plan had created tremendous pressure on the directors, had it not? Miller asked.

BOISI: "They felt under even more pressure [after the "Dear Hugh" letter was disclosed]. This was the first time true acknowledgment was made that they would be thrown out of their positions if they did not knuckle under to the proposal. There was a combination of outrage and a feeling of terrific pressure."

MILLER: "No further questions."

Boisi had obviously sufferd at the hands of Irv Terrell during his video deposition, but that could hardly have prepared him for what confronted him in the courtroom.

It's clear, after listening to his questioning style on video depositions, that Terrell comes alive in the courtroom with spectators, a judge, and especially a jury to catch his act. He unleashed his full complement of weapons on Boisi, but with almost an air of restraint.

While he had gone after Winokur with relish, no holds barred, he held back slightly with Boisi, recognizing that the man did not have a hard shell of self-defense like that Winokur had erected. In the end, it just meant that Terrell did not have to try quite as hard to savage the testimony of the man from Goldman, Sachs.

Terrell immediately jumped right in on the last point of Boisi's direct testimony, and asked increduously:

TERRELL: "Are you telling the jury that fifteen members of the board of directors, when they voted 15 to 1 on the evening of January 3rd, did not have the courage to stand up and say, 'No, you bullies, we won't do this,' that only Chauncey Medberry could muster the courage to ward off Pennzoil?"

BOISI: "I am not saying that."

TERRELL: "Okay."

As surely as the sun rises in the east, the subject of the November 11 board meeting came up. Boisi was on his guard; in fact, his evasive answers about the subject had been played during his video deposition. Of course, Terrell went right back into this subject when he had nervous Geoff sitting on the stand.

TERRELL: "You remember the November 11, 1983, board meeting in Houston, don't you?"

> BOISI: "I remember the November meeting. I don't remember the day."
>
> TERRELL: "And you remember it was in Houston?"
>
> BOISI: "I remember it was in Houston."

From his reactions, this courtroom in Texas was as foreign to Boisi as the Soviet Union. He ducked, dodged, and squirmed his way through minimal answers as Terrell had him confirm his role in coming through the back door with Winokur to advocate the lawsuit against Gordon.

Whether it was actually the back door or the side door or the front door or a hatch in the roof doesn't change the duplicity of the tactics used on Gordon Getty. But if you hear somebody wanting to argue about which door it actually was, you have encountered a smokescreen, my friends.

The real damage to Texaco from Boisi's testimony came on the subject of the invitation to bid that Boisi had offered to DeCrane on the morning of January 4. The circumstances of the call had been duly recorded by DeCrane on page 1 of his notes. When asked the critical question, however, Boisi waffled.

Conscious of the Getty bylaws that required the consent of fourteen of the sixteen directors to take any major action—like trying to sell the company—Boisi said that he did not solicit a bid from DeCrane on January 4. He had called him as a "courtesy," since Texaco had indicated a desire to be kept abreast of developments. Boisi made a number of calls to other potential bidders that morning which he now characterized the same way:

> TERRELL: "So your evidence is that when you talked to Mr. DeCrane on the 4th, you did not solicit an offer, it was merely a courtesy call?"

Boisi wouldn't give a direct answer to the question, but dragged out a string of equivocation.

> TERRELL: "Was the answer to my question yes or no?"
>
> BOISI: "Well . . . "
>
> TERRELL: "Yes or no?"
>
> BOISI: "I would say no."

The harm this exchange did to Texaco's theory of the case was compounded by Terrell's skillful follow-up questions that gave the lie to the defense Miller had been building since Day One.

TERRELL: "Did the board vote at any time on January 2nd and 3rd to authorize you to go out and sell the company?"

BOISI: "No, not in a formal sense."

The answer to that question was a foregone conclusion. There had been no mention of such a vote in the sixty-five pages of the Copley notes. No authorization for Boisi to sell the company had even been sought by Chairman Petersen.

Every member of the jury was familiar with the bylaws imposed on the Getty board by the Museum and the Trust under the Consent. Gordon only had to attract two votes to block any board action.

This was the true legacy of the back-door board meeting. Did Pennzoil relentlessly harp on this issue? Of course it did, and to great effect. It *was* relevant to the facts of this case. Texaco was now sponsoring the authors of that duplicity, and it was heavy baggage to carry. When these men came to court to testify for them, it was only natural that Pennzoil would impeach them with their participation in that treachery. The direct connection here was that by thwarting the Pennzoil deal, they had again defeated the ambitions of Gordon.

The irony of the back-door meeting was that the Getty board's efforts to hamstring Gordon directly led to drastic limits on the board's authority and, ultimately, the demise of the company.

There may be a lesson there somewhere . . .

For Texaco, the case was not going particularly well. Its first two live witnesses had not been well received in the jury box.

Winokur had brought some compelling evidence for Texaco's case, but it had been wrapped in a poisonous package. As we have just seen, Boisi started out all right, but on cross-examination he gave Pennzoil some vital testimony. While protecting Goldman, Sachs' flanks, he left Texaco open for a full frontal attack.

The next day, Texaco would seek safe harbor in the pages of a transcribed deposition, the first of twelve that would be read to the jury during Texaco's defense. While the reading of these depositions was a long, slow process, there was little chance of the harmful surprises Texaco had encountered thus far with its live witnesses.

27

Richard Keeton stood before the bench. "This is Mr. Martin Lipton of New York, Your Honor," he drawled.

Lipton, immaculate in a dark-blue pin-striped suit, took the witness stand. He peered out from behind thick glasses as Keeton began his examination. If Lipton was at all nervous, his manner and expression did nothing to betray it.

KEETON: "Mr. Lipton, this jury has heard your name quite a bit so far in these many weeks . . ."

With this typical understatement, Keeton began the questioning of Lipton. In fact, there had been very few witnesses in the trial who had not mentioned his name. Lipton had been active in virtually every battle of the Six Day War except the first, and his New Year's Day party had served as sort of a rear guard staging area for that one.

Hugh Liedtke had left little doubt of his animosity for Lipton. Gordon had described Lipton's central role in convincing him he was forced to sell to Texaco, but the DeCrane and Lynch notes had perhaps said it best: "KEY PERSON—MARTY."

So Keeton was on fairly safe ground when he said the jury had heard the man's name a time or two.

Listening to Lipton talk about financial dealings reminded me of hearing Henry Kissinger give a lengthy discourse on some obscure aspect of foreign policy. Although Lipton didn't have a German accent, he was vaguely reminiscent of Kissinger.

After the trial, I read an article in *American Lawyer* about the blood in the water phase of this story that described Lipton's efforts in obtaining the ill-fated Standstill Agreement as "shuttle diplomacy" and quoted an unidentified participant as describing Lipton as "running around like some kind of goddamn Henry Kissinger."

Well, maybe it wasn't an original thought, but as Gordon Getty had said about another matter, "Maybe we both thought of it independently. It was my idea; I'm not saying it was not their idea also."

Keeton's questions covered what was by now well-worn ground for the jury. Indeed, we had heard so much about Lipton from all sides that initially there were very few surprises in his testimony.

When the story turned to the night of January 5, when Texaco was coming to call even as the Pennzoil-Getty final merger documents were being prepared, Keeton moved to defuse the attack on Lipton's double dealing on that evening, specifically, the charge that he had kept Vlahakis away from the negotiating session with the Pennzoil lawyers.

KEETON: "Did you prevent her from going to the meeting or ask her not to go to the meeting in order to in some way delay or frustrate the negotiations with Pennzoil?"

LIPTON: "No."

JAMAIL: "Your Honor, I object. The answer he's going to give is obvious. It speaks for itself. He told the lady not to go to the meeting and now to explain or alibi or make some self-serving statement by this witness as to why he did it, I object."

KEETON: "Your Honor, the motivation which Pennzoil has put in issue on this question is certainly one that the man can state and that's the question I asked."

JUDGE FARRIS: "Mr. Jamail, you will, of course, have the opportunity on cross to correct that if you feel that it needs correcting. The objection is overruled."

JAMAIL: "And I will."

Lipton would have to deal with Joe Jamail when his direct examination was over, and Jamail had just served notice that it would be no day at the beach if he had anything to say about it.

While the jury was on a break, Judge Farris inquired how long Jamail thought his cross-examination would take.

JAMAIL: "You've heard me, Your Honor, and I think I will be a full three days with this man."

JUDGE FARRIS: "What man?"

JAMAIL: "Lipton."

MILLER: "That's okay with us, the longer the better."

JAMAIL: "That's right. You think that."

There was no let up in the war of nerves between Miller and Jamail. If Miller had any trepidation about putting Lipton on the stand (and he did), the last person in the world he would tell was Jamail.

The Texaco meeting with Gordon at the Pierre on January 5 was brought up in due course. Lipton had been there, of course, and had played a significant role in the outcome, according to Gordon. Texaco's defense said that wasn't true, but why was Lipton there?

KEETON: "All right. Did you at some point that evening receive a phone call inviting you or asking you to go to the Pierre Hotel?"

LIPTON: "Yes."

KEETON: "And from whom did you receive that phone call?"

LIPTON: "Either from Martin Siegel or Tom Woodhouse. . . . There had been a meeting between Gordon Getty and John McKinley that had not gone well, and they thought it might be helpful if I came to the Pierre Hotel."

Tom Woodhouse was the chief lawyer representing Gordon at the Pierre that night. His partner, Charles Cohler, you will remember, was in California trying to get the court there to lift an injunction so Gordon could sign the final merger document with Pennzoil.

Lipton's arrival at the Pierre that night had indeed proved helpful to Texaco, as Gordon had testified. By giving this testimony at the trial, however, he seemed to be validating Pennzoil's theory of the case.

As with many witnesses, the questions jumped back and forth from event to event. Earlier that same evening, Lipton had met in his offices with the Texaco contingent sans McKinley. The Lipton agenda, an exhibit in this case, had contained a list of demands to present to Texaco; prominently featured was the indemnity.

KEETON: "And did the Texaco people make any comments when you made that request?"

LIPTON: "Yes, they did not want to give that indemnity and they wanted an explanation of all of the background . . . they questioned me closely as to whether I believed the Museum was in a legal position to contract with Texaco, whether they were breaching or inducing the breach of a contract."

JAMAIL: "Your Honor, I object to any legal opinions he may have given. This is totally and solely for the jury."

KEETON: "Your Honor, the witness said what the Texaco people were questioning him about. He's not giving a legal opinion."

JAMAIL: "The witness has just said, 'They were asking me to give them an interpretation of whether they were breaching the contract or whether Texaco had induced the breach.'

"Now, he's been an all-purpose type of witness but we can't put him in the jury box. This jury is going to make that determination, not Mr. Lipton."

Richard Keeton had done a fairly efficient job of leading Lipton through his story, hitting the high points that bolstered Texaco's case. To my mind, Texaco hadn't really had much choice in the matter. Lipton had played such a prominent role in the case that his absence would have itself been grounds for suspicion.

There was an element of risk in all this, however.

Joe Jamail rose to begin the cross-examination of Martin Lipton. Although always combative, Jamail seemed to have raised his intensity level a peg for this encounter. He stood at the Pennzoil table and asked his first question.

JAMAIL: "Mr. Lipton, you seem to be a man who uses his words precisely. Do you?"

LIPTON: "I try to."

JAMAIL: "Did I understand you earlier to testify that not one of your clients ever intended to be bound [prior to] signing a definitive merger agreement?"

LIPTON: "That's correct."

JAMAIL: "And you meant it when you said it?"

LIPTON: "Yes."

JAMAIL: "Were you involved in the acquisition by Esmark of Norton Simon?"

LIPTON: "Yes."

Jamail picked up a document from the table and showed it to Lipton.

JAMAIL: ". . . There was an acquisition agreement that was entered into by the parties and there was later going to be a definitive merger agreement. . . . I'd like you to look what I have now marked as Plaintiff's Exhibit 445 and ask you to take a look at it. . . . Do you recognize it, Mr. Lipton?''

LIPTON: "Yes."

JAMAIL: "Now, Mr. Lipton, isn't it true that the parties, Esmark [and] Norton Simon intended to be bound by an acquisition agreement, this two-and-a-half-page document that you have in front of you?''

LIPTON: "Yes."

JAMAIL: "And there was to be a definitive merger agreement, was there not?''

LIPTON: "Yes."

JAMAIL: "So it is not really precise, is it Mr. Lipton, to say that not one of your clients ever intended to be bound without the signing of a definitive merger agreement?''

LIPTON: "No, I don't agree with that."

JAMAIL: "All right, sir . . . You say that . . . You are uncomfortable with that even though this agreement is not a definitive merger agreement?''

With that first question, Jamail had nailed Lipton to the wall with his own words. It was an auspicious beginning for an eventful cross-examination.

  Jamail immediately moved to the next subject, Lipton crony Martin Siegel.

JAMAIL: "You say Mr. Siegel is a man who used words precisely, correct?''

LIPTON: "Yes."

JAMAIL: "And Mr. Siegel is a knowledgeable man in the merger and acquisition trade?''

LIPTON: "Yes."

JAMAIL: "And a man who would not use words foolishly if under oath especially, would he?''

LIPTON: "Yes."

  Jamail had Lipton read aloud the affidavit confirming the Pennzoil-Getty deal that Siegel had telexed to the California court the day Texaco made its move.

JAMAIL: "Mr. Lipton, he used the words, 'transaction now agreed upon,' didn't he?"

LIPTON: "Yes, he did."

JAMAIL: "Well, do you believe he knew what he was saying when he said that?"

LIPTON: "No."

JAMAIL: "You just disagree, is that it?"

LIPTON: "I disagree."

JAMAIL: "Yet you find him to be a man precise, who was under oath and he stated this as late as January the 5th. You've never been shown this, have you?"

LIPTON: "I have not."

Score another point for Jamail. The jury, of course, had long been familiar with the Siegel affidavit. Lipton's all-encompassing knowledge had fallen short of one of the critical pieces of evidence in the case.

Jamail has set the tone for his cross-examination in those first few minutes, but it would go on from there. If Lipton was on his guard, he didn't show it in some of the answers he would offer the next morning.

Jamail took Lipton through almost all the defenses raised by Texaco. The idea that Pennzoil had applied undue pressure to try to force its deal on the Getty board was confronted head on.

JAMAIL: "When did you first become aware of the Pennzoil tender offer, Mr. Lipton? Did you feel threatened and under pressure at that time?"

LIPTON: "Yes, I did."

JAMAIL: "Just a mere tender offer put you under pressure? Does it cause you to act irrationally?"

LIPTON: "Well, I sure hope not."

JAMAIL: "You do know how to react under pressure, don't you?"

LIPTON: "I hope so."

JAMAIL: "Well, you're paid large amounts of money to help people under pressure learn how to react, to do what they have to do to ward off tender offers, and to do all these things you've told us about. Isn't that correct?"

LIPTON: "Well, my professional life is basically one of advising targets of tender offers how to respond to them. That's correct."

JAMAIL: "People under pressure? You're paid large amounts of money to go and advise these people as to how to handle these things, aren't you?"

LIPTON: "That's correct."

JAMAIL: "To make sure they don't act unreasonably under pressure, right?"

LIPTON: "That's correct."

JAMAIL: "To make sure they don't succumb to pressure?"

LIPTON: "That's correct."

JAMAIL: "You felt terribly pressured?"

LIPTON: "Yes, I did."

JAMAIL: "You didn't crater, did you?"

LIPTON: "I hope not . . ."

JAMAIL: ". . . Are you here telling this jury that you, the leading takeover lawyer in the country, could not prevent that board from just cratering because of this assault by Pennzoil?"

KEETON: "Your Honor, Mr. Jamail's introduction of Mr. Lipton's expertise may be correct, but there has been no testimony from this witness that he is the leading takeover expert in this country, even if he be that."

JAMAIL: "I did not hear Mr. Lipton deny it, but I will withdraw it."

Finally, in response to the last barrage of questioning, Lipton did allow that he didn't think that pressure from Pennzoil was what motivated the directors, that in fact the pressure had existed before Pennzoil came into the picture.

Another Lipton contention that backfired was the idea that there could be no binding agreement because the deal had not involved lawyers who were oil and gas specialists.

LIPTON: ". . . And they would need expert advice in the oil and gas business to work out a real agreement with respect to that."

JAMAIL: "You mean they can't agree without hiring a bunch of experts to tell them what they think they want to do?"

LIPTON: "I think that the Getty Trust and Kidder, Peabody and Mr. Siegel would need considerable expert help in working out any agreement along those lines, yes."

JAMAIL: "Mr. Lipton, I want to be real sure that I understand this.
            "Are you saying that two people cannot agree unless they
            hire a bunch of lawyers to tell them they've agreed?"

LIPTON: "No . . . I'm not saying that at all, Mr. Jamail.
            "I'm saying that two people who are contemplating an
            agreement with respect to a ten-billion-dollar transaction
            would be awfully foolish to do it on the basis of an outline
            and the absence of experts' advice."

JAMAIL: "Mr. Lipton, are you saying that you have some distinction
            between just us ordinary people making contracts with each
            other and whether or not it's a ten-billion-dollar deal? It's a
            different standard in your mind?"

LIPTON: "Yes, indeed."

At that statement, my jaw dropped. Lipton had just affirmed Pennzoil's whole case, as I saw it. Of course, there was more.

JAMAIL: "I see. So if it wasn't a bunch of money involved in this Getty-
            Pennzoil thing, it could be an agreement?"

LIPTON: "Well, if there was five or ten dollars involved, I guess you
            might say that . . .
            "All I'm saying is that in a transaction of that kind, I would
            expect that Mr. Getty would have other lawyers. Every time
            I have been involved in a transaction that involves the oil and
            gas business and operating agreements and so on, I've con-
            sulted with experts in that field."

JAMAIL: "I'm sure the legal profession would be very grateful to you
            for creating all this business, but that is not the point. The
            point is that people can agree without lawyers, can they
            not?"

LIPTON: "Yes."

It would later be suggested that Lipton was referring to a legal principle codified in 17th Century England known as the Statute of Frauds,
which held that complex transactions of great size must be evidenced in
writing. Despite this contention, the Statute of Frauds was *never* at issue in this trial.

Jamail's questions came back around to the Memorandum of Agreement, which Lipton had attempted to disavow, even though his client
had signed it. Jamail confronted Lipton with the signature.

JAMAIL: "Mr. Williams signed that, did he not?"

LIPTON: "Yes, sir."

JAMAIL: "And you didn't tell him not to sign it, did you?"

LIPTON: "No, I did not tell him not to sign it."

JAMAIL: "You advised him to sign it, didn't you?"

LIPTON: "I advised him to sign it, yes, sir."

JAMAIL: "And you had no reservations at that time about signing it?"

LIPTON: "No, I had no reservations about his signing it."

JAMAIL: "You weren't trying to trick Gordon Getty into some compla-
            cent posture, were you?"

LIPTON: "No."

JAMAIL: "You meant it when you had Mr. Williams sign it, didn't
            you?"

LIPTON: "Yes."

JAMAIL: "And he meant it?"

LIPTON: "He certainly meant to sign it, yes."

JAMAIL: "Well, did he mean to honor his signature?"

LIPTON: "Well, I can't answer that question because I don't know
            what you're referring to. I'm sure Mr. Williams always hon-
            ors his signature."

JAMAIL: *"That's exactly what I'm referring to.* He had no pressure ap-
            plied to him by Pennzoil, you've already told us . . .
                "Well, I want to look at this Memorandum of Agreement
            with you for just a minute."

Jamail reviewed the signature page of the document.

JAMAIL: "And it is signed by Mr. Harold Williams who is president of
            the Museum?"

LIPTON: "Yes."

JAMAIL: "Now, he's written in some language. Or did you write that
            language in for him, please?"

LIPTON: "I wrote the language."

JAMAIL: "We want to come back to that and we'll talk about it in just
            a moment."

Jamail detoured briefly to touch on the Museum escrow agreement,
negotiated by Pennzoil with Patricia Vlahakis on behalf of the Museum. Under that agreement, Pennzoil had amassed $1 billion in cash
overnight to put the Museum shares in escrow.

JAMAIL: "This escrow agreement prepared by your partner would ensure that you had some advantage over the other shareholders, financial advantage?"

LIPTON: "You said Ms. Vlahakis was my partner. She's my *associate.*

JAMAIL: "Excuse me. I strike that.

           "The rest of the statement is true, is it not?"

Lipton's statement was not as flat as those words sitting on the page might indicate. He seemed almost indignant that Jamail would elevate Patricia Vlahakis to partner status. Jamail realized it too, and referred to this distinction Lipton had drawn on several occasions. A minor point, but it seemed like a definite slap at someone the jury had to admire for having some guts, unlike some of the other high-powered professionals we had seen.

Again and again, however, Jamail hammered at the essence of the lawsuit, the deal. Lipton wouldn't give in, but his repeated invoking of technical, white-is-black, black-is-white reasoning found no takers in the jury box.

JAMAIL: "Now, Mr. Lipton, you have told this jury that you attempt at all times to be a very precise person, correct?"

LIPTON: "Yes."

JAMAIL: "And if you want to make something clear, you make it clear, don't you?"

LIPTON: "I try to, yes."

JAMAIL: "You wrote this, printed in some language at the bottom of this Plaintiff's Exhibit 2.

           "Would you mind reading it aloud? The jury has read it, but I would like to focus on it. It's right under the signature of the J. Paul Getty Museum that Mr. Williams signed."

LIPTON: "It says, 'Upon condition that if not approved at the January 2 board meeting referred to above, the Museum will not be bound in any way by this plan and will have no liability or obligation to anyone hereunder.'"

JAMAIL: "You understand how the construction of the English language is, do you not, sir?"

LIPTON: "I hope so."

JAMAIL: "Well, sir, is this just a game that was going on up there for two days? Nobody is bound but Pennzoil?"

LIPTON: "No, absolutely not. It was not a game at all."

JAMAIL: "Mrs. Vlahakis is your colleague, as you put it, associate."

LIPTON: "Yes, sir."

JAMAIL: "You find her competent?"

LIPTON: "Yes."

JAMAIL: "You find her a precise person?"

LIPTON: "Yes."

Jamail then read Lipton her testimony that the converse of what he had written under the Williams signature line was that if the plan is approved, the Museum is bound. Lipton agreed that was the converse, but it did not mean the Museum was bound.

Later, he seemed to be trying to distance the Museum from the Memorandum of Agreement even further.

LIPTON: "No, this whole plan was not something that the Museum wanted at all."

JAMAIL: "Well, did somebody have a gun at Mr. Williams's head?"

LIPTON: "No."

JAMAIL: "Did somebody force him to make the motion to adopt the Pennzoil proposal?"

LIPTON: "No."

JAMAIL: "He did it, didn't he?"

LIPTON: "No."

JAMAIL: "He didn't move to adopt the Pennzoil proposal?"

LIPTON: "What Mr. Williams did was move to approve going forward with discussions and negotiations with Pennzoil on the basis of a hundred and ten in cash and a five-dollar note, but he certainly didn't move to approve what you are referring to as the Pennzoil proposal, no.

JAMAIL: "Was Mr. Copley sober when he was taking these notes?"

LIPTON: "I don't know, sir. I don't know."

JAMAIL: "Would you look at page 65? And look at that last paragraph. Well, just read it aloud."

LIPTON: "'Mr. Williams then moved that the board accept the Pennzoil proposal . . .'"

JAMAIL: "Well, am I missing something or did Mr. Williams move the board accept the Pennzoil proposal?"

LIPTON: "The answer is yes, and I think that the answer needs to be explained as to what the proposal was."

Lipton's explanations were destined to fall far short of the mark. I don't know what their expectations were, but by the time Marty Lipton left the stand, he had driven the last nail in Texaco's coffin. I first wrote that sentence two days after the trial ended, and it is as true today as it was then.

28

Martin Lipton's second day on the stand was a Friday, and the Jamail-Lipton clash then in progress had been the most memorable battle since Miller and Liedtke had locked horns.

When we returned to the courthouse on Monday for resumption of the duel, it was with a sense of anticipation that had been missing for much of the trial.

Getting off the elevator on the fifth floor with a group of jurors, we were met by a worried looking Carl Shaw. "Go home, the Judge is sick. We'll call you later and tell you when to come back," the bailiff said.

Although Judge Farris was tougher on him than anyone else, Carl Shaw was obviously devoted to his boss. The grave nature of his concern was written all over his face, and we wondered what he knew that he wasn't telling. By then, we had all long since become aware of Judge Farris's heart problems of recent years and had observed the deterioration in his health in recent weeks. He was obviously not a well man. The week before, we had noticed him turning blue around the mouth as he sat up on the bench.

On the way home that night, I wondered if he would survive the trial. I sometimes thought of this trial as an eternal beast, formless and without end, consuming those who had haplessly wandered in its path.

Cliff Bennett, the court clerk, called the next day and said the trial would resume in one week. No, nothing was known about the Judge's condition.

When we returned, the case had been assigned to a new judge. Judge Farris had been forced to step down. Although you couldn't have convinced me of it a month before, I was going to miss the gruff old lion. Later, we would learn that the problem was not his heart, but a recurrence of a previous bout with cancer.

* * *

Judge Solomon C. Casseb was a retired state district judge from San Antonio who was pressed into service after the trial had run fifteen weeks. Visiting judges are used all over Texas, especially in large cities where one lengthy trial can crowd out the other cases on a judge's docket. In such instances, visiting judges are appointed to hear these cases. Although it had been well over a decade since he had occupied a permanent seat on the bench, Judge Casseb had recently returned to the courtroom as a visiting judge.

His arrival ten days after Judge Farris had stepped down brought big changes to the way the trial of *Pennzoil Co.* v. *Texaco, Inc.* was conducted.

A courtly jurist, Judge Casseb nevertheless served notice the first day that the operation of the case would be tightened up considerably. Immediately, lunch breaks were slashed from ninety minutes to one hour. The new starting time would be 8:30, instead of 9:00. The court would not recess until 5:00 instead of 4:15. In addition, we would now be in session for a full day on Mondays and Fridays.

Judge Casseb also resolved to speed things up during the proceedings themselves. When an attorney would rise to make an objection, the typical exchange would go like this:

BROWN: "Your Honor, we object to—"

JUDGE CASSEB: "Sustained. Proceed, Counselor."

Protracted conferences at the bench all but became a thing of the past. Judge Farris had frequently liked to have the full give-and-take of discussion and argument from the attorneys before ruling on a motion. This encouraged counsel for both sides to launch into flights of rhetorical fancy at the slightest provocation.

Judge Casseb, on the other hand, wanted objections cited in a direct, concise manner and made his rulings accordingly.

Getting adjusted to the new order took a little time; the differences between the two judges were pronounced. The new schedule also worked a hardship on many of the jurors, who had already drastically altered their itineraries to meet the demands of the trial. In light of Judge Casseb's work ethic, the pressure began to build. It wasn't too long before it blew a gasket.

Two days after the jury returned to resume the trial was Halloween. Linda had previously arranged activities with her child, based on the schedule we had followed for almost four months. The unexpected change in hours created a conflict. She appealed to the judge for an accommodation, without success. As it turned out, Theresa had a similar problem that she approached the judge about later that same day.

In the face of this second request, Judge Casseb announced a one-time-only dispensation for that particular afternoon. Simple, right? Wrong.

Linda was black, Theresa was white. There was immediate suspicion on the part of two black jurors that Judge Casseb was prejudiced, since he had refused permission upon request from a black juror, then granted it when asked by a white juror.

When I returned from lunch, this tempest was brewing in our little jury teapot. "What's going on?" I asked innocently.

"The Judge is prejudiced," Ola flatly stated. Based on previous experience, I detected a fledgling protest movement in the works here.

"Please, tell me what happened," I said. "If that's true, let's go back to his chambers right now and we'll tell him it just won't do."

They were singularly unimpressed with my gesture of solidarity, but told me the story anyway. "Wait a minute," I interjected, "Don't you think the same thing would have happened if Theresa had gone in there first?"

Ola was still a little perturbed with Judge Casseb because the change in hours had wreaked havoc with her bus schedule, but Linda seemed to accept the logic of this rationale, which was undoubtedly true.

This first note of racial tension had an unsettling effect on my perceptions of our little group. Frankly, racial differences hadn't really been an issue to date because there was such a diversity of age, geography, employment, and experience among the jurors that race was somewhere down the list. As I've indicated, this group of sixteen (now fifteen) people would not have come together under any other circumstances imaginable. All in all, we had done remarkably well together to make the best out of what all agreed was a bad situation.

The length of the trial and the necessity of making the adjustment to the new judge had produced circumstances where a misunderstanding could blossom into an incident. Judge Casseb's innocently intended efforts had merely been the catalyst. The resolution of the matter was of no great satisfaction to anyone, but the trial took precedence over such concerns.

One of the best witnesses Texaco called to the stand was Henry Wendt, a member of the Getty Oil board of directors. The CEO of pharmaceutical giant SmithKline Beckman, Wendt was a participant in the January 2–3 board meeting and cast one of the fifteen votes for what the Copley notes defined as the Pennzoil proposal.

Like several other witnesses who had testified at the trial, Wendt was obviously a man who was used to giving orders instead of taking them. He was decisive in his answers to Bob Brown, who handled the direct examination for Texaco, and frequently abrupt with John Jeffers, who put him through a relatively intricate cross-examination.

Wendt testified forcefully that the 15-to-1 vote was for a price proposal only, and took exception to the description given in the Copley notes and the press release. He did have a piece of direct evidence, a handwritten notation made near the end of the board meeting that detailed the stub that brought the $110.00 price to $112.50. The Wendt note, as of course it became known, stopped there, however; it was not conclusive either way in shedding any more light on that critical vote.

The best thing about this testimony for Texaco was Wendt himself. Without a financial interest in the outcome of the deal, the authoritative corporate chieftain could be said to speak from a relatively disinterested position.

He testified that there was no binding agreement ever with Pennzoil and took a cue from Brown and roundly criticized Pennzoil's handling of the negotiations. He reserved special scorn for the "Dear Hugh" letter and expressed the belief that Gordon Getty was trying "to sell the other shareholders down the river."

On cross-examination, Jeffers reminded the jury that Wendt had been on the board throughout the long struggle between Petersen and Gordon, much to the irritation of the witness.

Jeffers handed him the Siegel and Cohler affidavits to the California court. Wendt branded them both as incorrect, saying he didn't agree with their "phraseology."

He couldn't make much out of the Garber note, dismissing it practically out of hand.

The reason for this line of questioning was not to review these documents for the jury yet again, but to demonstrate the sharp contrast between that evidence and the testimony of the witness. Jeffers never did manage to impeach the testimony of Wendt, but he beat on the documentary evidence hard enough to prevent anyone from seizing the inconclusive Wendt note as some sort of definitive signpost.

On the balance, though, this round went to Texaco on points. No knockdowns, just a series of solid jabs.

While Wendt had been on the Getty board for several years before the troubles of 1983, Chauncey Medberry III was a long-time board member put in by J. Paul Getty himself. Medberry, the retired chairman of the Bank of America, had cast the sole dissenting vote against the Pennzoil proposal approved at the end of the board meeting.

Marty Lipton had described Medberry as "a special case," and indeed, the wily curmudgeon was one of the most colorful witnesses in the trial. Medberry was a hard-liner who expressed a series of opinions that went beyond the contentions Texaco had made in this case. His attitude toward Gordon made it clear he considered him to be just this side of an idiot.

Pennzoil was a villain of the worst sort:

MEDBERRY: "Here is a buyer, sensing an opportunity for a quick kill, hovering around, pressing. They had encamped two of the major shareholders who, for different reasons, would have favored a quick deal with Pennzoil, and the other directors were sort of handcuffed . . ."

Not Chauncey Medberry, however. He described his lone holdout on that vote by thumping himself on the chest and saying "Medberry was going on the record."

He even contributed to the rampant distrust of Lipton by making it clear that he disdained the actions of the man he called "Mr. New York lawyer."

During his brief time on the stand, Medberry did deliver the message Miller had brought him from California to give:

MEDBERRY: "You would not sell a ten-billion-dollar corporation based on a discussion that takes place in the middle of the night. There would have to be a document with all terms agreed upon, hammered out, negotiated. That's the way business is conducted in this country."

When John Jeffers began his cross-examination, he wasn't starting from scratch. The jury was well aware that the co-trustee the lawsuit authorized in the back-door board meeting sought to appoint was the Bank of America, whose former chairman was now on the witness stand. Jeffers gently extracted those details from Medberry, who overreacted to many of his questions.

In addition, the jury had seen a blushing Gordon recount how Arthur Liman had referred to Medberry as a "duplicitous prick," which I suppose was the flip side of Lipton's description.

When Jeffers showed him the Memorandum of Agreement, Medberry blurted out, "I know what you want me to say, and I'm not going to say it."

Jeffers stopped and raised his eyebrows, frankly appraising the witness. "Mr. Medberry, I'm not trying to make you say anything, I just wanted to ask you . . ."

Medberry was so intent on not being tricked into confirming Pennzoil's deal that he was caught totally off guard when Jeffers deftly changed direction.

JEFFERS: "Bank of America makes money financing unsolicited tender offers from time to time, doesn't it?"

MEDBERRY: "It has in the past . . . it reviews them very carefully before it does it."

JEFFERS: "You don't have any ethical or moral objections to the bank making some money on unsolicited tender offers?"

MEDBERRY: "Well, now, let's not be quite so cavalier about it . . ."

Medberry went into a little speech to the effect that *his* bank didn't lend money to just any back alley snatch 'n grab takeover effort.

JEFFERS: "Did you know that the Bank of America was one of the banks that was providing the financing for Pennzoil's tender offer for Getty?"

Jeffers had again neatly underplayed what was a devastating blow to this witness's heretofore unflappable corporate Sunday School lesson.

There was no point in a protracted effort on redirect for Texaco, because the jury had gotten the message loud and clear from Medberry. He had been the only Getty director to vote against the Pennzoil-Getty deal, and nearly two years later he was still voting against it in every way he could.

The final score on Medberry's testimony was at best a draw for Texaco, which appeared to be desperately in need of a knockout punch.

29

Judge Casseb had indeed expedited the progress of the trial, but it was hard to get any real sense of momentum.

Between live witnesses, Texaco would offer portions of depositions that would be read into the record, usually by two of the lawyers. The depositions were long and drawn-out because, as with the video depositions, both sides would make designations of material to be presented to the jury.

Although Dick Miller had conducted most of the depositions, he would get on the witness stand and read the part of the witness; one of his partners would take the part of Miller himself.

At every answer that could be considered remotely meaningful, Miller would have his voice swell slightly and end with a pregnant pause when he looked at the jury. "Isn't that some significant stuff?" his look seemed to say. At many such moments, I could not help but look over at Jamail, who would roll his eyes skyward if he weren't already counting the holes in the ceiling tile.

The deposition testimony was by no means uniformly helpful to Texaco's case. When it made an offering from the deposition of Harold Williams, Pennzoil designated a series of questions that showed he knew the price of the Texaco offer for the Museum before Lipton went to rendezvous with the Texaco contingent at Gordon's suite on the night of January 5. Back in California that day, Williams had received the information in a phone call from Lipton that was clearly made before he went to the Pierre.

Lipton's testimony on that point was still fresh in the minds of the jury, and he had apparently been contradicted by his own client. It was obvious by this time that Williams would not come to testify at this trial and, in truth, I'm not sure what more he could have added.

* * *

Texaco had another big gun, and it wheeled him out a few days later. Laurence Tisch was a well-known investor who was then in the news because he had taken a strong financial position in **CBS** stock in the wake of Ted Turner's much publicized (and ultimately unsuccessful) move on the network.

As one of the four "new" directors added to the Getty board in the wake of the Consent, Tisch had attended his first Getty Oil board meeting on January 2–3. He had voted for both the Pennzoil and, later, the Texaco deals, and walked away with $4 million profit on the shares he had bought when he became a director.

Bob Brown, who had been taking some of the burden off Miller, conducted the questioning of Tisch for the defense. After reviewing the obligatory background information and details of how he came to serve on the Getty Oil board of directors, Brown got right to the heart of Texaco's case.

BROWN: "Mr. Tisch, will you express to this jury as accurately as you can, please, sir, what proposal you understood that you voted on when your board meeting resumed on the afternoon of the 3rd?"

TISCH: "The proposal for Pennzoil to pay one hundred and ten dollars per share with a five-dollar stub that we estimated could be worth in the area of two-fifty or even three dollars a share."

BROWN: "Mr. Tisch, did you or did you not understand that the proposal included a vote on the Memorandum of Agreement?"

TISCH: "It did not include a vote on the Memorandum of Agreement."

BROWN: "How can you be certain of that, Mr. Tisch?"

TISCH: "I know what I voted on."

This declaration in no uncertain terms was as strong as the testimony got for Texaco on this point.

Like Wendt, Tisch was not riddled with self-interest, his $4 million profit notwithstanding. Unlike Wendt, Tisch had not been part of any of the earlier bloodletting at Getty Oil. He had been placed on the board at the behest of Gordon, so *that* familiar Pennzoil refrain would not work on this witness . . . or would it?

Before he passed the witness, Brown asked Tisch about the handshakes Arthur Liman had testified he made in the board room after the meeting broke up. Liman had testified warmly about Tisch being a long-standing friend of "myself and my family," and the jury watched to see the response Tisch would give. He was basically noncommittal to Brown's question, saying he couldn't recall seeing that.

Joe Jamail came forward to cross-examine the witness. Judge Casseb had reined in the more flamboyant impulses of the attorneys on both sides, but Jamail was fairly low key.

He did ask Tisch questions about his relationship with some of the other figures in the case. It was established that Lipton had represented Tisch in a number of transactions. On the subject of Gordon, Jamail was apparently trying to get Tisch to say something nice about the man, but a curious exchange followed:

JAMAIL: "You're not friends with Gordon Getty?"

TISCH: "No, sir."

JAMAIL: "Did Gordon Getty *think* you were his friend?"

TISCH: "Define friend and I'll answer the question."

JAMAIL: "Sir, I can't define the New York friendship . . ."

TISCH: "There's no difference between a friend in Texas and a friend in New York, sir . . . Mr. Getty is an acquaintance of mine."

Jamail returned again to the subject of Arthur Liman, having detected some reticence on the part of Tisch to join in the trashing of Liman.

JAMAIL: "You have told the jury that you have known Mr. Liman for twenty years?"

TISCH: "Maybe longer."

JAMAIL: "Has that acquaintanceship offered you an opportunity to assess his integrity?"

TISCH: "Yes, sir."

JAMAIL: "Would you tell us whether he's a man of integrity?"

TISCH: "Yes, sir."

JAMAIL: "High-principled man?"

TISCH: "Yes, sir."

JAMAIL: "Honorable man?"

TISCH: "Yes, sir."

JAMAIL: "Will tell the truth?"

TISCH: "Yes, sir."

JAMAIL: "When you answered in response to Mr. Brown's question

that you did not recall Mr. Liman coming into the directors'
room and shaking hands, are you saying that it didn't happen
or you just don't recall it?"

TISCH: "I'm saying that I don't recall it."

JAMAIL: "And if he remembered it happening, you would not dispute
that, I take it?"

TISCH: "His recollection may be different than mine."

JAMAIL: "You did not see him in the board room, did you not?"

TISCH: "Yes, sir."

JAMAIL: "And you did shake hands with him outside the board
room?"

TISCH: "Yes, sir."

JAMAIL: "And this was after the vote?"

TISCH: "Most likely, I don't recall if we did shake hands after the
vote. We may very well have."

JAMAIL: "And you were offered an opportunity to go to the Getty suite
to drink and sip champagne?"

TISCH: "Yes, sir."

JAMAIL: "To celebrate?"

TISCH: "Yes, sir."

JAMAIL: "What were you going to celebrate?"

TISCH: "I think it was the acceptance of a price that could lead to
the agreement on a contract."

Jamail had led Tisch back to some of the core ingredients of the
evidence that seemed to support Pennzoil's central claim in this case,
that it had a binding agreement with Getty Oil.

The champagne toast and the handshakes did not prove anything,
of course, but taken in concert with the documents and much of the
testimony, they began to add up. The Getty press release was further
indication of an agreement; it contained terms that Pennzoil claimed
it had a contract on, terms from the Memorandum of Agreement.

JAMAIL: "How can this [press release] be authentic unless the board
voted on it? You have told me earlier that the only way that
a board expresses itself is through directors' votes."

TISCH: "They didn't vote on this. The board voted, as you can see
from Mr. Copley's notes, which you referred to so frequently

that we voted on the price of the transaction. This never came up for a vote by the board. . . . This is not a contract."

JAMAIL: "Mr. Tisch, I thought you earlier said you were not going to express opinions about what is and is not a contract. You were going to leave that to the lawyers."

TISCH: "Well, this is very clear on what it says that it's not a contract."

JAMAIL: "Well, sir, that isn't my question."

Joe Jamail couldn't have been too unhappy about the testimony of Laurence Tisch, because he didn't really go after him. Tisch had not tried to distance himself from the Copley notes or the press release; he just maintained a different interpretation of those documents.

By this time, the jury was better acquainted by far with this evidence than Tisch could possibly have been, so we took that into account in evaluating his testimony.

Again, a good round for Texaco, but far from the home-run shot it was looking for.

After the reading of still more depositions, Texaco Vice-Chairman James Kinnear took the stand. The jury was at ease with Kinnear, who had been seated in the courtroom throughout the many long days of the trial. There was a certain bond among all those who had suffered through the tedium that marked the trial's most exhausting passages, a shared experience, if you will.

Kinnear had been a key player in Texaco's moves on January 5, going to the Wachtell, Lipton office after the board meeting to "MEET WITH MARTY." He knew every nuance of the testimony that had taken place in the more than four months before he took the stand, so he didn't steer blithely into the pitfalls that other witnesses had. Rather, his direct testimony was a careful affirmation of Texaco's case, with all the strengths and weaknesses that implied.

On cross-examination, Jamail took it easy on him. There was no point in roughing up someone the jury had become comfortable with, and there was really no need to. Kinnear had seen all too well the hazards of evasion and the resultant effect on the jury. As I listened to his testimony, I couldn't help but wonder what his assessment was of Texaco's case at that point.

On one point, Jamail elicited a memorable admission from Kinnear: "That night at the Wachtell, Lipton offices, when you saw that Lipton demanded an indemnity protecting the Museum against Pennzoil, by name, weren't you concerned?"

Kinnear answered that of course he was concerned, he was *very* concerned. After the four months of testimony about that very indemnity, concern might not seem like a strong enough term.

Texaco didn't fare as well with the testimony of William Weitzel, its general counsel. Along with president Al DeCrane, whom he succeeded as general counsel, Bill Weitzel was Texaco's top lawyer. Responsibility for the legal ramifications of the Getty deal would appear to be primarily on the shoulders of these two men.

Again, it was Bob Brown who handled the direct examination for Texaco. Weitzel was a forthcoming witness, and Brown's questions often seemed like a mere formality.

Weitzel had been part of the group that had gone to Gordon's suite at the Pierre on the night of January 5. Although that meeting had been described by at least seven of those present that night, Weitzel added some new details to the story.

BROWN: "In the first part of this meeting when you were with Mr. Getty, would you express to the jury your recollection about what Mr. McKinley said to Mr. Getty?"

WEITZEL: "Well, they started and they had some pleasantries. There was some talk about Texaco's sponsorship of the Metropolitan Opera.

"Then Mr. McKinley got into why we were there and he said that Texaco understood that the Getty entities were interested in receiving an offer from Texaco . . ."

Every witness had described this opening exchange of "pleasantries" as concerning McKinley's "mutual interest in the arts" with Gordon Getty. Weitzel revealed that the interest revolved around a radio program Texaco had sponsored since the early days of World War II.

The first part of the meeting was interesting for another reason. Gordon wouldn't indicate that he wanted to deal with Texaco. After this false start, the Texaco contingent repaired to the lobby. Shortly thereafter, first Lipton, then Tisch arrived. When the Texaco group was summoned back upstairs by the two late arrivals . . . Bob Brown had Weitzel pick up the action:

BROWN: "Recount for us, Mr. Weitzel, as best you can recall, what Mr. McKinley said to Mr. Getty to reopen the conversation."

WEITZEL: "Mr. McKinley said that he was prepared to make an offer to Mr. Getty, since Mr. Getty would be happy to receive one; that he had given a lot of thought as to what price that

Texaco could go for would be, that he had been thinking of a hundred and twenty-two dollars.

"And he sort of smiled and said, 'But I have gotten some indications here that there is another price that would be more agreeable to you, and so I am prepared to offer—' and before we could say anything further, Mr. Getty said, 'I accept. Oh, you are supposed to give the price first.'

"And Mr. McKinley said, 'Yes. I am prepared to go to a hundred and twenty-five.'

"And Mr. Getty said, 'Fine. Fine. Thanks. I accept. Thank you.'

"And so that's sort of how the price agreement was handled."

Weitzel had capped his direct testimony with this detailed, revealing anecdote that confirmed the stories told by all the other participants in that meeting, but with more color, precision, and detail. He smiled broadly as he concluded the vignette and beamed at the jury, as if to say, "What could be more friendly than that?"

For Pennzoil, John Jeffers had a few ideas about Weitzel's actions on January 5, and he asked the witness about them. First, he questioned Texaco's top lawyer why, if the deal was so friendly, it had abandoned its long-standing outside law firm to hire takeover specialists Skadden, Arps.

JEFFERS: "And the reason you hired these experts with the capability of moving swiftly in this field is because you and Mr. De-Crane and the others at Texaco did not want any repetition of the Conoco fiasco, did you?"

WEITZEL: "There was no Conoco fiasco. That question is loaded with so many misstatements that I don't know where to start in answering it."

JEFFERS: "And this was at a time that certainly the reserves of Getty represented an attractive and needed buy for Texaco, right?"

WEITZEL: "We thought it would be very desirable?"

JEFFERS: "I'll put it to you as a question. When you went to New York to see Mr. Lipton to make these inquiries you are talking about, at that point there was no way that you, Bill Weitzel, was going to turn back and say, 'Sorry, we've run into some contract principles here, sorry we rushed you up here, Board.'

"I'm saying your state of mind, Mr. Weitzel, when you left White Plains that night of the 5th to go see Mr. Lipton was that you were not going to turn back?"

WEITZEL: "That's *really wrong* because if I *ever* had the *slightest* idea that there was *any* contract, a binding contract between Pennzoil and Getty, Texaco would not have moved *one inch* because I know Mr. McKinley would not have moved against my legal advice, and he would have had the advice not to do it.

"So that's just *really wrong.*"

JEFFERS: "Haven't your resonsibilities been expanded since that time?"

WEITZEL: "Well, I've been pretty lucky. I've had some ups and downs, but I've had it expand pretty much over most of my career."

JEFFERS: "I gather from your job description this morning you have been promoted since the time of this acquisition?"

WEITZEL: "Yes."

Jeffers had touched a nerve with this last string of questions, because Weitzel had almost come out of his chair. This was easily the most excited reaction from the witness stand during the entire trial. Weitzel just burst out, practically shouting at times.

It was somehow fitting that the instigator had been John Jeffers, the quiet diplomat of what had been a rough and ready band of inquisitors for Pennzoil.

Weitzel's outburst had not deflected attention from what I truly felt was a perfunctory examination on his part of the detailed, involved dealings between Pennzoil and the Museum, the Trust, and the Getty Oil Company, the results of which had been announced to the public early on the morning of January 4.

Texaco President Al DeCrane followed Weitzel to the stand. Dick Miller came forward and took DeCrane on what amounted to a four-day review of the case.

He started off with some instructions for the witness, which reminded the jury that DeCrane might be president of the company, but he, Miller, ran the show for Texaco in this courtroom.

MILLER: "Mr. DeCrane, be conscious of the fact that it will be utterly useless for you to talk if nobody hears you, and keep your

voice up and still don't holler at us. Listen carefully to my
questions."

As Pennzoil had done with some of its witnesses, especially Kerr and
Liedtke, Miller had DeCrane examine and comment on virtually all the
critical documents in the case. At this late hour, none loomed so large
as the handwritten notes of DeCrane himself. Miller's thrust, after all
the evidence, seemed to be to use the notes to explain the case. This
was a risky strategy for Texaco, but infinitely preferable to leaving the
notes to speak for themselves.

Initially, Miller tried to deflect the bright light that had shone on the
notes for four months, as in, "Can you believe we're still talking about
some notes?"

DeCrane: "They were strictly for my purposes."

  Miller: "Did you ever have any idea they would be an exhibit in a
            lawsuit?"

DeCrane: "No."

  Miller: "Or that they would be the subject of court testimony?"

DeCrane: "No, that never entered my mind at the time I was making
            these notes."

  Miller: "Or that questions might be asked, great detailed ques-
            tions concerning the notes themselves, and what they
            meant?"

DeCrane: "No, not at all."

As he had from the beginning, DeCrane disavowed much of the contents of his notes and those of Lynch, which were taken at the same
meetings, as the product of the First Boston braintrust, principally
Bruce Wasserstein. He maintained that these notes did not reflect Texaco's strategy, that the details of how Gordon Getty could be squeezed
out had no relationship to what had transpired.

He said the references to Lipton, "KEY MAN—MARTY—MUSEUM,"
were just some more First Boston ideas, but since Texaco didn't make
its deal with the Museum until after Gordon signed up, that notation had
no relevance.

"ONLY HAVE AN ORAL AGREEMENT," a note that had haunted
Texaco throughout the trial, now at the eleventh hour is explained as
confirmation that Texaco believed Pennzoil did not have a deal with
Getty.

On and on it went, explanations and alibis for the DeCrane and
Lynch notes. To my mind, it was much too little, far too late.

When Irv Terrell came up for the cross-examination, he knew De-Crane was a tough nut to crack. But how great was the pressure on Irv Terrell to "crack" Al DeCrane? What had DeCrane actually testified to? What damage had he inflicted on Pennzoil's case?

Terrell, of course, went ahead and asked DeCrane a lot of long, detailed questions that DeCrane coolly deflected like the lawyer he was. Unlike Weitzel, DeCrane did not get excited. Hell, he barely even twitched.

Control was clearly DeCrane's strong suit, and it seemed to me that he had exercised great control throughout Texaco's entire Getty misadventure. While McKinley and Kinnear vacationed, it had been DeCrane who had taken the reins. Although the final decision had definitely been John McKinley's, DeCrane had everything in place by the time McKinley returned.

Did DeCrane know what it would take to obtain the blessing of McKinley and the board? Well, he had managed to get himself installed in the president's chair, so there's a clear indication that he had a pretty good idea.

From his rambling cross-examination of DeCrane, there was one exchange by Terrell that had electricity crackling in the courtroom. It concerned, what else, the notes.

TERRELL: "Do you live up to your oral agreements, Mr. DeCrane?"

DECRANE: "I live up to my contracts."

TERRELL: "Do you live up to your oral agreements?"

DECRANE: "I live up to my contracts and commitments."

TERRELL: "I want you to tell this jury right now whether you live up to your oral contracts. I want to know!"

DECRANE: "I live up to all of my contracts . . ."

DeCrane had sat implacably through Terrell's harangue; he may have blinked a couple of times, that was about it.

Miller came back for a last series of questions. The testimony of Patricia Vlahakis that the indemnity obliged the Museum to cooperate with Texaco in the preparation of its defense of this case had been repeatedly used by Pennzoil to discredit the string of Getty witnesses.

MILLER: One other thing. . . . 'Cooperation' doesn't mean lying under oath, does it?"

DECRANE: "It certainly doesn't."

Miller resumed his seat, but Terrell came back to pick up this last thread.

TERRELL: "And, Mr. DeCrane, if this jury believed that people had come in here and lied to it under oath, you'd want the full power of the Court to redress that, wouldn't you?"

DeCrane didn't answer; he just looked at Terrell. It didn't seem like a difficult question.

TERRELL: "Wouldn't you?"
DeCrane: "I don't believe anybody has lied that I'm aware of."
TERRELL: "If they have, and the jury believes they have, you would want the full power of the Court and the jury to put an end to that, wouldn't you?"
DeCrane: "I believe that justice and truth should be what we seek in this whole proceeding."

After Terrell sat down, Miller came back for one last question. He wasn't going to cede the moral high ground to anybody at the Pennzoil table.

MILLER: "Would that apply to Liedtke?"
DeCrane: "That applies to everyone."

Terrell wasn't going to let Miller get the last word.

TERRELL: "Would it apply to Mr. Lipton?"
DeCrane: "It applies to everyone."
TERRELL: "Mr. Boisi?"
DeCrane: "To everyone."

DeCrane left the stand and I wondered what was coming next. The jury never knew who the next witness would be.

After the break, Miller offered a portion of the deposition of Martin Siegel, an attempt to deflate the Siegel affidavit that we all knew by heart. Pennzoil found something in the Siegel deposition it wanted to read also. Months after Arthur Liman had testified about going into the board room to shake hands after the meeting broke up, Siegel's deposition placed Liman in the board room shaking hands, the first direct confirmation of the story Texaco had spent the past three months trying to discredit. Maybe it was an omen.

When the Siegel deposition was completed, Miller stood and turned to smile at the jury. "Counsel for Texaco rests," he said simply.

I was stunned by the announcement. That's it? I suddenly got a heavy feeling inside my rather substantial gut that I wasn't going to like what happened next.

PART X

"ARROGANCE, POWER, AND MONEY"

Arguments, Deliberations, and a Verdict

It is not what a lawyer tells me I *may* do; but what humanity, reason, and justice tell me I *ought* to do.

EDMUND BURKE

30

After Miller rested the case for the defendant, the jury received a rare day off so the attorneys and the judge could prepare the charge to the jury. On Thursday morning, we were back to hear the reading of the judge's charge and final arguments. Was the end really in sight?

The courtroom had been largely vacant for the majority of the trial, but word had obviously spread that the endgame was at hand; every seat was filled. Since the testimony had been concluded, the top executives from both companies were back in the courtroom. As witnesses, they had been barred from the trial, except for company observers Kerr and Kinnear.

Judge Casseb had allotted each side four hours for final arguments. As they had throughout the trial, Jeffers, Terrell, and Jamail would divide Pennzoil's time. Miller alone would make the argument for Texaco.

Jeffers and Terrell would lead off, followed by Miller. Jamail would deliver the closing argument. Since the burden of proof is always on the plaintiff, it has the first and last word.

There was an air of expectation as John Jeffers began the final arguments.

JEFFERS: "Good morning, ladies and gentlemen. I said in my opening statement that we had serious matters to attend to here, but I had no idea that it was going to take us quite so long to attend to them. We've been here four and a half months. Technically, we've had a change in seasons.

"I'm going to say one thing very quickly because you've had a job to do and you've done it. If you didn't do your job, the system wouldn't work. You know that without me saying it."

With that preamble, Jeffers turned his attention to the facts of the case.

JEFFERS: "I have here a lot of the transcript from this trial. Because of the time, I'm actually going to read [very little] of it to you this morning.

"I will represent to you that I have read it all yesterday and I believe what I am going to tell you it says is right. But it's here, and if the Texaco lawyers think I made a mistake, then [they can] look through it when I finish."

It was fortunate that Jeffers chose not to read very much of the testimony; he began his final argument at page 23,793 of the trial transcript.

JEFFERS: "I want to recall first for you the testimony of Mr. Tom Barrow, Pennzoil's expert. Mr. Barrow, you will recall, was the first person ever to receive a Ph.D. in oil and gas geology from Stanford University in California.

"He worked his way to the very top of the Exxon organization, was a director of Exxon for many years, [he] retired from Exxon and became chairman and chief executive officer of Kennecott Copper, the largest copper company in the world . . . I think an eminent man.

"He testified without pay. For one reason, he said he did not want to become a professional expert witness. For another, he said he believes that the principles here in this case were important ones.

"To him, an agreement in principle signifies a meeting of the minds on essential terms and carries with it a duty of good faith to proceed towards consummation."

Barrow's review of evidence detailing the proposal presented to the Getty board on January 2 and 3 and the subsequent actions of the board was recounted by Jeffers.

JEFFERS: "There was counteroffer and acceptance, in his experience a binding agreement which he thought was confirmed by the press release. A meeting of the minds on essential terms and intent to be bound and they are bound.

"Contrast that testimony, if you will, with the testimony of Alfred DeCrane, the president of Texaco. Mr. DeCrane said that an agreement in principle carries with it a business obligation to proceed to a final transaction but not, in his mind, a legal obligation; that Texaco lives up to its signed contracts as distinguished from its oral agreements."

Next, Jeffers read from the court's instructions that accompanied Special Issue No. 1 of the jury charge, which dealt with the Pennzoil's alleged contract with Getty Oil: "Where an agreement is fully or partially in writing, the law provides that persons may bind themselves to that agreement even though they do not sign it where their assent is otherwise indicated."

JEFFERS: "The question is not determined by the parties' secret, inward, or suggested intentions. Persons may intend to be bound to an agreement even though they plan to sign a more formal and detailed document at a later time.

"That is the law of this Court under which this case is to be decided. It is the law as stated, corroborating the experience of Tom Barrow.

"You have heard a group of mergers and acquisitions specialists who have said that an agreement in principle has no more value than the air between my fingers; that it is an agreement which announces to the world that there is no agreement; that it is an invitation for bids.

"The law is that where parties have reached agreement on essential terms and intend to be bound, they *are* bound. There is an opportunity here in this case to send a message which will preserve honor and integrity in the American business system."

Jeffers turned to the actions of the Getty Oil board of directors.

JEFFERS: "They tried to get away with selling the company twice. Maybe they didn't reflect on what they were doing so carefully because the eminent Mr. Martin Lipton said it was okay, he said, 'Sure, you can always withdraw a counteroffer after it's been accepted.'

"Remember what Mr. Barrow said about that when I asked him what you would do as a board member, as a lawyer, would you withdraw a counteroffer that had already been accepted? He said, 'I'd invite him out of the room.' But that's what Lipton did. He said, 'Let's withdraw this counteroffer that's already been accepted and let's sell the company twice and try to get away with it.'

"And then when Texaco's hand got called down here in Houston, I suppose, these Getty board members felt like it was their duty under the indemnification to assist in the defense."

This last point had been verified by Wachtell, Lipton lawyer Patricia Vlahakis.

JEFFERS: "Ms. Vlahakis testified that it was her duty under the Texaco indemnification to assist and testify in the case because they had helped start Texaco down this road, and [had done so] secretly. The stakes were very high.

"Now, what they testified, I suppose, cynical people could reconcile as being just a shading of the truth.

"What these board members did was' just shade some, [from] saying that 'Our main concern was price and that we approved the deal' to saying that 'All we talk[ed] about was price and that's all we approved.' It's that much shading in an effort to make a difference and distance themselves from the Memorandum of Agreement, because they were told that Memorandum of Agreement, as Mr. Winokur said, was an absolute agreement.

"It was written that way and it had essential terms of price and structure in it and would reflect an intention to be bound if it was approved, so they had to shade away from that, these witnesses, and that's what they did."

There was another contradiction, Jeffers said, with the testimony that what the board approved was only the price.

JEFFERS: "The Memorandum of Agreement provided that the company was going to buy the Museum shares. Now, it would be a problem if the board ratified that kind of provision at the end, along with the price, because it would show that the Memorandum of Agreement came right on through to the end [of the meeting].

"And, at page 63 of the Copley notes, Mr. Lipton describes the Pennzoil proposal of one hundred and ten dollars plus the three-dollar stub and he describes the terms that go right back to the Memorandum of Agreement, twenty-four million shares [to be purchased by Pennzoil], and the purchase by the company of the Museum shares, followed by a cash merger to take out the balance of the shares.

"So right there at the end of the Copley notes, you have this provision from the Memorandum of Agreement still there. That document is still before the board, clearly, with the addition of a price, and that's where the argument breaks down that they disconnected this Memorandum of Agreement and just talked about price.

"What did they do about that? Well, they have people testify, Winokur, Wendt, Medberry, Lipton that that was Mr. Lipton's proposal, but that's not what the board voted on.

"Mr. Lipton articulated the proposal; Mr. Williams made a motion [but their testimony is] that they weren't voting on Pennzoil's proposal.

"Now, that's a strange story and it broke down because Mr. Tisch testified that the proposal he voted on was the proposal on page 63. He said it was all price. Well, fine. If he wants to call everything at page 63 price, that's fine, but it is price and structure and it is the terms of the Memorandum of Agreement."

Getty Oil's investment banker, Geoffrey Boisi, also seemed to confirm this detail.

JEFFERS: "Mr. Boisi testified that the board approved the purchase of the Museum shares. And it's [in response to] a question by Mr. Terrell: 'Well, in fact, do you recall that the Getty board voted, what it approved by a 15-to-1 vote was that the company was going to buy the Museum shares?' Boisi's answer: 'I believe so.'

"Now, after a recess, that witness came back and was coached to change his testimony, and not in answer to a question, but in answer to a comment by Mr. Terrell in response to an objection. Mr. Terrell said, 'Mr. Boisi already testified that the board approved the company's buying the Museum shares' and Mr. Boisi said, 'Well, if I said that, I misspoke.'

"That is insulting and presumptuous, in my opinion, to the jury. I mean, this man who had made eighteen million dollars was asked this very clear question: 'Was the company going to buy the Museum shares?'

"'I believe so.' And just undertake, after a recess, to change that testimony?

"But you are the judges of the credibility."

In his polite fashion, Jeffers savaged the Texaco witnesses who said the January 4 Getty press release was wrong.

JEFFERS: "And I say the story about the press release being wrong broke down . . . Mr. Winokur and Mr. Copley for the company approved it. Martin Siegel testified that the press release was true and complete and Mr. Lipton had a problem because [his associate] Ms. Vlahakis approved it. Ms. Vlahakis said that the press release was very accurate but doesn't mean anything. That's the effect of what he [Lipton] said and that was because, I suppose, Ms. Vlahakis approved it.

            "So, what they come back to then after this story breaks down, after you have too many breaks in the dike, they say 'Well, this press release still says it's an agreement in principle.''

The meaning of agreement in principle was debated at great length, from the voir dire to the final argument. Despite this agony, it would not be the point on which this case would turn.

Another point Jeffers covered was the contention by Texaco's Getty witnesses that they never intended to be bound until a definitive merger agreement was signed. Jeffers reminded the jury that Lipton had suffered a dramatic loss of credibility when he was confronted by Jamail with a two-and-a-half-page agreement from the Esmark–Norton Simon deal, a binding agreement that clearly contemplated a subsequent definitive merger agreement. Lipton had said only seconds before that no client of his was ever bound before the definitive agreement.

Jeffers also pointed to the actions of Getty Oil attorney Winokur, who he said was "clearly an obstructionist in this deal." Jeffers recalled that Winokur had said the deal wasn't binding "because of all of these so-called open items that had to be negotiated"; the fact was the Copley notes recorded no mention of the open items at the board meeting.

JEFFERS: "So just to wrap it up, the story about how 'We approved price only, we detached the Memorandum of Agreement, the press release is wrong'; that story doesn't hold up because their witnesses can't all corroborate it. Some people slipped up, and the documentary evidence doesn't corroborate it."

Not content with simply refuting the Texaco defenses, Jeffers again set his sights on Winokur, this time on the drafting session for the press release.

JEFFERS: "Mr. Winokur, you will recall, scratched that provision out of the draft press release for Getty Oil Company to buy the Museum shares and that's what I say clearly shows that Winokur had got there to act as an obstructionist, to try to slow this deal down while Boisi shopped.

"At the same time, however, in his villainy, he did not want Pennzoil off the hook. He wanted [Pennzoil] to believe that the Getty Oil Company intended to be bound because, as Mr. Petersen said—remember the *Fortune* magazine article—Pennzoil was a bird in the hand.

"Why have a press release when nothing has been agreed? . . . Why not tell Pennzoil that if that was the truth? The reason is that that was not the truth and Winokur had no authorization from his board to tell Pennzoil there was no deal. He was just trying to slow it down and so he acted that way."

The methodical John Jeffers did not mince words as he concluded his argument.

JEFFERS: "In my humble opinion, the defenses to these issues have been an outrageous cover-up. If Texaco's witnesses can get away with the things they said here, if they can do what they did to Pennzoil in the marketplace, they can do it to someone else somewhere else.

"But they are here; we're four and a half months into it. It's time for an accounting.

"Thank you."

The second part of Pennzoil's final argument would come next, after the lunch break. If past performance was any indication, Irv Terrell would not be as gentlemanly as his partner.

Terrell delivered his argument the same way he had tried his part of the case, as if he was a criminal prosecutor who wanted to send Texaco to the penitentiary for a long, long time. He also reiterated the theme that this was a case of great significance.

TERRELL: "You can tell, as you could in the beginning, the importance of this case. This case is not just important because of Pennzoil; it's of tremendous importance to the people at Pennzoil, there's no question about that."

Terrell motioned toward the Texaco executives.

TERRELL: "These men at Texaco consider this case important. You've got the four big hitters of Texaco sitting right over there looking you right in the eye, and they should be.

"But there's more, and that's the business community of this country, because you're going to be able to do something that I, in my lifetime, will never be able to do as a lawyer.

"You will decide the ethics. You will decide whether you can go out and take somebody else's deal just because you're bigger and stronger, you've got a lot of muscle, you've got a lot of lawyers, you've got a lot of investment bankers.

"And, above all, what have you got? You've got indemnities.

"These people do these things and they have no personal responsibility. And I'm telling you that if you speak out, they will hear you. Not just these men—they'll hear you for sure—but so will the whole business community."

It would be imprudent to believe automatically that any emotion expressed by a lawyer in a trial is genuine, and not just manufactured for the occasion. Deception is the actor's art, as well as that of the trial lawyer and the confidence trickster. I am persuaded that at the critical moment, when an artist of the confidence trade is reeling in the pigeon with promises of wealth and glory, in his mind and in his heart at that moment he, too, believes.

Casting no doubt on the purity of Irv Terrell's motives and actions, I can assure you that as he delivered this frequently emotional final argument, he believed it fervently. That belief was transmitted to the jury box like a current down a wire.

TERRELL: "But if you don't care about the way corporations treat each other, and I know there are people who believe in that, then you don't have to do anything.

"But I'm telling you that our society is not just separated. If something can happen to a three-billion-dollar company like Pennzoil, it can happen to anybody.

"So that's what this case is all about, at least to me. And each of you will decide in your own hearts what it's about to you. That's up to you."

Terrell addressed Special Issue No. 8, which asked if the $112.50 Pennzoil offer was a fair price.

TERRELL: "You remember when I cross-examined Mr. Winokur, who
             was a difficult witness. He's a smart man. There is no
             doubt about that; there are no flies on Mr. Winokur. But I
             surprised him with his own document."

One of the strategies Getty Oil had planned to fend off the Pennzoil
tender offer was a self-tender, where it would buy up 16 million shares
of its own stock at $110 a share.

TERRELL: "That's about 40% of that public stock and they'd pay them
             a hundred and ten dollars and [then] send Boisi out to sell
             the company at anything over a hundred and ten, cash or
             securities.
                 "Well, Mr. Winokur and Mr. Boisi got stuck on that. Mr.
             Boisi said, 'Well, yeah, I knew about it but, you know, I
             didn't approve it.' I said, 'Well, did you disapprove it?' He
             said, 'No.'
                 "Mr. Winokur said he hoped it would be signed. Well,
             what happened? Why is this price of a hundred and twelve
             fifty not fair when the hundred and ten is fair for these
             wards, as Mr. Miller calls them, the wards of these direc-
             tors. Sixteen million of those shares."

Salomon Brothers for the Museum and Kidder, Peabody for the
Trust had advised the board that $112.50 was a fair price. The fact that
Boisi had refused to issue a fairness opinion on the $112.50 during the
board meeting is what kept this issue alive to be used as a Texaco defense almost two years later.

TERRELL: "They can't have it both ways. I suggest to you that if you
             have two investment banking firms, for whatever they're
             worth and I'll tell you I don't think they're worth much, for
             whatever they're worth you've got these two investment
             banking firms saying it's fair. Only Boisi, who says he's not
             prepared to comment.
                 "You look at the Copley notes. That says again that Salo-
             mon and Kidder, Peabody say the Pennzoil price is fair.
                 "You've got Tisch, who's got sixty-seven thousand
             shares. And if you look at Mr. Lynch's notes, it says 'LARRY
             TISCH, PUBLIC SHAREHOLDER. OPEN TO HIGHER OF-
             FER BUT THOUGHT $110 OKAY.' $110 okay. So you've got
             Mr. Tisch, the man who made four million dollars here; so
             it's fair.

"So where are we now? Well, we're down to coercion. We're down to the 'Dear Hugh' letter, about which you've heard so much. If there was ever more of a red herring, I don't know what it was. Because the "Dear Hugh" letter was a gate that was opened and could not be closed unless Lipton closed it. And Lipton got up here and said he wasn't going to close the gate."

The situation contemplated by the "Dear Hugh" letter could only have been effected by the Museum again joining with the Trust.

TERRELL: "Now, I'm not vouching for Mr. Lipton's testimony, because I don't believe Mr. Lipton about his evidence. And I don't think you do either.

"But extrinsic, separate from Mr. Lipton, it's clear as a matter of Delaware law that you can't do anything unless you get over 50%. So we couldn't swing the gate shut without Mr. Lipton.

"And who was offended, had we swung the gate? We had Chauncey Medberry, he was offended. We had Henry Wendt, he was offended. These are men who told you of their indignation, their outrage, how upset they were. I guess they know now how Gordon Getty felt on November 11th when they did it to him.

"What did Mr. Barrow say? Had he been coerced, what would he have done? He would have, at a minimum, abstained and he really believes he would have resigned.

"Well, not one person abstained in that vote, and I may say, it was a hard vote to get. Fifteen directors, Sidney Petersen voting for it. I mean, *remarkable*. And if you think Sidney Petersen, with his army of lawyers, thought he had been coerced or Pennzoil had done something wrong after he saw that "Dear Hugh" letter . . . Do you think he would have done something about it? *He didn't say a word*."

What did this issue have to do with whether or not Pennzoil had a binding contract? Terrell had an idea about that as well.

TERRELL: "This 'Dear Hugh' letter, let me tell you what this is. This is Texaco's way of trying to make up for the destruction of the Copley notes.

"Lawyers try cases many times looking for dirt to throw on the other side. And they particularly do that when they know that their face is covered with it because of these Copley notes and these indemnities."

If the jury found there was in fact a contract, the next step was to determine if Texaco interfered with that contract.

TERRELL: "Okay, let's talk about interference, Issue No. 2. And that's really where I'm going to spend most of my time; it is an absolute and fundamental issue.

"The key language in the issue is 'knowingly interfered': 'Do you find from a preponderance of the evidence that Texaco knowingly interfered with this agreement between Pennzoil and the Getty entities?'

"If you look at the instructions, it says, 'A party may interfere with an agreement by persuasion alone by offering better terms'—I submit that's the $12.50 premium (over Pennzoil's price)—or by giving an indemnity against damage claims.

"Boy, I think if we've got anything in this case, we've got that coming out of the ears."

Texaco's final witness had been its president, Al DeCrane.

TERRELL: "They get Mr. DeCrane, who I think is probably their smartest witness. They get their clean-up witness up here and put him on the stand and run through our documents. You remember that?

"The Copley notes: 'Okay, page 63 of the Copley notes is just the price.'

"Steadman Garber's note about 'agreement' and 'deal should be done': 'Oh, no, that just confirms me in my view.'

"Mr. Siegel's affidavit, the Cohler statement to the California court, plus the Cohler affidavit or declaration after Siegel had clued him in, that just confirms him in his view.

"What could I show Alfred DeCrane that would not confirm him in his view?"

Above all the Texaco executives, even the chairman, Terrell targeted DeCrane.

TERRELL: "I think that it's crystallized in Mr. DeCrane's testimony on two points. When I asked him about an agreement in principle, what is that? 'Well, it's a business obligation, but, you know, that's not a legal obligation.'

"He's got in his notes 'ORAL AGREEMENT.' He can't remember what it relates to. He can't really tell us what it is.

"By the way, they've got Perella and Wasserstein and Petrie on the witness list but you will never see them, the
guys who are talking [to Texaco] about this.

"Well, what does he say that means? Well, he's not sure.

"Okay. 'Do you live up to your oral agreements?' In the
contract [portion of the jury] charge, you can have an oral
agreement and it can be enforced. 'Do you live up to your
oral agreements, Mr. DeCrane?'

"How many times did I ask him that question? Four or
five, I think. And what did he come out with?

"'Mr. Terrell, I live up to my signed contracts.' That is
their whole defense. They're going to be picky. They're going to be technical. They're going to do everything they
can."

He characterized the nature of the testimony offered by the defense
witnesses, the Texaco officials, and their Getty indemnities.

TERRELL: "These guys, when you show them something, every one of
             their witnesses, when they get into a tight spot, and I sub-
             mit to you they all did, what do they do when you show
             them the documents? Steadman Garber's note, page 63 of
             the Copley notes, inconsistent testimony from some other
             witness, what do they do?

             "If it's a Getty witness, it's 'free to deal.' That's one of
             their buzz words.

             "And if it's a Texaco witness, it's 'no binding agreement.'
             Mr. DeCrane got off that. Finally, when he finished, it was
             'no binding contract.'"

The semantical game over the difference between 'agreement' and
'contract' was beaten to death by both sides, to little effect. The key word
was 'binding'; if the parties intended to be bound, the judge instructed,
they were bound.

TERRELL: "What we need to do, if you don't mind, is to look at what
             Texaco's strategy was and what their actions are and then
             we're going to look at their motive briefly.

             "You have to start with the DeCrane and Lynch notes. If
             there was ever a smoking gun, that's it. That's where it all
             is. That's the strategy.

             "You remember Mr. McKinley got up here on this witness stand and said that the key sections of those notes were never discussed with him. And then we get Mr. De-Crane on the stand and he says, 'Yes, I discussed it with him. I discussed those notes with him before the board meeting, but I didn't tell the board.' Mr. DeCrane contradicts Mr. McKinley.

"But both Mr. McKinley and Mr. DeCrane are solid on one thing: 'These notes are not strategy, this was not a strategy meeting. It just wasn't. We are friendly. We stayed up late, you know, but it wasn't a strategy meeting. We had our investment bankers there, we had our outside lawyers there, we had our planning group internally there, but it was not a strategy meeting.'"

This had been Texaco's explanation of what Terrell was now calling "the smoking gun."

TERRELL: "Unfortunately, Mr. Weitzel blew the whistle on that. He testified that it was. Of course it was. You can look at those notes and tell that. But that shows you how scared they are of these notes.

"After all, who was chairing that strategy session but none other than Alfred DeCrane? I mean, not a person who is a lightweight. He was there, holding forth with these people. He wanted to know what to do.

"Nevertheless, they maintained it was a friendly deal. They said that if Gordon Getty had just told them, 'Look, I don't want to sell' that they would have said, 'Okay, good night. It's been fun to see you.'

"And you know Wasserstein would have gone along with that because he only made ten million dollars out of this deal.

"And you know that Texaco really meant that because they really didn't need those reserves."

On the knowledge and interference issue, it was apparent that Pennzoil felt the DeCrane and Lynch notes were persuasive evidence. By that point, I was inclined to agree with them.

TERRELL: "You have seen all these notes. You have seen about 'STOP THE TRAIN.' You have seen 'TIMING 24 HOURS, MEET WITH MARTY AT NOON,' which we now know that McKinley did call him at noon and [said] 'Don't sign anything' . . .

"We have got 'KEY PARTY—MARTY—MUSEUM, CON-

TROL TIME.' We have got 'IDEAL—1. MUSEUM, 2. TRUST, 3. BALANCE.'

"And each of these fellows says, 'Well, that's just in there, but that's not what we planned to do. That wasn't our strategy.' Then you see it in the [Texaco] board resolutions, where it's in the same order.

"And now, they still deny it. You get them to admit, 'Well, but that's what you did.' And they say, 'Well, but no, that isn't what we wanted to do. We were looking for Gordon.'

"Where was Gordon? [They say], 'We were looking for Mr. Getty. On the 4th, our people talked to Lipton. On the 5th, McKinley talked to Lipton at noon and at 4:00.'

"Wasserstein, you can imagine what he was doing, running around trying to make sure this deal went through and he got his ten million dollars."

Terrell had cast Wasserstein as part of the unholy alliance since the beginning of the trial. The strong-arm tactics of First Boston were duly recorded by Lynch and DeCrane, so even though Wasserstein never appeared in the courtroom, his words rang out time and time again.

Texaco had gone to some length to establish that the Museum shares had not been committed at $125 before going to see Gordon Getty.

TERRELL: "They went to see Lipton, and they say they didn't make a deal with Lipton. He wanted a price, but 'Oh, no, we won't give it to you.'

"You know the whole story. But what proves the absolute lie to that? The testimony of Harold Williams. Here it is. In Mr. Jeffers's question of him: 'What was the status of discussions between the Museum and Texaco as of the time Mr. Lipton went to meet with Mr. Getty?'

"ANSWER: 'We at that point had an oral understanding that we would sell our stock to Texaco.'

"QUESTION: 'It was an oral understanding with the price?'

"ANSWER: 'One hundred and twenty-five dollars a share.'

"So all this business about somehow they talked about it down in the lobby, and how, 'Gee, Mr. Tisch [and] Lipton persuaded him and they went back up to Gordon Getty's place and McKinley just couldn't make up his mind but somehow he did in the hallway up there' is baloney. Williams knew about it going in."

This was Terrell's refutation of Texaco's claim that it didn't put the squeeze on Gordon.

TERRELL: "I thought what Mr. Weitzel did was particularly telling, when he said 'When we went back up there, Mr. McKinley said we would like to offer, oh, . . . ,' remember that? 'I accept, oh, I shouldn't do that.'

"Gordon Getty knew the price. He knew what the story was. Gordon, whatever you think of people who have been raised as he has, with a lot of money, and I suspect a lot of problems come with the money, he is not dumb.

"And Gordon got the message. If you doubt that, you know even after the man's indemnified, and I am sure embarrassed by this whole thing, he doesn't want any part of it.

"What did he say when Mr. Jeffers asked him, 'Would it be fair to say that when Texaco showed up at this meeting and you found out that the Museum was going to sell to Texaco, that you felt you had no choice but to do so?'

"ANSWER: 'I think that's a fair statement.'

"Well, sure. Siegel is not dumb either. He knew. They knew. He didn't have any choice.

"And Gordon Getty is a gentleman. He stuck with Hugh Liedtke two days in that board meeting; I am talking about with people who just didn't like him, who treated him like dirt at that earlier meeting.

"And do you think Gordon Getty, if he hadn't been threatened by Texaco, wouldn't have picked up the phone and called Mr. Liedtke on the night of the 5th unless he was forced [to sell]?

"I mean, I just don't think he would do that. I really don't.

"Well, Texaco got to him, and that was the end of that."

The strategy outlined by the Texaco notes spelled out the threat to Gordon and the Trust: if there was a tender offer for the whole company and Gordon didn't tender, he'd end up with "paper."

TERRELL: "By the way, I don't think I need to cover these notes with you again, about leaving Gordon with paper and all that.

"I think you know that. I mean, it's just plain as day, in Lynch's notes and DeCrane's notes.

"It's just there, and there is no way for them to explain it, try as, bless his heart, as much as Mr. Miller tried when he put DeCrane on."

It was more than slightly out of character for Terrell to wax nostal-
gic about Dick Miller, but there it is.

TERRELL: "You look at Mr. DeCrane's own notes and it talks about
            'OPTION TO PENNZOIL. WON'T GET IT TILL MERGER
            AGREEMENT SIGNED. COULD BE A WAY TO TAKE CARE
            OF LIEDTKE.'
               "Well, Mr. DeCrane tried to wriggle free of that by say-
            ing, you know, 'Look, they didn't have the option.' Well, if
            that's true, why write it out? It doesn't make any sense as
            an explanation."

Terrell then advanced the most speculative of his arguments, but
one that had a certain perverse logic. It concerned the night of January
5, when the Texaco contingent went to see first Lipton, then Gordon.
Where was DeCrane? He had gone to a midtown hotel to await a report
on the Lipton meeting. He then went to the Skadden, Arps office to
await a report on the meeting with Gordon.
   Irv Terrell had this theory.

TERRELL: "Now, I don't know that I could ever prove it to you, but in
            my heart of hearts, I believe that Mr. DeCrane had drawn
            the black bean. That in the event that Gordon Getty didn't
            get the message, was too dumb, or had the courage to call
            Liedtke, Mr. DeCrane was going to see Mr. Liedtke, and he
            was going to try to make a deal.
               "Well, it didn't turn out that way. He didn't have to
            take—I can tell you it would have been an unpleasant
            chore. He didn't have to do that. But to think that the man
            who did the merger agreement, the man who did Gordon
            Getty's agreement, the man who went out and quelled the
            beneficiaries, the man who I believe will run Texaco one
            day, to think that *that* man was just sitting around on this
            important night without an assignment defies logic."

It hadn't required a brilliant intuitive leap for Terrell to predict that
DeCrane would run Texaco one day; he was already president with
over a decade until retirement. Although the theory was not ultimately
relevant, it was a nice piece of speculation that underscored the damaging portion of the DeCrane notes that talked about "A WAY TO TAKE
CARE OF LIEDTKE."
   Returning to his favorite theme, Terrell pounded on the indemnities
again.

TERRELL: "We have seen what they planned and what they did. . . .
             You have to start with these indemnities, I mean, these
             people can talk until they are blue in the face about how
             these indemnities just aren't unusual, no big deal.

             "'Harold Williams and Mr. Lipton have got a psychologi-
             cal problem, that's the only reason they wanted them.' So
             Weitzel said.

             "DeCrane said he never heard about this psychological
             problem.

             "Lipton is a man who is not only proud of the fact that
             he got this higher price and he got protection for his client,
             this indemnity, but he wants to sit here and tell you that
             Arthur Liman's been his friend for twenty-five years. And
             I tell you, Liman thinks it . . . he spoke of Lipton as his
             friend.

             "But if ever somebody denied somebody three times,
             that was it. Lipton did, to Liman.

             "I mean, you take your pick between these two men. I
             know who I believe."

Geoff Boisi's shopping spree after the 15-to-1 board vote was again
recounted.

TERRELL: "I think Boisi was the most telling on this. You remember
             when he was up here? Here's the man that shopped this
             deal and made a ton of money, and even with his indemnity
             up here, he's worried. He knows he shouldn't have shopped
             it. He's afraid to admit it.

             "And what did he say he did on the 4th and the 5th? He
             made courtesy calls.

             "Now, if you believe that Geoff Boisi made courtesy calls
             on the 4th and the 5th, we need to talk about a car I'd like
             to sell you, because it's just incredible.

             "He knows he shouldn't have done that . . . Boisi couldn't
             shop the company without fourteen of sixteen directors.
             DeCrane knew that, it's in his notes! And yet he did, and
             here we are. I think that's pretty clear on both knowledge
             and interference, and what those indemnities mean. It
             shows what I would at least regard as lawless conduct on
             the part of Texaco."

Following a review of the testimony on damages, Terrell entered the
home stretch.

TERRELL: "What is this case really about? I mean, what have we got
here? Oh, it's a big contract case. It's a big interference
case, lots of tremendous harm to Pennzoil.

"Pennzoil would have become a completely different
company. It would have become a major oil company and
not just a mid-size oil company. It would have made a tremendous difference in the lives of all of the people associated with Pennzoil. We're not going to have Getty Oil Company, never will.

"This case is more than that. This case is about arrogance, power, and money, three rather common human
traits, that they're in many people in different degrees. But
you put all of these people together and you are to consider
not just Texaco but also the circumstances and, I suggest
to you, these Getty people. There was only one way Texaco
acquired Getty Oil Company. Indemnities. They were essential.

"These men over here have admitted they couldn't have
done it even with their premium price without those indemnities.

"Now, what you've got on the other side of this case are
people who have no personal responsibilities. They don't
have to answer for their acts, none of them. That's arrogance and that's power and that's money. And sitting right
over there are the men who did that."

It was impossible to keep from looking at the Texaco executives as
Terrell repeatedly invoked their names. Every one of them had been a
witness in this trial. He was talking about men we had seen and heard
at great length.

TERRELL: "If you want to do something, if you want to stand up and
say something to these men and to the business community
in this country about, 'Listen, people are going to act responsibly or they're going to pay the piper. They're not going to be permitted to just do whatever they want to do
without any possible punishment, they're just not going to
do that.' This is your time and you have the opportunity to
do it."

The case for punitive damages consisted of a broad attack on a variety of alleged Texaco actions.

TERRELL: "You know, we started out talking about the Copley notes on punitive damages in voir dire. I think that plays a part because Ralph—Mr. Copley, you saw him on TV.

"He's not a man that threw his notes away because it was his idea to throw those notes away. Mr. Copley just wouldn't do that. He's not dumb enough to do that and I don't think he's evil. I don't think he did that.

"And when he threw them away that Monday the 9th, where had he been the day before? In White Plains.

"I can't prove to you that Texaco did it. I can prove to you that Texaco didn't stop it. The Copley notes are an important ingredient of this.

"Both of those factors, I think, are important. If you permit the destruction of evidence after litigation, if you permit people to walk around with no responsibility, then it's just going to go on.

"And it's not just going to be companies. It's just going to be people because, you know, it just is. That's the way things work in this country."

Terrell had started out on a frankly emotional note, and finished the same way.

TERRELL: "Well, this is the last time I'll have a chance to get up and talk to you and I really enjoyed it. This has been certainly the most important thing I've ever done in my profesional life and maybe ever will do, to participate in the trial of this case.

"It's up to you now to determine this case. I would suggest this to you: that the law just sits in books and doesn't do anything for anybody unless juries make it come alive.

"It doesn't matter what is in this charge that you have in your hands unless you do something with it. It's yours to use or not to use.

"I suggest to you that you use it so that everybody, whether we're a big oil company or a medium-size oil company, whatever kind of person we are . . . that we all live by the same rules, that we're responsible for our actions and that's all I have to say.

"Just don't let them get away with these indemnities. Please don't."

It was 2:30 on that Thursday afternoon when Terrell finally sat down. After a fifteen-minute break, the ball would shift to Texaco's court.

31

Throughout the trial, Dick Miller had been backed up by a team of lawyers from his firm and Texaco. But unlike the trio of heavy hitters for Pennzoil, the Texaco defense was clearly a one-man show.

His partners, as he always called them, Richard Keeton and Bob Brown, had shouldered some of the load, but it was the redoubtable Miller alone who stood up for Texaco from the first day of voir dire to the final argument.

MILLER: "Ladies and gentlemen of the jury, I had the feeling that if I waited long enough I would finally get to talk to you. Sure enough, it's finally come to pass.

"It's 2:45 and I know you're tired. We're all a little bit tired, I suppose, but I'm going to talk to you for about an hour and fifteen minutes and then I'm going to quit.

"I'll try to keep you awake, and I hope Mr. Kronzer will stay awake now for the rest of the arguments."

Kronzer, the senior member of the Pennzoil team, had been a constant presence even though he never addressed the jury during the trial. In arguments at the bench or in chambers, however, Kronzer had been a particular thorn in Miller's side, providing reams of case law and precedents for every point of contention.

At the Pennzoil table, Jamail glowered at Miller's reference to Kronzer. This wasn't unusual; Jamail had glowered at nearly every joke Miller had cracked during the long trial.

MILLER: "Let me tell you what I conceive to be my obligation in this case.

"Most of these cases that get tried in court, the juries don't really get to know the lawyers because they're not around them long enough to get acquainted with them. And that's not true in this case. I have had a hard time not calling y'all by your first names because we've been together so long, it's kind of like being in a lifeboat. We've been through a horrible experience together and that makes me feel like I know you and I'm sure you know me, or think you know me.

"I don't conceive it to be my obligation to Texaco or to you or to this Court to fudge, or lie, or cheat to win this case. If we deserve to win this case, you're going to win it for us; and that's the way I want it."

Miller's argument would cover many specific points, but always returned to his central themes.

MILLER: "I've always been an admirer of Mr. Justice Holmes, who says that 'The wisdom of the law is the experience of mankind.' I like Mr. Justice Holmes. I like that thought.

"I like him for another reason. On his ninety-third birthday, as he was standing where they were having the party, a pretty girl walked outside the window and he looked at her and said, 'Oh, to be eighty again.' And I kind of like that.

"But he was a person who understood and knew the law and who appreciated the realities of the way people get along in the world.

"I'm here to speak to you about this case and the evidence in this case. I said to you that I don't think it's my obligation to lie, or cheat, or fudge.

"There's a story about that in the law. The young lawyer comes into the old lawyer's office and the old lawyer says, 'We've just been hired in a bad accident case. I want you to go out and talk to the witnesses.

"And the lawyer says, 'Great. What do you want them to say?'

"He said, 'Listen, you get the facts. I'll twist them.'

"And as I've listened to this case, that's the feeling I've had."

Pennzoil's twisting of the facts had to be the central theme of Miller's defense. If Pennzoil's "facts" were true, it was all over.

MILLER: "I don't know if you can recall for me, but when we started talking to you ladies and gentlemen on this jury panel four months ago and I came up with ten minutes given to me to talk to you after you had been listening to these lawyers for four days, I told you there's no way Texaco is going to lose this case if you understand it. No way.

"And I asked you to bring paper and pencil and to be prepared to take notes when this case started. . . . It's not my custom to thank people for doing their duty; that's what we're supposed to do.

"But I must say, if I ever get shipwrecked it's with one of you who's got the kind of devotion and responsibility to your duties that you have had for four and a half months, and that after four and a half months some of you are still taking notes. And I regard that as thoroughly remarkable."

Miller commented on the fact that the case ended up being tried in Houston.

MILLER: "When you see a complicated case like this, what's it doing being tried to people who have no background in the controversy, where you have to start out and explain tender offers and sharks and predators and targets, and all of the rules that companies like this have?

"And you could take the case to Delaware to some judge who has done this time, after time, after time. . . . You may say, 'Why didn't they file this case in Delaware?' They're both Delaware companies, 'Why didn't they file it in New York?' where all the witnesses are.

"I thank my lucky stars that they didn't and that they filed it in Texas where I would have an opportunity to participate in this case and to do what I could to present my side of the case to twelve people who have no axe to grind, who have no interest in the outcome but to do right.

"President Harry Truman, you know, grew up in Missouri. His mother lived to be eighty-eight or eighty-nine years old, and one time, when he was president and he came home, she said, 'Harry, always do right. It will please some and astound the rest.'

"And that's our obligation, as I see it, to do right."

His argument now confronted the case at hand.

MILLER: "Oh, I've listened to these attacks that have been made on the Texaco people and the charge that it's a big company. And it is. . . . They start right off in the middle of the case and appeal to your bias and prejudice, and they would do that in contravention of the instructions of the Court, and say 'We want you to vote against Texaco because it is a big company.'

"Golly, you would think this was some kind of filling station that's suing us, instead of another eight-billion-dollar company. I never saw such a bunch of crybabies."

Miller's scorn rang out in the courtroom. He went back to the fact the case was filed in Houston.

MILLER: "I thought they filed this case in Texas so they could bring their witnesses. They bring the chairman and the president and two middle-level people who don't know anything about the case basically, Mr. Fridge and Mr. Lewis. And their lawyer from New York, Mr. Liman, who's still representing them today, still on their team, who took all these depositions, who took the deposition of Mr. Wasserstein, took the deposition of John McKinley and then tried to pass him off to you as some disinterested witness? My eye!"

Pennzoil's missing witnesses was Miller's next theme.

MILLER: "Everybody in this case brought their investment banker to testify before you except Pennzoil."

"This is a key witness, Mr. James Glanville, who is a long-time personal friend of the chairman, who guided him through the transaction from start to finish, who was in on the transaction from the middle of December until it collapsed on him, has not appeared to testify in this case.

"Now, do they lack money to bring Mr. Glanville here to Houston? No, that's not it at all. One of the reasons he's not here is because he knows what an agreement in principle is. They can't afford to put Mr. Glanville on the stand because he's going to testify that an agreement in principle is not a contract.

"As I listened to this case and to the way these lawyers used these words, I think about a book that was one of my favorite books when I was a little boy which is some time ago, written in the 19th century by a man named Lewis Carroll, who wrote *Alice in Wonderland, Through the Looking Glass*, and *Humpty Dumpty* stories.

"In one of his stories, it's *Through the Looking Glass*, I think, Humpty Dumpty is talking to Alice, and they're having a conversation about the meaning of words. And Humpty Dumpty has used the word 'glory.' And he tells her that 'glory' means a big, good argument.

"She says, 'I never heard that word used that way.' He says, 'Well, words mean what I want them to mean.' She said, 'The question is can a word have so many meanings?' something like that.

"That's not the exact quote, but 'A word can have so many meanings.'"

In Miller's case, all roads lead to Hugh Liedtke.

MILLER: "When I listened to the lawyers talk about restructuring . . . when they make this tender offer to be what they say is 'a constructive force in the restructuring of the Getty Oil Company' . . .

"And that's almost an exact quotation in the hand of Mr. Liedtke himself, although he denied it when I took his deposition.

"I said, 'What does this restructing mean?' Alice, what does that mean?

"And when they say we want to be a constructive force in the restructuring of the Getty Oil Company, and they want the people who own shares in the Getty Oil Company to think that's what they have in mind, when their secret plan from start to finish, beginning to end, is to get 40% of the shares . . . so they can get 48% of the assets, dismantle the Getty Oil Company, leave 52% in the hands of Gordon Getty, and take the balance back to Houston at three dollars a barrel; that's what they call being constructive and that's what they call restructuring.

"So words mean what you want them to mean, don't they? You say I want this to mean that and that's what we're going to call it today.

"So in the final analysis, ladies and gentlemen, in many ways this case is a case about words. It's a case about companies, too, and the people who run them."

Liedtke the Hungry Predator was contrasted with the Friendly Men from Texaco.

MILLER: "I asked John McKinley and Jim Kinnear and Al DeCrane and Bill Weitzel to come to Houston to be here when you all were selected as jurors in this case. They didn't have to come; I didn't have to ask them.

"But it seemed to me that when people are faced with accusations that they're liars and cheats and frauds that you're entitled to look at them, that you're entitled to see them testify. . . . You have the opportunity to see them under pressure.

"You have the opportunity, if opportunity is the right word, to see them badgered, asked the same question over and over and over, and pushed and shoved and attacked. And you see how they handle it.

"So that's why I brought these people down here. They didn't all have to testify; we could have probably got away with just Mr. McKinley testifying. He's the head of the company.

"But it seemed to me that you ought to see them, and that's why I wonder, you see, why Dr. Howe, the president of Pennzoil, didn't testify.

"I had to read his deposition to you, and he couldn't even remember his own name. He managed to remember his name and that's about all I ever got out of him. And I think that's why they didn't bring him."

Miller had taken the measure of Liedtke and clearly found him lacking. Now, in the trial's final hours, he asked the jury to embrace that view. To effect this, he compared him head to head with McKinley.

MILLER: "You know, companies are like countries . . . they reflect the qualities and moralities of their leaders.

"You have one kind of country when Richard Nixon is president, and you have another kind of country when Harry Truman is president, and the third kind when Dwight Eisenhower is president, and the fourth when Ronald Reagan is president.

"So when you look at Texaco, you say Texaco is the kind of person that John McKinley is. And the kind of company that Pennzoil is the kind of a person that Hugh Liedtke is, because the people below those people reflect their qualities and their goals and their aspirations.

"So say what kind of a person is John McKinley? Well, you saw him on the witness stand. He's not an entrepreneur. He's not a wealthy man in the standards Mr. Liedtke is used to. He's a professional manager, and a fine one."

It was impossible not to look from McKinley to Liedtke and back again; all the players were present in the courtroom.

MILLER: "And every one of these men who represent Texaco came up
             through the organization. Every one of them started out at
             some job that wasn't too much, and now they've got a job
             that is something.

             "So when you look at them and you look at this company
             that's under attack, you say the company is these people.
             And that's why they're here.

             "When they say to you, to me, that Texaco has lied and
             cheated and stolen, they are talking about these people.

             "Mr. McKinley's deposition has been taken twice, once by
             Witness Liman and once by Mr. Jamail. And if there had
             been one lie told by Mr. McKinley, we'd have heard about it
             over and over and over.

             "As I understand what [Pennzoil's lawyers] have said,
             there is no witness presented by Texaco who has told the
             truth. Not one.

             "There is not one witness from Texaco who has not per-
             verted his testimony. Peverted and twisted by self-interest,
             by greed, by desire for some reason or another to get even
             with Pennzoil and could therefore come and testify."

This brought Miller to his most telling description of Pennzoil's case.

MILLER: "Why is it these people think that? Why is it that they cannot
             give credit to the proposition that some people would come
             to the Court and testify because it was the right thing to do?

             "Just like some people come and serve as jurors, because
             it's their obligation. So why don't they think that's possible?

             "The reason is that it's a reflection of themselves. It is a
             reflection of their own morality."

Within this framework, Miller responded to the attacks on the Getty
directors who testified.

MILLER: "Mr. Wendt comes from Philadelphia, and he has to take this
             kind of sly little remarks about being the chairman of one
             of the functions in Philadelphia, for the Pennzoil people have
             never run out of cheap shots, never.

             "This man, Mr. Wendt, owned two hundred shares of
             stock in the Getty Oil Company. Made three trips to Texas
             to testify. . . . Cross-examined like he was some criminal,
             treated like he was down here to do somebody in for some
             evil purpose, all to come in and tell what he knew.

             "Well, you tell me why he did it. You tell me why he came

down here and took that kind of personal abuse, and then came back twice.

"Somebody tell me. Was it for money? Jesus, God in Heaven, what kind of money would we be talking about? What, twenty-five hundred dollars he made on his stock?

"Was it because we have got some hold on him? He doesn't do any business with Texaco. He doesn't know any of us.

"Why would he come down here and say that all the Getty board voted on was the price, if that wasn't the truth? Somebody tell me the answer to that.

"And how about Mr. Medberry.... If one thing came across in his testimony it is that, 'Look, I am sixty-seven years old. I really don't care what you people think. I am going to tell you what I think.

"You want to pay attention to my testimony, fine. If you don't, it's okay with me. I'm going back to California and play golf.

"And how about Mr. Tisch? What's he doing down here? He hasn't got any business with us.

"Here is a guy who grew up on the streets of New York. Little fellow who has worked his fingers to the bone for sixty-three years. Has become one of the most successful people in the country, rich as Ben Gump.

"Maybe you all don't remember Ben Gump; I guess that dates me.

"Rich as he can be. Got no money to come down here and say a word ... and Mr. Jamail says, 'Well, I know what a friend is in New York, Mr. Tisch.

"If this case was in New York, he couldn't say that to Mr. Tisch, could he? He couldn't put that kind of a little shot on him and get away with it. And Mr. Lipton wouldn't have to take all the abuse he took."

Miller connected credibility with self-interest, contrasting these directors with Liedtke.

MILLER: "When I listened to Mr. Liedtke tell you ladies and gentlemen of the jury that he never read the tender offer, that he never saw the purpose clause, that they didn't plan this thing between Christmas and New Year's so it would make it easier to take over the Getty Oil Company, 'No, I don't know anything about that.'

"'No, I don't remember any papers being passed out at

the board meeting. What are you talking about? No, I never saw any studies. No, what, Mr. Miller? I don't think any of that happened.'

"So, well, Mr. Liedtke, where are the papers that were used at the December 19 board meeting of the Pennzoil Corporation? I would like for the ladies and gentlemen of the jury to see them.

"'Well, they are all gone. We were afraid some arbitrageurs would break into the building and find them.'"

Miller's tone shifted back and forth between scorn and amazement.

MILLER: "How have they got the nerve to criticize Mr. Copley? Finally have admitted it. It wasn't Mr. Copley. They say it was Texaco that caused Mr. Copley to destroy his notes.

"So what is it then that makes them think Texaco would do that? Or anybody would do that? Why? Where do they come by this mentality, that that's standard operating procedure?

"What you do when you are in a tight spot is you destroy evidence or you fix it, when there isn't any evidence, if that's a reason at all.

"Mr. Copley has got to take the abuse of being accused of being a destroyer of evidence from people who have destroyed more documents in this case, by far.

"Mr. Liedtke says to Mr. Fridge, 'I don't want to see that $120 pro forma walking around,' and it loses its legs. It's gone.

"So naturally, you would suspect somebody else destroying the papers, right? 'We are going to destroy our papers. That is naturally what you would do as well. . . . That's what smart businessmen do.'"

Miller's recasting of Pennzoil's actions through his own eyes continued with a key trial exhibit.

MILLER: "Not only that, we've got this 'Dear Hugh' letter that we needed to keep Gordon Getty in line . . . Mr. Liman says, 'No, it wasn't my idea. It was Mr. Siegel's idea.' Mr. Hertz said, 'No, it wasn't my idea. It was Mr. Liman's idea.' And nobody will claim that letter.

"As a matter of fact, they claimed in the deposition that it was Mr. Gordon Getty's idea. And why? Somebody tell me why?

"Because that letter is the most shameful example of business morality that you can imagine in any case, where they got Gordon Getty to agree to violate his fiduciary duties to the public shareholders . . . for the personal gain of the chairman and the president who owned between them around five hundred fifty thousand shares of stock in this Pennzoil Company and who stand personally to gain several hundred million dollars if they can collect on this oil gusher in this courtroom.

"And they have got the nerve and gall to call Geoff Boisi 'the eighteen-million-dollar man.'

"Now, how is that for nerve? He is going to make nine million dollars more and we are going to make several hundred million dollars more. And he is 'the eighteen-million-dollar man.' Except not to his face; only in his absence do they call him that. So much for corporate courage."

Miller said that in addition to evaluating the testimony of the witnesses, you had to evaluate the lawyers.

MILLER: "I remember Mr. Jeffers said when he commenced the opening statement that I have more tricks in my head than just about anybody you could imagine. That's what he said.

"Well, you and I, we have been around each other now for four years, and if you think that about me. I value your opinion. I guess I am like everybody else. I would prefer to be liked rather than disliked.

"But if you think that about me, that I have done tricky things and have mistreated witnesses and have beat on them and pilloried them, then you will hold that against me, and properly so."

In addition to Pennzoil's missing witnesses, Miller talked about Texaco's damage witnesses who also didn't appear.

MILLER: "The case lasted a long time. We haven't brought every witness we had, as has already been pointed out. We didn't bring these experts and I guess I sort of think you ought to consider that, but when you do consider that, just remember that no expert in this case knows as much about this case as you do, couldn't possibly know as much about it as you do.

"I looked at Mr. Barrow, and we've got no beef against Mr. Barrow except that I just have to ponder how he got involved in this case, if there isn't something there they don't know about. And I have to wonder whether or not the fact that he is on the board of directors of Texas Commerce Bank, along with the chairman of the board of the American General Insurance Company where Mr. Kerr serves as a director and Mr. Bill Harvin, who was then the managing people of the Baker Botts firm and himself a very fine lawyer, is on that board.

"Say, wait a minute now. What Mr. Justice Holmes said to Watson, 'Watson, I don't believe in coincidences.'"

In spite of Miller's mistaking Justice Oliver Wendell Holmes for Sherlock Holmes, he had locked in on an important point. Barrow had been a very effective witness for Pennzoil. The suggestion that there was some secret arrangement involving Barrow had been made during the voir dire; it reappeared in the final argument. Not much was said about it in the interim; certainly no evidence was offered.

Miller attacked the number of documents Pennzoil showed to Barrow on the issue of contract formation.

MILLER: "[Barrow] doesn't know one-fiftieth of what you know about this case, and they have got the nerve to suggest to you that you ought to give his opinion some weight.

"So, I have to say to myself, 'Well, I suppose they're entitled to do that. They're entitled to manipulate the testimony of that witness, I suppose, legally.' It's up to us to call their hand on it."

Miller's argument would conclude the next day. For the rest of this afternoon, he went back to Pennzoil's actions that triggered the Six Day War.

MILLER: "This case really began in November of 1983, when Mr. Liedtke and Mr. Kerr saw from the public press the death throes of the Getty Oil Company.

"These are two men who are able, intelligent, experienced, who know as much about tender offers as any executive in the country, and who are proud of the fact that they made the largest and first hostile tender offer ever made in this country. They brag about it."

This reference was to the 1965 Pennzoil tender offer for United Gas.

MILLER: "[They] have since then made five or six or seven other hostile tender offers . . . the growth of [their] company has come

> about from this business tactic that, I don't care what you say, is the most maligned tactic in the business community today and some day will be illegal."

There was some agreement in the jury box about the undesirability of hostile tender offers, but our oath as jurors was to decide the case on the evidence, not on our personal feelings about takeover tactics. Texaco's oft-repeated position that it engages only in friendly takeovers has not held up well. If there is any aspect of this case that is friendly, I have yet to see it.

MILLER: "[They] have made their company grow by taking over other companies, not only that way, but that's the principal source of their growth. It's a company run by lawyers. . . ."

While addressing the mechanics of tender offers, Miller recalled the testimony of Geoffrey Boisi of Goldman, Sachs.

MILLER: "I don't know what you think about Mr. Boisi. I can't believe that anybody who saw that guy, who saw him testify, could do anything but believe him. You know, I've seen a lot of witnesses—what could he have done better? How could he have been more straightforward with you, more honest with you, and more open with you? And more incensed, I suppose.

 "Yes, that's right, he did make more money. But I don't happen to think and I hope you don't think that that's the reason he came here to testify or that that motivated him to seek a decent offer for this company."

The testimony of Boisi was critical for Texaco. Miller's opening line of defense ("We didn't crash this party, we were invited") depended to a large degree on Boisi; he had issued the "invitation." On direct examination, he did come across as being fairly straightforward.

On cross-examination by Irv Terrell, however, he was evasive to an amazing degree. He said he didn't issue invitations to bid, he made "courtesy calls." Jeffers and Terrell had both savaged this and other Boisi testimony during their arguments earlier in the day.

Miller's attempts to rehabilitate Boisi at this late hour could not overcome this reality. He returned to his favorite theme, Pennzoil as villain.

MILLER: "If [Pennzoil] wanted to buy some Getty Oil Company oil, why didn't they just go to them and say, "Look, we'd like to

buy some of your oil. We want to be a big company. You've
got a lot of oil. You guys don't know how to run your business, anyhow. Your chairman's a boob. He bought this insurance company you are only going to make five hundred million dollars off of it. You dumb clucks can't run this
company. Sell some of this oil to us, but we only want to pay
three dollars a barrell.'

"You [Getty Oil] say, 'Well, gee. That's not fair.'

"'Well, by God, we'll take you over then.'

"[Pennzoil] didn't want to do anything without their
tender offer in force. . . . This purpose clause that's set out in
that tender offer is the most misleading, deliberately evasive
statement of purpose you can imagine. And their true purpose, their only purpose . . . was to go into partners with
Gordon Getty."

The true purpose of the tender offer, Miller maintained, was to
"scare the stuffing out of Gordon Getty."

MILLER: "So what would you all say is the most prevalent human
            emotion? What is it? It's fear, isn't it?
                "When you worry about not having enough money, that's
            fear. If you are worried about not having a place to live,
            that's fear. If you are worried about not having a job, wor-
            ried about 'What's going to happen to me?' worried about
            all those things. If it's not one of the most important human
            among emotions, it's certainly among the most.
                "And a skilled negotiator—and nobody denies this fits Mr.
            Liedtke—understands fear, understands how to use it, and
            what he did by that tender offer was to tighten Gordon Get-
            ty's gut up so tight with fear that he was willing to deal."

Miller noted the lateness of the hour and said he was going to ask
the judge to recess fifteen minutes early. He concluded for the day by
turning again to Pennzoil's real intentions toward Gordon Getty.

MILLER: "You tell me, do you really think that a man who has been
            chairman of the board of a major oil company for twenty
            years is going to participate in a five- or six-billion-dollar
            deal and let somebody else run it? Do you really think J.
            Hugh Liedtke was ever going to let Gordon Getty run the
            Getty Oil Company even if he owned 57% or 80? Not quite.
                "But what they did understand was his intense desire to
            be the head of his father's company so [Liedtke] said, 'You

are going to be the chairman of the board. I'm going to be the chief executive officer. Mr. Kerr is going to be head of the executive committee. You are going to get a big office and it's going to say on the door that you are the chairman, and you are going to decide what we do about the war in Bangladesh and how we stand on the international monetary policy and all those other important things that we're concerned about and we'll just run the rest of it.'

"Well, Mr. Getty is nobody's fool but he is in a real bind. He's got good advisors and he's like the rest of us, I suppose.

"I've got my weaknesses; you've got yours. People are often smart enough to spot what mine are. I always resent it a little bit when they do it, but they do it.

"And so, they had his number. They did. And you know how I know that? Because they made a deal with him in two hours. In two hours they talked Gordon Getty into that Memorandum of Agreement and he'd never seen them before in his life. Fantastic!"

Dick Miller still had nearly three hours in which to conclude his final argument, but the central difficulty that confronted him was readily apparent. Although the burden of proof was on Pennzoil, Miller still needed a persuasive scenario to replace the compelling tale spun by Jeffers, Terrell, and Jamail.

The defense did not have such a scenario, other than the aforementioned Liedtke the Hungry Predator theme. The strongest legal points in Texaco's favor were of a more technical nature. Taken one by one, each had merit to a degree. Taken as a whole, it could be argued that Miller had aroused reasonable doubt with these points.

That was not the standard of proof upon which this case would be decided. If Pennzoil's lawyers had not proven Texaco's guilt beyond a reasonable doubt, they had come close. The preponderance of the evidence pointed one way.

32

Early the next morning, the jury returned to the 151st District Court, which was again packed for this last day of final arguments.

Dick Miller would conclude his argument for Texaco in the next three hours. After lunch, Joe Jamail would take the final hour and a half for Pennzoil.

MILLER: "Last night my ever-vigilant partners pointed out to me that I had said that Mr. Justice Holmes had spoken to Watson, who was Sherlock [Holmes].

"They also pointed out to me that I ought to get to the point and I ought to quit acting like I know everything. So I'll try to do both of those things, and before I sit down we're going to have a little firm meeting because I don't want to hear any more of that when I'm through.

"Now, let's get on with the case."

Miller began by concentrating on the way the Memorandum of Agreement was presented to the Getty board.

MILLER: "They take this Memorandum of Agreement. Mind you, no director had seen this except Gordon Getty until 6:30 P.M. on the 2nd of January.

"Anyway, they take it there and flop it on that table and say, 'Vote. We want you to vote. You've got four hours, and if you don't, we're going to withdraw it, take it back, and then you directors are in this position: You had a chance to vote for one hundred ten dollars and if you don't vote for one hundred ten dollars and we go forward with our tender offer and the shareholders take one hundred dollars, you are going to get sued. You have violated your fiduciary duties.'

> "They say, 'No, we didn't put any pressure on these guys.
> What are you talking about? This is a friendly tender offer.
> All we want to do is cut your guts out. We are going to put
> you into a position where you get sued.' Very clever ploy.''

Miller argued the evidence showed that, even after the 15-to-1 vote,
the Getty directors did not intend to be bound until there was a formal
written agreement.

MILLER: "First and foremost, after you consider the suspicions and
          doubt and discord that existed between the parties, is that
          they never had any negotiations.
              "Mr. Wendt, I think you could tell from his attitude that
          he was very disappointed in the way this matter had been
          conducted by Pennzoil, and one of the last things he said on
          the witness stand is, 'We never knew what we were dealing
          with.' First one guy then another guy.
              "Nobody with any authority ever talked to anybody with
          any authority with the Getty Oil Company except that one
          time and then all they said was, 'We don't want to talk to
          you. We have got a deal with a 40% shareholder. Get lost.'
              "And they sat over there in the Waldorf Towers and
          talked to each other. And never negotiated, never said a
          word, never suggested any [meeting] times."

He turned his rhetorical powers on Pennzoil, which was made to
seem ridiculous by the image of Kerr and Liedtke sitting in the Waldorf Towers talking to each other.

MILLER: "I say this. I said this when we picked you on this jury and
          I am saying it now, and I am going to say it to you when I
          finish:
              "*Don't tell me that Texaco beat you. You beat yourself. You
          beat yourself, Pennzoil.*
              "*So arrogant, proud, so full of yourselves that you will not
          talk to anybody.*
              "And then what happened? See, now, if you take the most
          charitable view you possibly can take is that, I suppose, that
          there was a failure of communication."

Miller described the breakdown of the Pennzoil-Getty deal, concentrating on his contention that the 15-to-1 vote approved only the price.
Clearly, he felt that "no contract" was by far Texaco's best defense.

At the same time, he blasted the man at Pennzoil's table who would deliver the last argument when Miller finished.

MILLER: "The thing that amazes me about this whole case is when we picked this jury, Mr. Jamail spent twenty-five minutes talking to you about handshakes.

"He said, 'This is a handshake. This is a deal, I mean, you know, a deal is a deal,' whatever that means.

"He said that 'When I was a little boy, I was taught that when we shake hands, that's a deal.'

"Well, the only trouble is that these people from Pennzoil had to rely upon a designated handshake. Who? They never shook hands with anybody except themselves, as if it makes any difference whether you shake hands or not, and it doesn't.

"You don't have to shake, all you have to do is agree."

Jamail's handshake had been a flight of rhetorical fancy that had encapsulated his first presentation of the case. In that same sentence, however, Jamail had included "handshake" with "promise," "your word," and "contract."

Pennzoil's case did not rely on a handshake. Arthur Liman testified that he shook hands after the Getty board meeting on January 3. Miller had spent a lot of time ridiculing that contention. Witness after witness said that Liman never entered the board room, never shook hands with anybody. Tisch refused to join in the trashing of Liman, saying he wasn't sure *where* he shook hands with him. Martin Siegel, in a deposition read after the last live witness, said Liman *did*, in fact, come in the board room after the meeting and shake hands.

That testimony was fresh in the minds of the jury. Texaco had gotten a lot of mileage out of lampooning the handshakes, but by this stage of the trial, this worm, too, had turned.

Miller also contended that there was no contract because there had been no agreement on the essential terms.

MILLER: "The testimony that I can recall, about what the Pennzoil witnesses said the essential terms were, is in total conflict.

"Mr. Terrell said on the voir dire examination of the jury that the essential terms were the price and who was going to buy and who was going to sell, and the 4/7ths–3/7ths, three terms.

"Mr. Kerr gets on the witness stand and says the option was also or could be essential terms.

"Mr. Liedtke gets on the witness stand and gives another

list, and Mr. Liman gets on the witness stand and he's got eight items that he says are essential.

"Now, what kind of a contract are we talking about? Doesn't that just prove that they did not know what their own deal was? And if they can't come into this court after eighteen months and all of them tell the same story on the most critical issue in the case, well, I just don't buy that business that you've got a contract where you have agreed to essential terms and you don't even know what your side of the deal is."

Having dealt with the contract issue, Miller moved on to interference. He plainly stated that he did not believe the jury would get past Special Issue No. 1, but he offered some ideas on Pennzoil's contention that Texaco knowingly interfered with a binding agreement.

MILLER: "When these notes from Mr. Lynch and Mr. DeCrane say 'STOP THE TRAIN' . . . but you see, I thank God for these notes. They're one of the most compelling pieces of evidence we have in this case.

"The only thing I regret is they had the right to introduce their evidence first and put their Plaintiff's mark on them and they look damaging to us.

"But, in fact, what those notes conclusively prove they *did not* have a contract is what we believed.

"We are listening to people saying 'STOP THE TRAIN, STOP THE SIGNING.' On the 4th of January, there was nothing to sign. . . . The only way you would stop the train would be if you did not have a contract. And if that's not the most conclusive proof that you could present that we did not know or believe they had a contract, I don't know where we would be."

Miller turned to Pennzoil's battering ram, the indemnities.

MILLER: "An indemnity agreement is an insurance policy like you've got insurance on your car . . .

"Their whole theory of the indemnity rests upon the assumption that the person asking for the indemnity feels threatened, that he has done something wrong and that he wants protection.

"Does that mean that when you buy insurance for your car that you think you're going to have a wreck? No, it doesn't mean that at all. Means that you might have a wreck and you might be in the right and you might nevertheless get sued. And I can assure you that happens all the time."

Miller minimized the importance of the indemnities Texaco granted to the Getty entities.

MILLER: "Mr. DeCrane, Mr. Kinnear, and Mr. McKinley were advised by Mr. Weitzel that the risk of granting the indemnity . . . would be that they would incur some expenses for attorney's fees if groundless suits were brought.

"And the evidence in this case is quite clear. There have never been any suits filed against the directors of the Getty Oil Company, not one, or against the Museum trustees. Not a suit in the world.

"What are they talking about? What risks are they talking about? We haven't spent a dime on any judgments. We're not ever going to. They know that . . .

"What's the point of [using] the indemnities to say we knew we were violating an agreement because we were granting indemnities . . . and in the history of this dispute, Texaco has not paid out one red cent in judgments and is never going to."

Miller went on to address other evidence of interference that had been cited by Pennzoil, including Texaco's sworn statement that no one had read the January 5 *Wall Street Journal* article detailing the Pennzoil-Getty agreement.

He then sounded a familiar refrain.

MILLER: "I guess, to my mind, the most persuasive piece of evidence we have about the issue of tortious interference is the overwhelming persuasiveness of the fact that *we did not crash this party*. We were invited to the party. We didn't look them up. They looked us up [and said] 'Mr. Texaco, we want you to bid. We're in a squeeze and we think the offer we're getting is shameful.'

"When they say we interfered with their contract, in the face of this proof that we did nothing but respond, this has got to be the first case in history where the white knight got sued by the dragon. And that's exactly what happened."

Finally, Miller spoke to the record damages sought by Pennzoil.

MILLER: "Why have they sued for so much money, when they say it's by far the biggest suit in numbers that's ever been filed?

"It's to get publicity. If they were suing for fifty million dollars, there wouldn't be a reporter within fifty miles of this courthouse, and that's the truth.

"It's got to be one of the dullest cases that's been tried in the history of mankind, when the most exciting thing that happened was that I came into the courtroom with my fly unzipped . . . and that wasn't exciting to anybody but me.

"So what I am concerned about is somebody's going to say, well, you know, 'Let's give them a little bit.' I don't want you to give them anything, and if you don't understand that that's the way I feel by now, then I haven't been talking plainly.

"And what I am really afraid of is that somebody will say, well, 'What about this letter they wrote where they said they wanted their eight million shares of Getty stock at one hundred ten dollars a share? Let them make one hundred forty million dollars,' which would be an outrageous sum."

The reference here is to the 8-million-share option the DeCrane notes said "COULD BE A WAY TO TAKE CARE OF LIEDTKE." Was this really a back-handed suggestion that if the jury *did* find that Pennzoil had a binding agreement, the most it had coming was this "outrageous sum"?

The amount of damages sought by Pennzoil drew more fire and provoked a naked plea for reason.

MILLER: "Do you realize that today, eighteen months after we acquired the Getty Oil Company, we've got two hundred forty million shares of Texaco stock that're selling for less than forty dollars a share? Which means the whole company of Texaco, Incorporated is worth less than ten billion dollars.

"And that today, they are suing us for all our company is worth, plus all of what Getty is worth, plus half again?

"And that in the history of [Pennzoil], since 1952, twenty-three years, they have managed to get together and earn a total of a little over two billion dollars? And they have the nerve to come in here and sue for this outrageous sum of money? And I [tell] you, nobody in their right minds thinks they're entitled to that."

Miller tried to give some human scale to the kind of numbers he was talking about.

MILLER: "My little grandchild has got a little book that says 'How much is a billion?' And there is a little thing in the book, it says it would take you ninety-five years to count to a billion . . . But these people can sit out here and file this.

"Why did they sue for this monstrous sum of money?

"I think it's basically to get publicity, to satisfy the desire for revenge for getting beat out.

"These people can't bring themselves to admit they got beat; they want to think they got cheated.

"That's what they want to think. So that's the reason for this case.

"Now, if our courts are the place for people to seek vengeance and revenge and ego gratification and to bring these suits and put these people through this to get even, then I don't think that ought to go on. I don't think you ought to let that go on.

"I think you ought to tell them no, and you ought to tell them in a way they will understand it. . . . If you give them one thin dime, they're going to be encouraged. . ."

Miller briefly referred to the claim for an equal amount of punitive damages.

MILLER: "I am not going to waste my time or yours talking about this claim for seven billion dollars in fines.

"If we are in that bad a shape in this case, if that's the kind of shape we are in in this case after all this evidence, then there is no hope for us."

This seems almost poignant now. As I listened to Dick Miller that morning, the verdict was still in doubt. I knew how I felt, but I had no idea if this argument had found fertile ground with the rest of the jury.

MILLER: "Now, I'm about finished. I want to say a couple more things to you and I'm going to sit down."

He briefly looked at the crowded table where Keeton and Brown sat with the Texaco contingent; then he held up a little pile of yellow sheets of paper.

MILLER: "You see all these notes I'm getting from my partners? I'm going to put them in my pocket. . . . When you get down to it, the most important thing to realize in the case is that these people beat themselves.

"Mr. Liman says, 'Mr. Liedtke does not talk to agents,' that he does not deal with agents, and I guess if one thing has been proved in this case, that's true.

'They don't have an Exhibit A. They don't have a contract. I think you know they don't have a contract.''

Miller's Oklahoma drawl softened as he reflected on the trial that was nearing the end.

MILLER: "This has been a tough case. The fact of this case being in Texas has been a very lucky and fortunate thing for me, because I have got a chance to meet these people I represent and become acquainted with them, and I feel very comfortable vouching for their integrity and for their conduct in this case, so that I'm very confident about this case.

"I believe we're right. I believe you think we're right. I believe you're going to find 'No' to that first issue and you're going to answer these other issues like I've indicated. . .''

His part of the case complete, Miller was almost serene. He had to be physically exhausted as well. His defense in a nineteen-week trial he described as "a horrible experience" had ended with him talking for four hours, most of it this morning.

MILLER: "I hope you will decide this case. Remember what Harry Truman's mother said to him. You do right and the case will come out like it ought to come out."

Dick Miller thanked the jury and sat down. Although the audience had swelled to capacity and beyond for these final arguments, there remained a curious bond between those who had shared the long days and weeks and months of this trial.

I thought of Judge Farris who wasn't there for the conclusion of the voyage he had captained most of the way. Liedtke and McKinley and all of the top executives from both companies were present, but I thought of Jim Kinnear and Baine Kerr, who had made the whole long, strange trip.

This case was headed for the jury. I wasn't starting to wax nostalgiac (okay, maybe I was), but I knew this case was not quite over.

After lunch, Joe Jamail would address the jury for an hour and a half. He's a man who had been known to put a word or two together during the trial of this case.

* * *

From the first day of voir dire, the jury had heard the impassioned account of how Hugh Liedtke and Pennzoil had been done wrong.

The trial would end as it began, with a oratorical tour de force from the lead counsel for the plaintiff. Despite the methodical exposition by the cerebral Jeffers, despite the slashing rhetoric and probing analysis by the exuberant Terrell, Joe Jamail had always been the moral compass of Pennzoil's case. As such, he had consistently pointed the same way; there was only one right thing to do.

He rose to deliver what would indeed be the final argument in the case of *Pennzoil Co. v. Texaco, Inc.*

JAMAIL: "You have listened for the last four hours to the reason for our rules of law: What lawyers tell you cannot be taken as evidence.

"The evidence [is what] you heard from the witness stand, and you're going to have to decide the case based on what you heard.

"There were so many misstatements in what Mr. Miller has told you that I'm not going to try to answer all of them; it would take all of the time.

"He starts out by giving you four hours of excuses for Texaco's conduct and he ends it up by saying to you, 'Don't take our company because if something went wrong, it was Getty that did it.'

"They bought and paid for this lawsuit when they gave the indemnities.

"They knew right then that Pennzoil had been cheated out of its agreement, but they were willing to do it, pay any cost, because to them it was cheap.

"You heard it. You heard what they saved. You know what their motive was.

"And, for Mr. Miller to stand here and attempt to change the court's charge, the wording in it, by inserting things that he likes, is typical.

"Judge Casseb is one of the most honored and brilliant judges in America. If he had wanted to say 'contract,' he would have said it. I have no quarrel with it, but it's his charge, it's not Mr. Miller's."

The conflict over the words "contract" and "agreement" continued, but I felt then and now that it was truly a nonissue. Look up "agreement" in the dictionary; one of the first definitions listed is "contract." Still, the semantics battle raged on.

Joe Jamail was unrelenting.

JAMAIL: "The reason for that is this subtle way, the attempt to try to get you, the jury, to impose a far harsher burden on Pennzoil than the law does.

"You're not going to do that. You told me [that] when we talked the very first time. I have no reason to doubt that you're going to do what is just. You're not going to do it because I stand here and shout at you, and I hope I don't shout at you.

"He's told you a lot of old lawyer stories. I hope I don't take up your time doing that. I will tell you a lawyer's truism though: 'When the law is on the lawyer's side, he talks about the law. When the facts are on the lawyer's side he talks about the facts. When he has neither, he just talks' . . . and I guess tells Alice-in-Wonderland stories."

His tone was almost matter-of-fact, but only almost. Jamail talked directly to the jury, but the whole courtroom hung on his every word.

JAMAIL: "Mr. Miller spent a great deal of time trying to attack Pennzoil through its chairman, Mr. Liedtke, and Mr. Kerr. What did they do?

"Mr. Miller, if he has his way, will introduce into the American concept of business a new concept, that is, 'Pennzoil is not allowed to purchase the Getty shares at a bargain or a fair price to them.' Now, they need to arrest everybody that goes to Foley's [department store] then, if that's something we're going to put up with.

"The timing of the tender offer is meaningless. The proof is replete. Pennzoil could not purchase any of that stock through that tender offer until January 17th. [If the Getty directors] were pressured, all they had to do was go home and that deal was over and they would have thwarted Gordon Getty again . . . Gordon had been treated like a nincompoop and that's what they did."

Jamail scoffed at virtually every point raised in Texaco's final argument.

JAMAIL: "We got some criticism from Mr. Miller about not having our witnesses all have the very same pat story, couldn't get their lines in the same little channel.

"He didn't have the same trouble. Every witness they brought . . . reminds me of one of those squeeze dolls. You squeeze it, it says, 'Free to deal, free to deal.' That's all we got out of them. But, were they really [free to deal]?

"In their ostrich-like hunt to find out whether or not there was an agreement between Pennzoil and the Getty interests, what did they do?

"They didn't ask for the Copley notes. They know that boards of directors when they meet take notes, keep minutes. . . . Mr. Kinnear testified that getting those notes beforehand would have been diligent. That's his testimony.

"Why didn't they want them? You know very well why they didn't want them."

He then delineated a crucial point for the plaintiff's case.

JAMAIL: "Now Mr. Miller has told you, and it's right, that you can't take testimony after the fact and make it apply to this case. What was in the intent or in the minds of those parties at the time is what the case needs to be decided on, and you're admonished to do that.

"And yet, every witness that he relied on from that board of directors, Mr. Tisch, Mr. Wendt, Mr. Medberry, are after-the-fact witnesses. 'Well, we didn't intend' . . . but what is the proof? You know what the proof is. The proof is the Memorandum of Agreement.

"And what happened? Now, the company, in its ongoing war with Gordon Getty, had one thing in mind . . . to prevent Gordon Getty from fulfilling the wishes of the Trust, his family, his father, and himself and that is to regain control of that company and they were at war with him.

"They made a decision on their own that Gordon's father didn't know what he was talking about when he said he wanted it to stay in the family. He was dead, he couldn't defend himself."

Jamail addressed an issue that had not been raised by Texaco until the final argument.

JAMAIL: "You heard from Mr. Miller so many misstatements. Let me correct one of them now. He stood here and, in a plea of sympathy, which you are admonished in Judge Casseb's instructions not to let influence you, [said] that 'We only have ten billion dollars in our company.'

"That's not so. The assets of Texaco are in excess of forty-eight billion dollars. That's what their assets are."

Feeling as if he had disposed of that bit of confusion, Jamail moved to deflect the rest of Miller's barrage.

JAMAIL: "He tells you that Mr. Barrow's testimony should be tainted for some reason. What [reason]? He came here without pay because, as he put it, 'An important principle is at stake here.' And make no mistake about it, it is.

"He said that Mr. Kerr was on the board of directors at American General with Mr. Barrow. That is false. Mr. Kerr is not now, never has been on that board and was never asked to be on that board.

"Why he would try to tell you that is beyond me, except it's the 'win at any cost' philosophy of Texaco. Because 'Even if we lose here, we've won. We got Getty and you can't take that away from us.'

"So, the remedy that the law provides for us is to be here before you, but don't make any mistake about it, they don't want this. That's *our* remedy.

"And they ask, you know, 'You don't see Getty in here, you don't see the Museum in here, you don't see the Trust in here.' For what purpose? For what purpose, jury?

"After they've guaranteed them and protected them from this very contingency, Pennzoil's lawsuit.

"Well, why would there be any question [about] having them here, or for what reason?

"But that's another smokescreen. 'Anybody but Texaco, anybody but Texaco,' and listen, our fall-back position at Texaco is this: 'This is a revenge case. This is frivolous.'

'I don't believe that that should be dignified with any answer. I will make a statement, however, that if that were so, do you think that Judge Farris in a year and a half, and Judge Casseb lately, would have allowed us to progress this far?

"No. We have met every test the law as provided. Every one.

"The law says 'Go.' And now it's up to you."

While refuting another point from Miller's argument, Jamail continued to bolster his case.

JAMAIL: "He says to you, 'We weren't alone. Excuse us. Excuse us. There was Socal bidding. There was Mobil bidding, and there was the Saudi government bidding.' The old Saudi connection that Texaco has.

"But the problem they've got with this is nobody followed through but Texaco. Nobody. Everybody else backed off.

"And you remember Gordon Getty's testimony [about] when he talked to the Saudi government on January the 4th. [He talked to] Mr. Yamani or Mr. Shaheen, and the words were 'It's a done deal.'

"Did Gordon Getty think he had a binding agreement? Did he think he had an agreement with Pennzoil when he said to those people 'It's a done deal,' when he sends Mr. Siegel's affidavit to California to tell that Court 'transaction now agreed upon'?

"[Miller says] 'Don't believe that affidavit. You know lawyers make those things up.' He embraces Mr. Siegel part of the time and denounces him at other times. Can he have it both ways?

"We have nothing to gain from Mr. Siegel saying that. But Mr. Siegel testified by deposition that what he said in that affidavit on January 5th was true and correct then. Those are his words. What changed?

In other words, what transpired to make Siegel's "true and correct" affidavit inoperative?

JAMAIL: "The higher price, the indemnities, the limited warranties. And Texaco still continuing its very selective investigation of the facts say, well, 'Listen jury, if you don't believe that, we talked to all these people, Mr. Boisi, Mr. Petersen, Mr. Lipton, these people, and they told us free to deal, free to deal, free to deal. And, oh, by the way, they all stood to make millions on it if we upped the price . . .'

"That's good enough for Mr. McKinley, who recognizes no good-faith dealings. Only fair, not good faith. The Judge asked you about good faith."

This blast at McKinley was part and parcel of Jamail's attack on Texaco and its Getty indemnities.

JAMAIL: "Finally, they say these people, who have these millions to make on this deal, told us 'Okay to deal.'

"But when it came time for those very people to put in writing the words they had been telling [Texaco] or said they've been telling them, they balked. They wouldn't do it.

"As a matter of fact, they did the reverse. They said no. Does that sound like free to deal?

One by one, Jamail compared the facts of the case to the idea that Getty was "free to deal" after that 15-to-1 board vote approving the Pennzoil deal.

JAMAIL: "The specific indemnity against any claim of any kind against you made by Pennzoil? Does that sound like free to deal?

"Wouldn't that alert somebody? 'And in addition to that, we're going to require, on behalf of the Museum and on behalf of the Trust, that you take this stock with this reservation on it. We own it, except we make no reservation as to Pennzoil's ownership of this stock.'

"Wouldn't that alert somebody? Well, it would me. And it should have them and it did Mr. McKinley.

"The guaranteed price of a hundred and twelve fifty, is that a coincidence? No. The Museum knew and understood that it had a firm deal with Pennzoil at that price. They were not going to let that bird get out of hand.... That didn't come out of *Alice in Wonderland*. That came to you from this witness stand and from these documents.

"And the Copley notes? The third edition of the Copley notes? Page 63, which they want to gloss over, tells us a story about what was voted on by that board.

"But you are entitled to ask yourselves, 'I wonder what those originals looked like before they went to work on them in those Dechert, Price law firm offices? I wonder if they're just even more favorable to Pennzoil's position?'

"Mr. Copley goes to the Texaco offices ... and the first official act he did when he got back was to go over to the Dechert, Price offices, Mr. Winokur's law firm, and destroy them after editing them.

"Why did he have to go to White Plains and them come back and do that? You're entitled to ask yourself these kinds of questions.

"And I wonder what they truly would have shown. But they couldn't get enough out of there, because on page 63 when Mr. Petersen asked Mr. Lipton to restate the proposal, he talks about the structure of the company.

"You've read it. You don't need me to tell you about it again. It's there."

Jamail was not just answering the points raised in Miller argument, although he was surely doing that. His recitation illustrated the sharp contrast between Pennzoil's case and Texaco's defense.

JAMAIL: "They fall back on that old line of defense, fiduciary relation-
             ship.
                 "Well, I want to talk to you about fiduciary relationship
             for a moment. The Judge instructs you as to that, and there's
             testimony that you have heard, that says there's a time when
             that relationship ends, once you make an agreement.
                 "I suppose when Texaco talks to you about good faith and
             fiduciary, we have to start with the Getty board's lawyer that
             has adopted the Texaco position, and their witnesses, [who]
             came on their behalf."

   In talking about the Getty witnesses, Jamail recalled the testimony
of Lipton's associate, Patricia Vlahakis, an attorney for the Museum.
She said that "because of the full and complete indemnities that were
given by Texaco, it was incumbent upon and necessary that they have
to come and testify, to help them defend the case. . ."
   Since the other Getty entities were also given indemnities, the logical extension of this argument is that they, too, were obliged to assist
in Texaco's defense.

JAMAIL: "They had no choice but to come and give you some sugges-
             tive after-the-fact opinion. That's what they did.
                 "I suppose we have to start with Mr. Bart Winokur and
             Mr. Boisi back in Houston. You say, 'What has that got to
             do with it?' Well, it [gives you] some idea how Texaco views
             fiduciary relationship and good faith, because they have
             adopted [Winokur] and they vouched for him when they put
             him on.
                 "You remember him, 'Back-door Bart'? Snuck in the back
             door when they got Gordon Getty out of the room? Kicked
             him off or tried to instigate a lawsuit to kick him off the
             board of trust of his own family trust.
                 "That's really good faith and that's a lot of fiduciary. They
             really laid a whole lot of fiduciary on Gordon Getty.
                 "That's what you are dealing with, and he stands here and
             wants to talk about morality? Not today."

   The Texaco response to the interrogatory about the January 5 *Wall
Street Journal* article was not a critical part of the evidence, but it appeared to be part of a pattern of deception. Jamail skillfully exploited
this fact, inadvertently provoking laughter in the courtroom for the first
time that afternoon.

JAMAIL: "That's the only day in history that nobody in Texaco saw or read *The Wall Street Journal*, and they swear to that.

"If it's not important to them, why do they deny having seen it? You are entitled to ask yourselves that.

"Why this big denial under oath? 'They didn't see it.' They didn't see it because they didn't want to see it.

"And the excuse or alibi regarding the indemnity. Now, that was a strange excuse. It came from Mr. DeCrane, I believe. He said that Mr. Williams's paranoia affected his psyche so terribly that he needed this indemnity for peace of mind. But Mr. Williams and Mr. Lipton never said that."

As told by Jamail, even Texaco's choice of legal counsel during the Getty deal was grounds for suspicion.

JAMAIL: "I want to talk to you now about the friendly Texaco, if I may.

"They paint this picture of this great big friendly teddy bear that had these law offices and law firms represent them everywhere. They have one that they've used for years in New York, but when it came down to the time for this acquisition, did they call them? Nope.

"Why not? 'Well, we called Skadden, Arps.' They are the biggest takeover lawyers in the country, right ahead of Lipton's group.

"Why did they hire Skadden, Arps? Why didn't they go to their own people? As you remember, Lipton suggested that they hire Skadden, Arps; Mr. Lipton, the lawyer for the Museum, the lawyer for Mr. Tisch, and the lawyer for Mr. McKinley and Texaco.

"He was the engineer, and they got to him and they stopped the train. It's just that simple, and they did it, and [Lipton] slicked them with indemnities."

Lipton had indeed represented many of the players in this saga at one time or another. Whatever you think of the case, there is no doubt Lipton had a central role.

Even as his associate Vlahakis was working on the final documents with Pennzoil lawyers, Lipton struck a deal to deliver the Museum shares to Texaco. This gave Texaco a club it could use to beat Gordon into submission.

Lipton even showed up at the Pierre when the Texaco contingent was there, to make sure a recalcitrant Gordon got the message.

JAMAIL: "They said, well, 'That isn't what the intent was.' You know, we're dealing with people who supposedly know how to read and write.

"You have to take them at their word at the time, not their alibis later in the courtroom. The last thing they wanted was to be in Texas before a Texas jury, but the law provided that and that's why we're here."

The claim that Pennzoil exerted undue pressure on the Getty board was also addressed.

JAMAIL: "They say, finally, 'Pressure made us do it.' Only Mr. Medberry had enough courage to stand up to the pressure.

"Well, that got out and then it comes back and Mr. Lipton says, 'Mr. Medberry is just a strange case. You wouldn't really understand.' I guess I don't.

"But if that thing was meaningless, that Pennzoil proposal, and the board approval of it was meaningless, is Mr. Medberry just some stubborn old coot that, by golly, he's not going to vote yes to anything?"

I remember thinking this was a pretty fair assessment of Medberry, but that wasn't Jamail's point. He read a passage from the transcript where Lipton had said the pressure on the board was not caused by Pennzoil, but by the threat that the Museum would again act in concert with the Trust, as it had with the Consent less than a month earlier.

JAMAIL: "Now, that's his testimony. The pressure was put on by he, Lipton . . .

"Along that line of questioning, I thought it was really strange when Mr. Lipton blurted out something about an agreement, a binding agreement. And I think it's worth looking at. 'There's a different standard in my mind between ordinary people making contracts with each other and whether or not it's a ten-billion-dollar deal.'

"To him, money is the object. That's what it is. That's what he said, that this could be an agreement, this Pennzoil thing.

"People can agree without lawyers, but in his mind the amount of money makes the difference."

While ostensibly recounting the Lipton testimony that supported his contentions, Jamail in fact vividly re-created the scene that had occurred in the courtroom weeks earlier, when the self-described "soreback" lawyer had confronted the merger maestro.

Jamail had effectively cast Lipton as the epitome of what John Jeffers had referred to in his opening statement back in July when he said that "one of the things that ought to be done in this case is to bring that circle of people square with the law."

JAMAIL: "That's that specialized group you had to deal with that would have injected themselves into our business community and into our law, calling themselves mergers and acquisitions lawyers, carving up corporate America to their liking and nobody loses.

"Those investment bankers and the lawyers that comprise this mergers and acquisitions field, they get together. They're all within a couple of miles of each other. One wins one day and another wins another day, and the investment bankers win all the days.

"And they win with these indemnities. Do you think for a minute that Marty Lipton or that Mr. Siegel, the advisor, or Woodhouse for the Trust would have for one minute, *one minute,* agreed to this Texaco deal without those indemnities?

"I know you've heard the testimony. Mr. McKinley says it was an essential part, and you heard Lipton on it."

Jamail deftly wrapped Texaco and the mergers and acquisitions community into one big conspiratorial package. Some have subsequently argued that this was grossly unfair to Texaco, but consider this point: the Texaco purchase of Getty was facilitated by this same group; if Texaco did not want to be held accountable for them, it should never have bestowed these blanket indemnities. Of course, as Lipton and others testified, there could have been no Texaco takeover of Getty Oil at any price without those indemnities.

Jamail also reminded the jury that he had said in voir dire that "it was not going to be possible for me to extract a confession on the stand Perry Mason–like from these people" and that the law did not require that of him or of Pennzoil.

Getting back to the central issue, whether the Pennzoil-Getty deal was a binding agreement, Jamail read from the judge's charge.

JAMAIL: "Now, you turn to Special Issue No. 1, and you can read it as well as I can, 'Do you find from a preponderance of the evidence that these Getty entities and Pennzoil intended to bind themselves to an agreement' . . . an *agreement.*

"And again, if Judge Casseb wanted to say 'written contract,' he knows how to do that. He didn't."

"The law requires no such harsh burden of us; would be impossible.

"And you go back to the Copley notes, and you go back to the press release, and you go back to the Memorandum of Agreement, and you go back to the Garber note, and you go back to the January 5th newspaper release in which it confirms that an agreement has been reached; the turmoil is over."

This was yet another reference to that *Wall Street Journal* article.

JAMAIL: "It talks about an agreement no less than fifteen times. What can we do? What more can we bring you?

"Well, there was one other thing. Mr. Petersen, brilliant one moment, experienced one moment and not knowing what he was saying the next moment, after the *Fortune* magazine article appeared. He just didn't know what he was saying when he said, 'We approved the deal but we didn't like it. We approved it because it was a bird-in-the-handish situation.'"

"What did that mean? Mr. McKinley doesn't know. He doesn't know. Perhaps by your verdict you can tell him. Everybody else knew. Mr. Tisch knew. "It means the same to everybody,' Mr. Tisch said.

"You know, we didn't give that interview. It's unrefuted. Uncorrected. He didn't do anything. There it is, glaring at us. You know, we didn't have a gun at Mr. Petersen's head when he told those people at *Fortune* magazine what he told them.

"And they ask you to ignore that. They ask you to ignore all of this compelling evidence, much as they ignored Pennzoil's rights the same way. As they ignored asking for any information that would give them concrete evidence as to whether or not these people were free to deal.

"They avoided it like the plague. They knew it was there. They knew those notes were there. They knew the indemnities were there.

Like Miller, Jamail spoke extemporaneously; neither man consulted a prepared text. Unlike Miller, however, Jamail had a mountain of contemporaneous evidence supporting his argument. Miller frequently had to rely on testimony that was, Jamail continually reminded, after-the-fact and self-serving.

JAMAIL: "You go back to Mr. DeCrane's notes, where he says 'ONLY AN ORAL AGREEMENT.'

  "Well, the Judge tells us you can have an oral agreement. That's our law. It's been ingrained in our law. Otherwise, you couldn't deal. You don't need a lawyer there and a stenographer taking down what you're doing with somebody. That happens afterwards."

Jamail did not claim that the Pennzoil-Getty deal was 'only an oral agreement.' He was speaking here to Texaco's knowledge of that deal, as reflected in the notes of its president.

JAMAIL: "One other important thing about that: If there's any sanity, dignity, and morality left in the business community, one provision is that you cannot deal with me and harbor in your heart some secret that you're going to turn around and stick a knife in me. Which is what they claim they did: 'Well, we never told Pennzoil that we didn't intend to be bound, but the board secretly had in mind shopping the deal.'

  "Judge Casseb says you can't do that. If that's their defense, we've got a binding agreement on these points. You can read it."

Competition was another Texaco defense, but Jamail had a few thoughts on that subject as well.

JAMAIL: "If Texaco really wanted to bid fairly and compete, nothing prevented them from going out and making a tender offer for one hundred twenty, one hundred twenty-five. Pennzoil could be no threat at one hundred. Who is going to send in their stock to Pennzoil at that price?

  "That's the way you compete. You don't compete as Mr. DeCrane's notes tell: 'CREATE CONCERN OF ECONOMIC LOSS TO GORDON GETTY. THE PROBLEM IS WE CAN'T GET GORDON GETTY ON BASE FIRST, so let's manufacture some real work for Gordon after smearing him with some of this fiduciary stuff we did in Houston. Let's do some of that to him.' That's Texaco morality.

  "And they say it's just an accident when it says in these notes that the ideal way to consummate the deal is to get the Museum first. 'MARTY LIPTON THE KEY.' 'That doesn't mean anything. I don't want you to think about that, Jury, because if you do you might have some independent thoughts that don't coincide with Texaco's wishes.'

"Mr. Miller told you you couldn't find a contract. I tell you, I plead with you that under the evidence you should. That's all I can do."

Like Terrell, Jamail returned again and again to the notes of De-Crane and Lynch. In his final argument, Miller had tried to put the best face on it, even going so far as to say 'I thank God for those notes.' But in light of all the other evidence, those notes were absolutely devastating to Texaco; they could not be explained away.

JAMAIL: "This 'STOP THE TRAIN, STOP THE SIGNING'; we've got Mr. McKinley's admission. 'I told him I hope you don't sign this until you hear my offer.' This is Mr. McKinley talking to his one-time lawyer, Mr. Lipton.
           "Did they follow that script [from] that strategy meeting? He stopped the signing, but he did it 'nicely.' He says 'I hope you won't sign anything until you consider my substantially higher offer.'
           "I guess that's called nice stop the signing. 'STOP THE TRAIN'; they didn't stop it, they derailed it, blew it up.
           "You know, this didn't come out of *Alice in Wonderland*. It came from right here. They don't want you to accept it that way because, you see, Jury, the way they've got it figured you shouldn't look at the clear evidence. 'Don't believe what you see. Don't believe what you hear. We know best. We know best. Believe what we tell you, and then America will be strong.' That's really wonderful. I just don't think it will happen that way."

The notes also contained the Wassersteinian strategy Texaco ultimately used to get Gordon to sell.

JAMAIL: "They wanted to deny to you that they pressured Gordon Getty. But what does Gordon Getty have to say about it? That Marty Lipton left no doubt that the Museum was selling [to Texaco].
           "Where would that leave Gordon Getty? Except then in a minority position with one owner, Texaco, at their total mercy; at that point, squeezed out, 'leaving him with paper.'
           "Now they can't explain their own notes. They used the words. They didn't quite understand what 'leaving him with paper' means. Nobody understands that.
           "I believe you do. I believe I do.
           "So that's the one way they could get Gordon Getty to

abandon his dream of being in control of his father's company, and they did it. They did it, but they haven't gotten away with it yet.

"So, they get the Museum to sell, and that's the last thing Gordon Getty told you: 'I had no choice but to sell.'

"I thought that was sad. I really did. You know, we build for our children. Fellow builds a company for his children and wants his family to operate it and to run it. And along comes some people he has made, has put in their positions, and they turn on his family when he dies. It happens."

If anybody could make the distant, remote J. Paul Getty seem like a fount of paternal love, it was Joe Jamail.

He continued to dismiss Texaco's defenses and strengthen Pennzoil's case.

JAMAIL: "They made a big deal about Mr. Liman not being in that board room. And the last deposition you heard [had] Mr. Siegel saying 'I remember he came in after the board meeting.' Mr. Tisch never denied it: 'I shook hands with him . . . maybe it was just outside the board room. I can't remember.'

"And Mr. Tisch, along with Mr. Liedtke, got invited up to the Getty suite at the Pierre Hotel to drink champagne to celebrate the agreement.

"Did Gordon Getty think at that time he had an agreement? Did Pennzoil? Did the Museum? It's not important what Texaco told you they thought. It's what they thought.

"The board meeting had nothing to do with those two agreements between [Pennzoil and] the Museum and the Trust, except to approve or not approve. If not approved, it's gone and they had two more weeks [before Pennzoil could purchase stock under the tender offer]. There is no pressure to find another buyer.

"No, they went out secretly shopping this thing . . . and there's no board authorization anywhere that you will find that gives authority to Boisi or anybody to go and shop this deal of Pennzoil's. None."

Jamail prepared to close out his argument on the interference issue, asking the jury to look at the language in the judge's charge.

JAMAIL: "You can read it as well as I can. The evidence is replete.
"Mr. McKinley admitted from the witness stand that the

only reason the Pennzoil agreement was not finalized was the action of Texaco.

"Do you need more than that? I just can't—interfere, stop the train, don't sign up yet, hear from me, give indemnities, give special warranties.

"[McKinley] knew what he was going to do when he went to that Texaco board meeting and got that Texaco board of directors to indemnify him and Mr. Kinnear and Mr. Weitzel and Mr. DeCrane from any actions they might take in acquiring Getty.

"That's what he did; it's in their resolution. And if it wasn't on their minds to do this underhanded thing, what was on their mind?"

The question of damages was the only issue that remained for Jamail to address.

JAMAIL: "There has been no dispute about the damages in this case. None. You saw the models. It's easy to figure.

"They listed three damage experts to come here and tell you what their idea about damages were. They didn't bring any of them.

"They criticized us for not bringing some people, but I got those depositions somewhere and they want me to read that pile of depositions to you."

His tone indicated that Jamail thought those depositions were a pile of something else entirely. As he spoke, he kicked a cardboard box overflowing with bound transcripts that was under the defense table.

JAMAIL: "Now, do you want to hear that? I thought you were going to throw up listening to all the depositions you had to hear.

"And they criticized us for that. They were here if they wanted them. We produced those people for them, but they didn't produce their own experts.

"So unrefuted in this testimony is the damage model given to you . . . finding and development costs, which is the reasonable one, seven and a half billion dollars.

"Texaco, through its lawyers, do not think you are big enough to assess those kinds of damages. They think that you are not big enough to do that. I think you are.

"I questioned you very carefully about this, as I recall, when we talked the first time we met. You said nothing would deter you from that if the evidence preponderates in that way.

"Well, it not only preponderates; it's *all* the evidence. *There is no other evidence.*

"So to suggest to you subtly that you are going to give something—he didn't say it, this is what he was meaning: 'This difference there in stock is what you are going to do.'"

This last garbled sentence was his reply to Miller's comment about the stock option, which would have given some $124 million to Pennzoil.

Jamail disdained the idea that Pennzoil's damages consisted of the stock price differential on 8 million shares.

JAMAIL: "There is no evidence that this is the measure of damages, and that would be insulting to you and to us. It would not be following our oaths.

"Now, I can't force any of you to do anything, and I am not that way anyway. Wouldn't try to. It's dumb.

"But at this moment in time, you need to understand something. There cannot be half justice without there being half injustice. Pennzoil is entitled to full and complete compensation for its damages. Full and complete, that's what Judge Casseb tells us."

He concluded his argument with a demonstration of why he was alone at the top of his trade. Jamail's words alone cannot reveal the level of communication he achieved with the jurors, even those who were not particularly receptive to his message. It was a cumulative thing that had started back on the first day of the trial and swelled to a crescendo as he came to the conclusion. With few exceptions, the courtroom had been quiet throughout the arguments, but there was an expectant hush as Jamail neared the end. It was as if people instinctively knew they were witnessing a rare and beautiful thing.

JAMAIL: "I plead with you, if you are going to right this wrong, it has to be done here. This will never happen again, because it's so expensive to bring one of these kinds of cases, so expensive.

"Hugh Liedtke is my friend, Baine Kerr is my friend, and I can't divorce myself from that. But I can't divorce myself from these facts, either.

"They honestly endeavored to build this company. Hugh Liedtke, who got started with the help of J. Paul Getty. You heard the testimony, the man who came to his wedding.

"A stranger, interloper, going to rape the company? He got his start from that family. Believed he could work with them, and could have—if it had not been sabotaged.

"You people here, you jury, are the conscience, not only of this community now in this hour, but of this country. What you decide is going to set the standard of morality in business in America for years and years to come.

"Now, you can turn your back on Pennzoil and say 'Okay, that's fine. We like that kind of deal. That's slick stuff. Go on out and do this kind of thing. Take the company, fire the employees, loot the pension fund. You can do a deal that's already been done.'

"Or you can say 'No. Hold it, hold it, hold it now. That's not going to happen.'

"'I have got a chance. Me. Juror. I have got me a chance. I can stop this. And I am going to stop it.'

"'And you might pull this on somebody else, but you are not going to run it through me and tell me to wash it for you. I am not going to clean up that dirty mess for you.'

"It's you. Nobody else but you. Not me. I am not big enough. Not Liedtke, not Kerr, not anybody. Not the Judge. Only you in our system can do that. Do not let this opportunity pass you. Do not.

"The evidence is clear. Punitive damages are meant for one reason, to stop this kind of conduct. If we were not entitled to ask you for punitive damages, we would not be able to stand before you now and request it. . . . [In] certain cases only can we ask for and receive punitive damages. This is one of them.

"And the reason is that you can send a message to corporate America, the business world, because it's just people who make up those things.

"It isn't as though we are numbers and robots. We are people.

"And you can tell them that 'You are not going to get away with this. The law says you should be punished if you did this and it's not only to punish you for this that you took, it is to deter others from doing the same thing.'

"I ask you to remember that you are in a once-in-a-lifetime situation. It won't happen again; it just won't happen. You have a chance to right a wrong, a grievous wrong, a serious wrong.

"It's going to take some courage. You got that.

"It's going to take a logical reexamination of what went

on. You don't have to read everything. You remember better than anybody what happened in this courtroom.

"If you come to the conclusion, as I hope and believe that you will under all the evidence in this case, that a grievous wrong has been done to Pennzoil . . . then no verdict less than seven and a half billion dollars actual and seven and a half billion dollars punitive is enough to dent them, because they saved from forty to sixty billion dollars by their own statement by taking Getty and for every billion dollars you assess, it costs them 43 cents more per barrel than the purchase price of Getty.

"I know you are going to do the right thing. You are people of morality and conscience and strength.

"Don't let this opportunity pass you by."

With a nod to the jury, Jamail returned to his seat. The hollow feeling in the pit of my stomach hadn't left me all day; I knew our time to sit and listen was over.

Judge Casseb then said that he would read the order from Judge Farris designating the twelve jurors who will deliberate the case. Although I started the trial as an alternate, I knew I had become the twelfth juror. Judge Casseb said Judge Farris wanted to extend the gratitude of the court for the patient giving of the time of the three alternate jurors, who would now be dismissed.

JUDGE CASSEB: "I want to add my voice of gratitude to that. You may say, well 'Is all of my time wasted, that I have been here for all this time and now, when it comes to the crucial moment, I cannot go into that jury room and deliberate and become part of the verdict?

"I will admit to you that our law is not perfect, but unless we are able to perform under the law and within the law, our system would crumble. I don't know of any American who wants to give up that right."

Judge Casseb asked the alternates to take comfort from the fact that the whole trial had almost been lost after the illness of Judge Farris during the trial's fourth month.

He recalled his last stint as a visiting judge in Harris County over twenty years before, and spoke of the difficulty of his task, stepping into this case after so much testimony and evidence had already been presented.

JUDGE CASSEB: "I didn't have to take this assignment. It wasn't easy
for me to come in after four and a half months and
then try to grapple with these tremendously capable
lawyers and come in with the charge we have here,
try to follow the law of New York, Delaware, and
Texas.

"But I did it, and thank God it wasn't for the
money. I did it because I knew it was an opportunity
for me to advance the continuation of this tremendous system that we have in the administration of justice, American-style.

"I want y'all to remember that and I want you to
know that. And if I had seventy more years to give of
my life, I'd like to devote it to maintain the type of
administration of justice that we have in this wonderful country of ours.

"I will call the names of the twelve jurors and ask
you then to go with the bailiff to the jury room where
you will select a presiding juror.

"Ms. Lillie Futch, Juanita Suarez, Susan Fleming,
Richard Lawler, Israel Jackson, Ola Guy, Fred Dan-
iels, James Shannon, Velinda Allen, Shirley Wall,
Laura Johnson, Theresa Ladig.

One by one, we walked from the jury box to the side door. The events
of the day had left me emotionally limp, but the prospect of the upcoming deliberations filled me with a disquieting exhilaration tinged with
dread.

I thought I had become cynical and jaded during the long months of
the trial; the ability of Dick Miller and Joe Jamail to generate that level
of feeling had left me astounded on this momentous day. Like the dispute itself, these emotions were complex and overlapping. More than
at any time in the trial, I thought I glimpsed the essence of these two
men. It did not profoundly affect my view of the case, however; my
surprise when Miller had rested his case earlier in the week had been
genuine.

Until that point, I had awaited the appearance of something, anything that might mitigate the powerful case that Pennzoil had built. As
it developed, though, the more witnesses Texaco called, the stronger
Pennzoil's case became, at least to my way of thinking. When Miller
rested Texaco's defense, I knew no such mitigation would be forthcoming.

I wasn't measuring Hugh Liedtke for sainthood, not by a longshot, but I knew that he and his stockholders had taken a fierce beating under brutal circumstances.

The final arguments had sharpened this view. Miller could have wandered for forty days and forty nights in the evidentiary desert of Texaco's case and achieved little more than he had in those four heroic hours. Jamail had come in and worked his magic, after Jeffers and Terrell had set the table. His rhetoric had been relatively modest, and his argument all the more powerful for the restraint he had shown.

The inexorable wheels of our system of justice were turning again; the matter had been placed in the hands of twelve citizens. I wondered how the other eleven felt.

33

We filed out of the courtroom into the long, narrow hallway. For the first time, our number was twelve. The three alternates who had been part of this group for almost five months had been cleaved from our bosom. Nervously, we waited for Carl the Bailiff to "take charge of the jury." The jury room for the 151st District Court had been filled with a team of headset-wearing typists since the trial began, part of a system that produced the transcript on the same day as the testimony. We would be escorted to the jury room of a nearby court to begin our deliberations.

As we waited in that hallway, a thundering ovation sounded from inside the packed courtroom. Out of our hearing, Judge Casseb had commended the alternate jurors for their diligent service and told the spectators they could express their approval with a round of applause. It was an emotional moment inside the courtroom that was also felt out in the hall.

At last Carl appeared and we walked in formation across the fifth floor. The small room where the case of *Pennzoil Co.* v. *Texaco, Inc.* would be decided was no different from most of the jury rooms in the courthouse complex, rooms where other difficult decisions had been made, some literally on matters of life and death.

The first big issue the jury tackled that day was one on the minds of many Americans: smoking or nonsmoking? Thankfully, the non-smoking contingent prevailed, with a boost from health-conscious juror Laura Johnson, R.N.

The actual first order of business was the election of a presiding juror. To say that we were familiar with each other after nineteen weeks would be an understatement. Prohibited from discussing the one thing we had in common (the case), we had long since exhausted sports, politics, and local gossip as topics of conversation, leaving only such perennial Houston favorites as traffic and the weather. As of to-day, the shackles were removed. But the talk was strangely subdued. The trial had put a great deal of pressure on the members of the jury in a number of ways; job and family upheavals and other personal concerns had been necessarily subordinated to the task at hand.

We had sat in silence throughout the testimony and final arguments. During the trial, we also spent many long hours out in the hall while the judge and the attorneys argued the law. That was all over now; the decision was on the table in this room. But who would preside over our deliberations? It was not a subject that had merited much thought during the trial.

Although I was certain that I was the first alternate selected (and now the twelfth juror), I was not sure who the other alternates were. Gilbert Starkweather had been respected and well liked by everyone, and seemed like a natural selection for the top spot. As an alternate, he was not available.

As I looked around the table, the choice was immediately obvious to me. I nominated Fred Daniels for presiding juror. The nomination was quickly seconded, and Fred was elected by a unanimous vote. This was comforting for two reasons. Fred, a 41-year-old letter carrier, was a man of quiet dignity and unmistakable intelligence, clearly among the most qualified. He was also one of four blacks in the final twelve. Although he had not been involved in the misunderstanding that resulted in the suggestion Judge Casseb was prejudiced, he was sensitive to the feelings of those who thought it might be true.

I, too, was sensitive to those feelings. In the evenings, Shirley Wall and I would sometimes await our rides on the same patch of court-house lawn. We were both concerned that the fallout from this might pose problems when the case went to the jury. The makeup of our microcosm could be devastated by the injection of unwarranted fear and suspicion. My nomination of Fred was not a cynical move, but it *was* calculated to help build the consensus we would need to do justice to our task, which was justice itself.

Alas, Fred declined to serve. Knowing the members of the jury, he no doubt anticipated the turmoil that lay ahead. To ride herd on a stampede went contrary to his nature.

Susan Fleming then nominated Shirley Wall, who had taken voluminous notes throughout the trial. She was elected by acclamation and called the meeting to order. Almost immediately, she asked if the presiding juror had to read the verdict aloud in court. When told this was indeed a possibility, she resigned from office.

Again, the floor was thrown open for nominations. Rick Lawler, army brat turned heavy-equipment salesman, became the third presiding juror to be unanimously elected. Fortunately, he was not shy about speaking out, as the questions he had posed to John McKinley had shown. Further, it was apparent that any point of contention raised would have Lawler in the thick of the fray on one side or the other. As the deliberations progressed, Lawler gained strength and kept the discussion on a relatively even keel. Shirley sat next to him that first day and served as an unofficial vice-foreman.

Rick turned out to be an excellent foreman, ensuring that everyone had a chance to speak.

His view of the case was not at all clear to me at that point (how could it have been?), so I harbored some doubts about what would happen next.

We had been told to leave all papers in the courtroom, including our copies of the exhibits and the judge's charge. Carl was stationed right outside the jury room to ensure that we weren't disturbed. I guess he was also supposed to make sure we didn't leave. There were restroom facilities and a water cooler in the jury room, so there was no reason to leave.

Rick proposed starting the deliberations by reading the judge's charge to the jury and the accompanying instructions. Carl was dispatched to the courtroom to get the individual jurors' copies of the charge. He returned with all but one. Rick indicated he was ready to begin, but Ola Guy wanted to read along on her copy. This was not an unreasonable request; a recurring phenomenon during the trial was an attorney or witness attempting to read aloud from a document and inadvertently reading it incorrectly. We would wait for Ola's copy of the charge to arrive. Minutes dragged on but her copy did not arrive. Finally, Rick stuck his head out the door and made himself understood; Carl quickly produced the document.

With this fitful start we began. Rick read Special Issue No. 1 aloud. This was the contract issue. The instructions were specific about the meaning of the terms used to frame the issue. Rick read the instructions aloud also.

"Should we take a vote on No. 1?" asked Rick. Heads nodded. The moment of truth was finally at hand. This first issue asked the essential question of the case: Did Pennzoil have a binding agreement with Getty?

Slips of paper were passed around; it would be a secret ballot. Shirley counted the votes as Rick unfolded them. The result was 7 to 5, Pennzoil had a contract. I was shocked that Texaco's arguments had attracted 5 votes. What case had those five been watching?

That first vote surprised a lot of people sitting around the table. Rick looked up for help. Some one suggested that we go around the table, with each person in turn telling how he or she had voted and why.

This process was alarmingly brief, with statements such as "I voted no because I don't think Pennzoil had a contract" and "I voted yes because I think Pennzoil did have a contract." I mumbled something about the persuasiveness of the Getty directors' 15-to-1 vote for Pennzoil, regardless of how they later characterized that vote.

Susan said she voted "No" because of the trial testimony of those Getty directors about that same 15-to-1 vote, mentioning Wendt and Tisch.

Israel Jackson said he voted "No" because he didn't think Pennzoil had a contract. He said his mind was made up, wouldn't change, and he didn't feel like talking about it. I commented that this phase was called "deliberations" because we were *supposed* to talk about it. "Otherwise, they just would've given us an absentee ballot to mail in when we got home."

Fred Daniels said he voted "No" because the Memorandum of Agreement was not signed by all the parties.

When the "go-round" was completed, it became apparent that Susan had been joined by all four black jurors in voting "No." Had that misunderstanding with Judge Casseb helped to form a voting bloc against Pennzoil?

I didn't think so, in spite of the initial appearances of the split. Susan was not part of any protest movement; her arguments were clear and reasoned. I also believed that they were at total odds with the evidence, but resolving such differences are the reason for deliberations.

Fred's difficulty with the missing signature on the Memorandum of Agreement was more problematic. A major Texaco argument refuting Pennzoil's characterization of the Getty board vote was the missing Getty Oil signature on the last page. In this discussion, however, Fred pointed to the blank space on page 4 where Hugh Liedtke's signature belonged. Gordon Getty and Harold Williams had signed this copy; Liedtke's signature was on another version.

Shirley recalled how the issue had been addressed during the trial: counterparts of the same document can be signed by the contracting parties. As an example, she recalled a business transaction she and her husband had concluded during this trial. A counterpart to the document he signed was brought to the courthouse, where she signed it. The legal effect was that they had both signed the document.

We looked through the trial documents and found the version of the Memorandum of Agreement signed by Liedtke and Gordon Getty. Gordon had also signed the version signed by Harold Williams. The exhibit was passed around the table, and we took another ballot. Fred changed his vote, and the count stood at 8 to 4.

Unlike juries in criminal cases, we could deliver a verdict on a split vote of 10 to 2. As Friday afternoon wound down, however, 10 votes seemed an impossible distance away.

Although we had been deliberating for about an hour and a half, it seemed much longer. Judge Casseb gave us the option of deciding when we would conduct our deliberations. Nobody wanted to come in over the weekend. We quickly agreed to resume on Monday morning. This case had started in early July; we were now staring Thanksgiving in the face. This matter would surely keep until Monday.

The brief euphoria produced by the case finally going to the jury was long forgotten by the end of the day; I had a pounding headache as I waited for my wife to pick me up. Shirley and I shared a moment's speculation about how many weeks of deliberations the Court would insist on before calling it a mistrial. After the trial, I would read stories written that same day quoting observers who predicted that the deliberations would last several weeks. This dovetailed with my thinking after the first ninety minutes in the jury room. With that cheerful prospect in mind, I went home fervently wishing I were still an alternate.

Over the weekend, my spirits did not perceptibly improve. I searched in vain for an eloquent argument to break the gridlock in the jury room.

In truth, I had not anticipated the kind of dead-end discussion that ensued on the first afternoon of deliberations. I knew the jury would want to assess the case carefully before rendering a verdict after a nineteen-week trial, but the evidence was so one-sided I hadn't thought the end result was ever in doubt. Now, a mistrial loomed as a very real possibility. Of the 4 votes for no contract, 3 seemed rock solid. Besides Susan, Israel had already announced that he wouldn't change his mind. Ola didn't seem to care too much for Pennzoil's case either. Velinda Allen had yet to articulate any arguments of her own, so I marked her down as the wild card among the four.

When we reassembled on Monday morning, however, the mood had changed. Two days of peace and rest appeared to have lifted everyone else's spirits; the grueling days of final arguments had been almost as exhausting for the jury as they had been for the attorneys who delivered them.

All the participants were present when Judge Casseb reconvened the proceedings and immediately dispatched us to the jury room to resume our deliberations.

As we settled in our chairs around the table, I thought about my previous experience as a juror. I had formulated the theory that on a jury, as in "real" life, there are two kinds of people: shepherds and sheep. The problem on a jury was to figure out who was what.

An outspoken, opinionated juror who seemed to be taking up the shepherd's crook might, in fact, just be a very vocal sheep, challenging a real shepherd to point which way to turn.

On the other hand, someone who seemed to fit the sheep profile might turn out to be a silent motivator with a fully developed flow chart for flock movement. This turned out to be the case on the Pennzoil-Texaco jury.

When the deliberations got untracked that morning, we again utilized the go-round method of discussion. Rick would frame the issue, and each person in turn, would give his or her response and elaborate on the reason for it. Some of the jurors who sat in silence during the free-form shouting matches that marked the less structured portion of deliberations would come alive during the go-round.

Juanita Suarez was not a big talker, but when her turn arrived she opened her mouth and clear, reasoned arguments came forth. She hadn't missed a thing in the courtroom; her sense of the case was as fully articulated as my own.

Unfortunately, verbal restraint has never been my strong suit; I antagonized some of the jurors (okay, maybe all of them) with my persistent recitation of my view of the evidence. Like John McKinley, who had not been able to answer many questions without reciting the basic elements of Texaco's defense, I couldn't make a point without listing the sheaf of documents that supported it. As Terrell had said of Al De-Crane, you couldn't show me anything that would not confirm my view. I was not a strong candidate for Most Admired Juror.

A constant feature of every jury I have served on was the way the discussion in the jury room tracked the arguments of the attorneys. That was most apparent in this case in my exchanges with Susan Fleming. When I would echo a Terrell argument on a specific point, she would recount Miller's view of the evidence that mitigated against it. Susan had accepted the role of articulate advocate for the defense almost by default. There was no shortage of proponents on the other side of case. Shirley was outspoken on all the issues, with contributions from Laura. The normally taciturn Lillie Futch would crack me up with her wryly understated remarks about various Texas defenses that were raised. It was clear that she thought even less of their case than I did.

Rick, on the other hand, was wearing two hats. As presiding juror, he was responsible for making sure everyone had his or her say. He also had his own point of view. Although he had voted for Pennzoil's contract on the first ballot Friday afternoon, Rick returned on Monday with the fervor of a born-again plaintiff's juror. In his work on the sales force of Briggs-Weaver, Rick undoubtedly had to use the professional salesman's powers of persuasion. As I heard him methodically repeat the arguments on a particular point to a recalcitrant juror, I thought I was listening to a motivational tape on "Closing the Sale." When I called him on it, he smiled sheepishly as he realized that he was doing just that.

The sticking point as the deliberations resumed Monday remained Special Issue No. 1. Until we could put 10 votes together on that issue, there we would stay.

The discussion focused on what the Getty board approved by that 15-to-1 vote. If it was the price only, as Texaco argued, there was no binding agreement. If it was the Memorandum of Agreement with a modified price, it was a "done deal."

The Getty directors Texaco had produced in the trial had all testified that they voted on the price only. As Miller had argued, they ought to know what they voted on. Susan relied primarily on this testimony in arguing the "price-only" theory.

On the other hand, these were all after-the-fact declarations. The Pennzoil attorneys had argued that the contemporaneous evidence (documents and statements made at the time) strongly indicated that the board had approved the Memorandum of Agreement at $112.50.

One by one we examined the documents that pointed to a binding contract between Pennzoil and the Getty entities:

The Memorandum of Agreement itself, which even Winokur had admitted was written as an absolute agreement.

Page 63–65 of the Copley notes, which recorded Harold William's motion and the subsequent 15-to-1 vote approving the "the Pennzoil proposal." In the pages immediately following, the Copley notes also recorded the board's transmittal of that approval to Pennzoil, as well as Pennzoil's immediate acceptance.

The January 4 Getty press release, which announced the Pennzoil-Getty deal, containing many of the essential terms of the Memorandum of Agreement. If the board did not approve these terms, what were they doing in the press release? This release was approved by Copley and Winokur for Getty Oil, Cohler for the Trust, Vlahakis for the Museum, and Goodrum for Pennzoil. Except for Goodrum, they were all present at the board meeting.

The Garber note, Getty Oil treasurer Steadman Garber's note recording the telephone report he received after the board meeting concluded. He believed that Copley was the source of the information. Captioned "Agreement," this note also contained references to the 15-to-1 vote and terms from the Memorandum of Agreement.

The "Dear Hugh" letter, which demonstrated Gordon Getty's commitment to the Pennzoil deal. Gordon had kept the promises he made in the "Dear Hugh" letter throughout that long, difficult board meeting.

The Siegel and Cohler affidavits to the California court, which affirmed the Pennzoil-Getty deal as "a transaction now agreed upon." These affidavits were filed on January 5, even as Texaco was making its move on Getty Oil.

This wasn't all of the evidence by a long shot. Goldman, Sachs had submitted its fee letter the day after the board meeting concluded. Tisch testified that was standard practice—if the deal was completed.

The weight of this evidence chipped away at the resistance of two of the jurors voting "No contract." Between the go-round and Rick's closing-the-sale persuasion, we got 9 votes, then 10. Ola Guy was the tenth juror to vote affirmatively on Special Issue No. 1. Of course, she also said that she didn't want to give Pennzoil any money.

Susan and Velinda held firm at "No contract." Velinda had finally articulated her position: "I think what Susan says is right."

In the jury room, it was 10 to 2. Pennzoil's contract had finally been recognized by somebody other than their own lawyers.

The knowledge that we wouldn't end up as a hung jury was like a tonic. With renewed energy, we moved to tackle the rest of the charge.

Special Issue No. 2 asked, "Do you find from a preponderance of the evidence that Texaco knowingly interfered with the agreement between Pennzoil and the Getty entities, [if] you have so found?"

With Texaco having staked its case to the claim of "no contract," there really wasn't anywhere else for the jury to go on this issue. Jamail had elicited a response from John McKinley that he didn't know of any reason the Pennzoil-Getty deal would not have been consummated, other than Texaco. The first ballot on this issue produced a 10-to-2 vote, again enough to render a verdict.

Moving on to Special Issue No. 3, the jury was seized by the enormous implications of our decision. "What sum of money, if any, do you find from a preponderance of evidence would compensate Pennzoil for it actual damages, if any, suffered as a direct and natural result of Texaco's knowingly interfering with the agreement between Pennzoil and the Getty entities, if any?"

Pennzoil had asked for $7.53 billion, to replace the 1.008 billion barrels of proved oil reserves that would have accrued to the firm as 43 percent owner of Getty Oil. The "divorce clause" in the Memorandum of Agreement had stipulated that the New Getty Oil would be divided after one year if Gordon and Pennzoil could not agree on a plan for restructuring. This meant that, at worst, Pennzoil would walk away with those 1.008 billion barrels.

Based on the testimony in the trial, I really believed that Pennzoil and the Trust would have had a happy, prosperous partnership. No matter what you thought about Hugh Liedtke, it was an indisputable fact that he made money for his partners.

That was not the issue to be decided here. The question was what sum of money was necessary to restore Pennzoil's bargain *based on a preponderance of the evidence*? Well, what was the evidence?

There was the testimony of Dr. Ron Lewis and Clifton Fridge, two Pennzoil executives who delivered publicly documented numbers about the finding and development costs and the Getty assets that established the replacement model of damages. Taken by itself, their testimony would not have been conclusive; they worked for Pennzoil. Texaco did not choose to put on any witnesses to rebut Lewis and Fridge.

They may have survived that strategic oversight if it hadn't been for Thomas Barrow. His forthright, direct testimony had been a powerful argument for the validity of the replacement model. His description of how two different methods of calculating damages he ran had placed $7.53 billion squarely in the range was fresh in the minds of the jury, having been forcefully recalled by Jeffers in final arguments.

It was more than that, however. Barrow's background and reputation and his independence had not been diminished at all by Keeton's cross-examination; he exuded competence and an appealingly dispassionate logic. Miller's innuendos about secret compensation and some conspiracy theory involving Texas Commerce Bank had never borne fruit.

The cross-examination of the Pennzoil damage witnesses could not really be considered as any sort of mitigation for Texaco; what Richard Keeton or Dick Miller thought about Thomas Barrow was clearly not evidence.

Some of the jurors started trying to find a way to justify awarding Pennzoil a lesser sum. "Pennzoil's bad, too," the irrepressible Ola Guy contended.

These mental gymnastics were doomed to failure, for one simple reason: Texaco had presented no evidence, not one shred, that argued against the Pennzoil replacement model of damages. It had experts on its witness list, but it never brought them to court. Again, Texaco's whole case could be summed up in two words: "No contract." When we found there *was* a contract, our duty was clear.

Judge Casseb's instructions were very explicit: "You will not consider or discuss anything that is not represented by the evidence in this case." The grasping for an alternate way to figure Pennzoil's damages was not supported by any evidence that had been placed before the jury. I read that instruction to the jurors at the table, and added, "Look, it isn't our place to fabricate a case for Texaco. We have to decide the case on the evidence presented in court."

Rick called for a vote. With a show of hands, we were 10 to 2 for awarding Pennzoil the full damages of $7.53 billion.

For me, trying to imagine exactly how much money $7.53 billion is was as difficult as trying to imagine 1 billion barrels of oil. Wars have been fought for much, much less. But one fact was abundantly clear: Pennzoil's bargain would not be restored by a damage award of pennies per barrel.

The replacement model of damages has subsequently been called "outrageous," "without foundation," and "bearing no relationship to reality." In fact, Texaco's post-trial legal maneuvers would include several ways of calculating actual damages that came to around $500 million. These included a stock price differential on 3/7ths of Getty Oil (the difference between $112.50 and the $128.00 a share paid by Texaco multiplied by 32 million shares) and an exploration program spread over a number of years discounted to present cash value.

There is no doubt that some arguments can be made for either of these alternative valuations; however, I am still not persuaded that either is the correct measure of damages.

The point is *none* of these alternatives was explored at the trial. The Texaco strategy from the beginning was "no contract, no contract, no contract."

Texaco didn't hire Dick Miller to give away any settlements. Offering the jury a compromise value on damages would perhaps have been a way to have the jury settle the case. I can say without reservation that I would have strongly resisted any such attempts; the testimony of Barrow was a bulwark on which the replacement model was grounded.

After consulting our personal copies of the key documents throughout the trial, we were provided with the actual exhibits in the jury room. Whenever the discussion bogged down, I would examine these original documents whose contents we had come to know so well.

A copy of a typewritten page looks about the same as the original. The four pages from a yellow legal pad that contained the DeCrane notes were a different story; the tight, precise handwriting and the stars and arrows that highlighted the text belied their author's contention that the notes were meaningless. Pennzoil's success at establishing a strong link between the strategy outlined in the notes and what Texaco actually did made this an extremely persuasive piece of evidence on the knowledge and interference issues.

Turn back to the section of this book labeled *Critical Documents 2: Evidence of Knowledge and Interference.* Examine the first exhibit, the DeCrane notes. See the detailed information about the status of the Pennzoil-Getty dealings that flowed into Texaco. Note the strategy outlined by Wasserstein and Perella of First Boston. Review the Lipton agenda and the language of the Museum indemnity. Recall the actions of Texaco in the acquisition of Getty.

In that context, the DeCrane notes scream out one uncontestable fact: Texaco *knew.*

Special Issue No. 4 asked, "Do you find from a preponderance of the evidence that Texaco's actions, if any, were willful, intentional and in wanton disregard of the rights of Pennzoil, if any?" This was the question that had to be answered "Yes" before punitive damages could be awarded. Special Issue No. 5 asked, "What sum of money, if any, was Pennzoil entitled to receive from Texaco as punitive damages?"

There were real problems with both of these issues that threatened to divide the jury yet again. We simultaneously considered these Special Issues. Susan was concerned that, having already voted 10 to 2 to award Pennzoil $7.53 billion, we were considering giving them even more money.

She recalled Miller's statement about how Texaco was being sued for "everything we're worth, plus Getty, plus half again as much." I shot back with Jamail's assertion that Texaco's assets totaled over $48 billion. "Where's that annual report?" someone asked. I looked through the hundreds of pages of exhibits, but couldn't find it.

The meaning of "wanton disregard" was questioned by several jurors. After a few unsuccessful attempts to spell out an acceptable definition of "wanton," it was decided to send a note to the judge asking for clarification. Rick wrote out the note.

"Ask him if we can have a copy of Texaco's annual report," I requested. Rick added it to the note. The document had never been passed to the jury, but several people around the table recalled seeing it when it had been used in the questioning of some of the witnesses.

The note was passed to Carl, and we waited for a response from Judge Casseb. After a period of time, Carl knocked on the door and handed the judge's response to Rick.

When read to the assembled jury, the answer to our questions were brief: "Wanton is used in its ordinary sense and is synonymous with reckless or heedless disregard of the rights of others." Also, Texaco's annual report was not in evidence.

After tossing these responses around for a while, we decided to take a break for lunch. The fact that Texaco's annual report was not available to us meant that we could not explore the wildly disparate claims of Miller and Jamail about the worth of Texaco. I posed the question to the rest of the panel: If the figure cited by Miller could stand up to scrutiny, wouldn't he have mentioned it before the last day of final arguments? We would take the matter up after lunch.

When we regrouped after lunch, the debate again turned to the issue of punitive damages. With $7.53 billion on the table, there was a genuine reluctance on the part of a few to award anything in punitive damages at all. On the other hand, there was also strong sentiment at the table to award the entire amount of punitive damages sought by Pennzoil, another $7.53 billion. The brazen acts of the Texaco officials and the Getty parties they indemnified had robbed Pennzoil of the benefits of what this jury had already found was a binding agreement.

Texaco had brought three of the men we felt were the principal bad actors on the Getty side to court and vouched for their testimony. The evidence presented in the case led the jurors to believe that Geoffrey Boisi had shopped the Pennzoil deal without authorization from the Getty board and then come to court with his Goldman, Sachs lawyer and denied he had shopped it.

Barton Winokur had been a major obstructionist in the Pennzoil deal and had encouraged the unauthorized Boisi shopping spree. He also suffered from serious credibility problems stemming from his earlier actions against Gordon and the Trust. Although those actions were not at issue here and certainly didn't involve Texaco, the fact they relied on the testimony of these witnesses was a burden that Pennzoil properly exploited.

Winokur's unbridled arrogance on the witness stand had altered the intended effect of his testimony. When he should have been scoring points for the defendant, his antipathy for the plaintiff was so apparent it colored his testimony.

Last but not least, the ubiquitous Martin Lipton had been deeply involved in the offer-counteroffer-acceptance ritual that preceded the announcement of the Pennzoil-Getty deal. Two days later, he told Texaco that the Museum was free to deal, a fact that was belied by the blanket indemnity he obtained that specifically protected the Museum from Pennzoil. Similarly, he had refused to warrant to Texaco that Pennzoil had no claim on the Museum stock.

What had he done? Pennzoil had proved to me beyond a reasonable doubt, beyond any doubt at all, that Lipton had been the device Texaco used to put the squeeze on Gordon.

The long list of hypertechnical distinctions Lipton had drawn during his testimony had done little to endear him to the jury, but it was his actions during the Six Day War that sealed the case.

In the end, Lipton could not deny the basic scenario presented by Pennzoil. He quibbled about nearly every detail, but on the central issue he was without a place to stand. And his accomplice, his partner, the beneficiary of, if not the inspiration for, his actions? His former clients, the friendly guys from Texaco.

The trio of Lipton, Boisi, and Winokur had formed what many jurors felt was an unholy alliance with Texaco, whose protestations of innocence were dashed on the rocks of the DeCrane and Lynch notes.

But wait, Texaco was the only one on trial here. Is it responsible for Lipton, Boisi, and Winokur? How, in this situation, to assess punitive damages that went to the heart of the conduct that needed to be punished? On this point, the deliberations bogged down yet again.

Our steadfast foreman wrote out another note to the judge in an attempt to resolve this thorny question.

Rick's note, the wording of which was agreed to by the jury, read, "To what extent is Texaco liable for the actions of Lipton, Winokur, and Boisi?"

There was not as long of a wait this time, but the answer from Judge Casseb wasn't very satisfying either: "The Court is unable to in any way specifically answer your question other than to advise you that you are to follow the instructions of the Court, the Special Issues as encompassed in the Court's Charge, and the evidence as received in open court."

After considering this for a few minutes, there was another vote. We were 10 to 2 on Special Issue No. 4, Texaco's actions were in "wanton disregard" to the rights of Pennzoil.

When we moved on to vote on the amount of punitive damages, I realized there were grave risks to the consensus we had built. If the punitive damages awarded were excessive, there was the frightening possibility Ola might join with the two holdouts and upset the issues we had already decided.

I filled out my ballot with a question mark. When the votes were tallied, there were 6 votes for $7.53 billion punitive, 4 votes for "?" or "Don't Know," and still the 2 holdouts voting zero.

Then it got crazy. Rick tried to bridge the gap between the "Don't Know" faction and the half of the jury set on awarding the full damages allowed under the law. "What if we divide the punitive figure in half? In thirds?" "If she'll go down $2 billion, will you come up $1.5 billion?" "Can we give them more?" it went on and on.

As the afternoon wore on, two things became increasingly apparent. First, we *would* ultimately be able to reach some sort of agreement on punitive damages among the ten jurors voting together on the first three issues.

Second, that agreement was not going to occur this afternoon.

Rick sent yet another note to Judge Casseb. We wanted to go home and resume our deliberations in the morning. Although it was left unsaid in the jury room, the feeling was strong that tomorrow would be the day.

34

November 19, 1985, was a Tuesday, more than nineteen weeks after the trial began. The dog days of summer had finally given way to the moderate temperatures of November, when Houston's climate seems almost hospitable.

That morning, as I dressed to make another in a series of seemingly endless treks to the courthouse, my mind was on other things. I was convinced that today a verdict would be handed down in the case of *Pennzoil Co. v. Texaco, Inc.*

The notes sent to Judge Casseb on the previous day had undoubtedly given the attorneys a clue to the progress of our deliberations. The request for a definition of "wanton disregard" was a clear indication that we were considering punitive damages.

If that wasn't enough of a tipoff, I arrived at the courthouse in a suit for the first time in the trial. Of course, I didn't wear a necktie. There was a sort of gentlemen's agreement among the male jurors that we would not dress up for the trial; comfort was a prime consideration in enduring the lengthy proceedings. The only exception was Rick, who would sometimes show up in a tie when the trial was only scheduled to run half a day. He used those occasions to try to catch up on his backlog of clients. Fred usually wore a sportcoat, but he, too, disdained neckties.

The courtroom was again packed. Since final arguments, there had been a consistently large crowd of onlookers. The almost uniform deadpan expressions of the jurors offered little clue to the impending verdict.

When again seated in the privacy of the jury room, we immediately tackled the issue of punitive damages.

That morning, the seating arrangement was scrambled. I took a chair next to Susan, the intellectual bulwark of the loyal opposition. Locked at 10 to 2 with little hope for a unanimous verdict, I decided to make a last attempt at building a consensus.

Pennzoil's contract ultimately turned on one issue: What had the Getty board approved on that oft-mentioned 15-to-1 vote. Despite the testimony of the Getty witnesses that testified for Texaco, all outward manifestations strongly indicated that the vote had indeed approved the Memorandum of Agreement with a price modification.

Shortly after deliberations resumed, I took one of the legal pads that had been provided for our use and drew two columns down the right-hand side of the page. On the left, I listed the terms of the Memorandum of Agreement. Taking them one by one, I compared these terms with those in the press release, and placed a check mark in the appropriate column. At the end, every term I had recorded from the Memorandum of Agreement was also in the press release.

Seated next to me, Susan had silently followed my scribbling. I turned to her and said, "If the board didn't approve the Memorandum of Agreement, how did all these terms wind up in the press release? Where else could they have come from? Remember, Copley and Winokur both approved this release for Getty Oil."

I dug out the Lipton agenda for his January 5 meeting with Texaco, showing that he had demanded the indemnity upfront and a guarantee of the $112.50 the Museum would receive under the Pennzoil deal. The Representations and Warranties clause of the Museum's agreement with Texaco said it owned the stock free and clear, except for any claims arising out of "the Pennzoil agreement." The indemnity specifically required Texaco to protect the Museum against any claim by Pennzoil.

This evidence confirmed both the contract and Texaco's knowledge of it. Why it proceeded in light of these demands from Lipton was subject to debate, but the incontrovertible fact remained that it did. Not only that, Pennzoil's name was all over that Museum agreement. Had Texaco thought it was there for ornamentation?

The matter of how to set the amount of punitive damages to award in this case had caused some trepidation and confusion in the jury room, even among many of those who felt strongly that substantial punitives were definitely in order.

The night before I had spent a lot of time contemplating our verdict. Some $7.53 billion in actual damages was already on the table, a decision that I suspected would arouse great controversy. When adding the punitive damages that I knew would be awarded, the total would be somewhere in the range of $9 to $15 billion.

The prospect of awarding punitive damages on the basis of a number pulled out of a hat was totally unacceptable to me. Looking at the case objectively, what would punitive damages be designed to punish? They were also known as explemplary damages. What kind of example would we be trying to set?

Slowly, a plan had begun to take shape in my mind. Irv Terrell's final argument had pounded away at the indemnities. Kinnear and Weitzel had both testified that Texaco could not have done the deal unless they had granted the indemnities. Before he sat down, the last thing Terrell had said was, "Just don't let them get away with these indemnities. Please don't."

For Texaco actually to pay out under these indemnities, Pennzoil would have to obtain a judgment against the Getty entities. That did not reduce their significance in *this* case, however. The terrible wrong to Pennzoil was made possible by these indemnities, which absolved everyone of responsibility for their actions . . . except Texaco.

If punitive damages were tied directly to the indemnities, it meant the figure would not be arrived at in an arbitrary or capricious manner. This also addressed the matter of Lipton, Boisi, and Winokur, whose plotting would never have amounted to anything without this legal cloak Texaco so willingly draped over their shoulders.

I spoke for a minute about the indemnities, recalling those final words of Terrell, and then said, "I think we should award $3 billion in punitive damages: $1 billion each for the indemnities given to the Trust, the Museum, and the company."

This would bring the total verdict to $10.53 billion, which seemed somehow less ambiguous than, say, $9.9 billion. It was happened to fall roughly in the middle of the range that had been the matter of much debate over the past two days.

To my surprise, no arguments broke out when I offered the suggestion; nor was it ignored like some of my admittedly overwrought rhetoric of the day before.

The final result was one I could never have anticipated. Susan, who was still sitting at "No contract" and "No damages," looked up and simply said, "Okay."

I wasn't sure what she meant until she asked Rick to take another vote on Special Issue No. 1. Susan changed her vote, and we stood at 11 to 1. This left Velinda in somewhat of a predicament, since her holdout had been tied to her indication that she believed "what Susan said" was right.

With apparent reluctance, Velinda indicated that she, too would change her vote. One last ballot made it unanimous at 12 to 0: Pennzoil's contract had been validated by every member of the jury that had heard nearly five months of testimony about this bizarre tangle of events.

My feelings at this latest turn of events were mixed. A divided verdict would have added a measure of doubt to what was sure to be a controversial verdict. My exhilaration at unexpectedly arriving at a unanimous verdict was equaled by a deep and sincere respect for Susan Fleming, who had made it possible.

Susan's initial advocacy of Texaco's position had been completely sincere, in my view. She wasn't playing devil's advocate in any traditional sense of the term; she definitely wasn't playing. In her own mind, however, she had reconciled the facts of the case. Any venality on the part of Pennzoil was so heavily outweighed by the cupidity of Texaco that justice cried out for this affirmative verdict.

When we moved through the other Special Issues, we came up at 12 to 0 on every issue. Until we got to the amount of punitive damages, where we again stalled out at 11 to 1. Lillie stood firm at $7.53 billion in punitives, the figure that had attracted 6 votes the day before.

Rick was not content to let this vote stand at 11 to 1, and again became the senator from the Great State of Compromise. I repeated the rationale tying the punitive damages to the indemnities, but Lillie indicated she felt strongly that the conduct of Texaco had merited no discount. This was a difficult position to argue against, because I shared her belief that Texaco's actions in this case had been an outrageous violation of a legal contract which it continued to disdain.

Finally, and with great reluctance, Lillie acceded to the wishes of the group. The vote stood at 12 to 0 on all issues. I had a feeling in the pit of my stomach similar to that I experienced as a child when I stood on the edge of the 10-meter platform at the Glenbrook pool. Once you climb those stairs, it's time to jump. My problem was that I could never stop myself from looking down first.

Rick wrote out what would be his last note, telling Judge Casseb that the jury had reached a verdict. Moments later, Carl Shaw and several uniformed deputies escorted the jury through the chaos of television lights and cameras and spectators who couldn't get one of the precious few seats in the courtroom. Filing into the jury box for what I knew would be the last time, any feelings of nostalgia were put aside. This group of twelve citizens, who would have never come together for any other conceivable purpose, was about to become the instrument of justice.

Carl took the jury charge on which Rick had recorded our verdict and handed it to Judge Casseb. The courtroom was as quiet as a tomb as he read Special Issue No. 1.

JUDGE CASSEB: "'Do you find from a preponderance of the evidence that at the end of the Getty Oil board meeting January 3, 1984, Pennzoil and each of the Getty entities, to wit, the Getty Oil Company, the Sarah C. Getty Trust, and

J. Paul Getty Museum, intended to bind themselves to an agreement that included the following the terms:'

"'A. All Getty shareholders except Pennzoil and the Sarah C. Getty Trust were to receive $110 per share plus the right to receive a deferred cash consideration from the sale of ERC Corporation of at least $5 per share within five years;'

"'B. Pennzoil was to own three-sevenths of the stock of Getty Oil and the Sarah C. Getty Trust was to own the remaining four-sevenths of the stock of Getty Oil;'"

As Judge Casseb read the terms outlined, I looked first at Miller, then at Jamail. They both looked right at the jury with a seriousness of expression befitting the circumstances, as if they were in church. In a way, I suppose the courtroom *is* their church.

An electric charge crackled through the air as the Judge steadily continued reading.

JUDGE CASSEB: "'and C. Pennzoil and the Sarah C. Getty Trust were to endeavor in good faith to agree upon a plan for re-structuring Getty Oil on or before December 31, 1984, and if they were unable to reach such agreement, then they would divide the assets of Getty Oil between them also on a three-sevenths:four-sevenths basis.'

"The answer is: 'We do.'"

With this response, the silence was broken; a low rumble began to filter through the room. The judge turned to the jury.

JUDGE CASSEB: "Is that the answer each one made?"

We affirmed our verdict by raising our right hands; twelve hands were raised.

Judge Casseb continued on to read Special Issue No. 2, which asked if we found that Texaco had interfered in the binding agreement referred to in No. 1. Again, he read, "The answer is: 'We do.'" The low rumble perceptibly increased.

He next read Special Issue No. 3, which asked what sum of money we awarded for actual damages.

JUDGE CASSEB: "The answer is 'Seven point five three billion dollars.'"

With that, the rumble soared far beyond the level of background noise, prompting the judge to call for order. On through each of the Special Issues he read, with each affirmative answer followed by twelve hands raised in the air.

When the issue on punitive damages was read, I looked at Jamail. The $3 billion figure was less than half of what they had requested, but he nodded his head; his expression indicated to me that he accepted our judgment, as well he might . . .

When the reading of the verdict was complete, pandemonium broke out in the courtroom. Dick Miller looked stricken, and my heart went out to him. I recalled his last words to the jury: "You do right and the case will come out like it ought to come out." His words proved true, and it had taken a heavy toll on him.

We had done our duty, but none of the twelve would have described it as a happy task. We filed out of the 151st District Court, never to return.

Our service was over, but the spotlight was on what was now being called "the $10 Billion Jury."

PART XI

A BITTER AFTERMATH

The Battle Moves to the Court of Public Opinion

That a lie which is all a lie may be met and
fought with outright,
But a lie which is part a truth is a harder matter
to fight.

TENNYSON

35

Reagan and Gorbachev were having a summit meeting that November Tuesday, but the news from the courtroom in Houston came like a bolt from the blue.

After Judge Casseb had dismissed the jury and gaveled the trial to a close, Joe Jamail waved his arms amid the confusion and asked the jurors if we would step into the court reporter's room for a minute.

The jurors struggled through the mass of bodies and took refuge in the side hall. Jamail came in alone with a grin on his face and closed the door. Suddenly, he was serious again. He solemnly thanked the jurors for the verdict, not just for himself, but for Pennzoil. There had to be a few misgivings about the verdict among the twelve citizens in that room, but there was also a sense of completion, of release from what had been a taxing ordeal by any definition.

The road from the juror assembly room to that back hallway in the 151st District Court had touched the lives of everyone in that room in ways that could not have been imagined nineteen weeks before.

Jamail had another purpose in wanting to talk to us. He told the members of the jury that lawyers from Texaco would be contacting us to ask questions about the jury's deliberations and what went on among the jurors during the trial. "You don't have to talk to them at all; just tell them you don't want to talk when they call," he advised. "If you do want to talk to them, that's fine; I'm just asking you to please give me an opportunity to be present when you do. You've seen what they'll do, and really, if I were you, I'd just say, 'No, I don't wish to discuss it.' "

Somebody asked, "Do you really think they'll call us?" Jamail nodded and said we could count on it. Then he said, "I talked to Judge Farris yesterday. He's doing better, and it would really mean a lot if y'all could give him a call."

With that, he opened the door and we exited into the chaos outside. The fifth floor corridor was littered with television cameras, lights, reporters, lawyers, and what seemed like half the people who worked in the courthouse complex.

I recognized Tom Curtis, a reporter from the *Dallas Times Herald*, whom I had known slightly when he worked for one of the Houston dailies in the early 1970s. He asked me for a comment about what the verdict meant. I came out with something to the effect that this was a signal to American business that "the idea that 'anything goes' is dead."

As we talked, I saw Patricia Manson, who had covered the trial for the *Houston Post*, scribbling down the conversation. That same quote, in a slightly different form, would appear on the front page of the next morning's edition of her paper.

Manson smiled when she saw me looking at her and said, "Hi." This was the first word of many we would exchange over the next eighteen months. By this time, we had already been in the same courtroom every day for months, but the rules of the Court forbade even casual conversation between jurors and the press.

Rick Lawler was the only other member of the jury who stopped to comment to reporters that day. "We won't tolerate this sort of thing in corporate America," he was quoted as saying.

Susie and I took the elevator down to the courthouse lobby where I ran into Dick Miller. He still looked stunned, but came forward and offered me his hand. I grabbed it and looked Miller in the eye, like I had for the entire trial.

"Did I blow the case?" he asked me. I shook my head and said simply "No." He indicated he would like to talk to me about it (as Jamail had predicted ten minutes before) and I immediately agreed. "Why don't you give me a call?" I told him. There were many questions that raced through my mind, but I was uncomfortable with having an open discussion in a crowded courthouse lobby.

Jim Kinnear walked up just then. He, too, offered his hand and asked, "Did the jury think everybody at Texaco had tails and horns?"

I said no, but told him the indemnities they had handed out had sunk their case. It would later be reported that Kinnear felt this meant we had found against Texaco under the indemnities, which is not what I said and was not the case by any means.

David Luttinger, a senior Texaco attorney who, like Kinnear, had sat in on the whole trial, started telling me about the evidence the jury wasn't allowed to hear, a fascinating subject that led to an immediate question and answer session that grew intense as the pent-up emotions of the trial started to pour out. I would have other conversations with Luttinger, too.

This conversation was getting crowded, however. Susan Roehm and Paul Yetter from Baker & Botts had come right up behind me to monitor the discussion. A few words were exchanged and the heat started rising again.

Most of the jurors were making plans to meet at a restaurant for a post-verdict dinner immediately after the trial, but I was emotionally spent. With my wife, I retreated to the privacy of our home and savored my newly acquired freedom.

Reading through the news accounts of the trial that Susie had saved for me, I was surprised at the relative sparseness of the coverage. When the next day's papers came in, I started to understand something of the impact this verdict was having. This hardly prepared me for the onslaught of negative publicity that soon followed.

The attention now being paid to the case revived an idea that had lay dormant during the months that the trial had dominated my daily existence.

I resolved to put some of the things I had seen and heard down on paper. When I had a rough idea of what I might write, I called both Houston newspapers. The *Chronicle* indicated that it would consider a substantially longer opinion piece than would its cross-town rival, the *Post*.

Over the next day, I scrawled out a rough draft of the proposed opinion piece. Arriving early at work the following day, I typed out the second draft, which was essentially the story that was finally published.

When I brought it to the *Chronicle* building, I was shown to the office of David Langworthy, who worked on the Op-Ed page. He read the ten or so pages, looked up and said, "Can't we cut out some of this out? We don't need that stuff about the lawyers. I'd like to read more about what happened inside the jury room."

While explaining that I would be uncomfortable revealing specific details about the deliberations, I also refused to consider cutting the guts out of the piece. He tried to prevail upon me to cut the length, to no avail. Finally, he picked up the piece and said, "Wait here."

About five minutes later, he returned and said, "We'll run it."

He sat down at the composer, we worked out a new lead paragraph and that was it.

That Sunday, only five days after the verdict, the *Houston Chronicle* carried the following article. I thought that writing it would help to get this case out of my system. As I've indicated, that did not prove to be the case.

PENNZOIL-TEXACO:

Inside the Jury Room*

BY JIM SHANNON

Since the verdict in the Pennzoil-Texaco case was announced last Tuesday there has been a lot of criticism of the amount of damages awarded.

Actual damages of $7.53 billion; $3 billion punitive. That's more money than an average person can imagine. I know it's more money than I can imagine. And I remember that several times during the four-and-a-half month trial it seemed like it was more than the lawyers for both sides could imagine. They misspoke themselves, frequently saying "millions" when they meant "billions," and vice versa.

I wish those who criticized the award could have sat through this trial the way I did as a juror. They would have known that Texaco never disputed Pennzoil's actual damage claim of $7.53 billion. And as for the punitive damages, many of the jurors were willing to award the full $7.53 billion in punitive damages sought by Pennzoil and allowed under the law because of the outrageous conduct of Texaco and the Getty parties they indemnified. We reached our verdict based on the evidence presented in court and free of outside influence.

I personally never had anything against Texaco. I still don't. For those four and a half months we sat in the same rooms, rode up and down the same elevators and patronized the same snack bar as Jim Kinnear, the Texaco vice chairman. I believe he's a nice man.

After the verdict was announced, Kinnear came up to me and asked if we on the jury thought he and the other Texaco executives had horns and tails. Texaco's lead attorney, Dick Miller, wanted to know if I thought he had "blown the case." To both men I honestly answered, "No." I told Miller I thought he was a good attorney . . . he just had a bad case . . .

* * *

When I opened my mailbox in June and found a juror summons from the sheriff of Harris County, I wasn't too surprised. It had been almost five years since I was last called and I was still a registered voter. But I had no idea that fulfilling this routine civic duty would lead to 19 weeks in a ringside seat at the Battle of the Corporate Giants, a.k.a. Case No. 84-05905, Pennzoil vs. Texaco.

On the way to the juror assembly room, I heard over my headphone radio that the Texas Supreme Court had denied Texaco's motion to disqualify the law firm of Baker & Botts from defending Pennzoil. I knew that Baker & Botts was a big downtown law firm and that Pennzoil was suing Texaco for some incredible amount of money over the acquisition of Getty Oil. I didn't really know what it was all about.

At the assembly room we were seated according to the number on the little card Sheriff Johnny Klevenhagen sent us. When we heard the next jury panel would be 100 instead of the standard 30 my seat mate turned to me and said. "Either we're

---

going to a death penalty case or it's Pennzoil-Texaco." With this cheerful thought in mind, we were herded across the street to the old courthouse built in 1900, where Houston's last public hanging took place. Now it's strictly used to hear civil cases. It's a busy building.

In the courthouse we were again seated by number to begin jury selection. They call it "voir dire," which means to speak the truth. Presumably, this is when prospective jurors are suppose to truthfully answer all questions put to them by the attorneys. In fact, both sides took a full week between them to present their respective cases in great detail, complete with 30-by-40-inch blow ups of key documents. Occasional questions were posed to the jury panel as a whole, asking for a show of hands.

On some questions, individual panel members huddled to air their views one at a time at the bench with the judge and the attorneys. As it turned out, most of the final 12 jurors and four alternates rarely, if ever, spoke during "voir dire." On the whole, there was very little questioning of individual panel members. Finally, Court Clerk Cliff Bennett read the names of those selected. Much to my surprise and chagrin, my name was called. I looked in astonishment at Joe Jamail and Richard Miller, lead counsel for the two warring parties. They each smiled and no doubt wondered if they had made a mistake taking me. The next morning the trial began in the 151st District Court.

Judge Anthony J. P. Farris, a self-described "64-1/2-year old ex-Marine," was presiding judge, an imposing figure on the bench. More than a dozen attorneys crowded the counsel tables along with Baine Kerr, recently re-

tired president of Pennzoil and James Kinnear, vice chairman of Texaco. They were the designated observers, each would also testify during the trial. All other witnesses were barred from the courtroom during testimony.

An impressive array of legal talent was assembled in the courtroom to try the case.

For Pennzoil, lead counsel Joe Jamail is best known as one of the top personal-injury lawyers in America. In this case, his client was a large corporation he felt had suffered a grievous wrong. His demeanor left no doubt he intended to correct this wrong. The burning intensity of his presentation let us know from the beginning that this was serious business involving important moral and ethical issues.

His principal co-counsel were John Jeffers and Irv Terrell of Baker & Bolts, both experienced trial lawyers with different styles. Jeffers was methodical and low key, like a boxer who quietly builds up his points and at the end of the match you realize he has won the fight. His deadpan manner contrasted with that of equally effective Terrell, who often responded to testimony by smirking broadly, smiling, frowning, folding and unfolding his long frame, even laying his forehead on the table. He was fun to watch, even as he elicited damaging admissions from defense witnesses through adroit questioning.

They were assisted by Baker & Botts lawyers Randall Hopkins, Brian Wunder, Paul Yetter and Susan Roehm, the only female attorney who appeared before the bench during the lengthy trial.

The final member of the Pennzoil team was Jim Kronzer, who has practiced law for more than 40 years. Although his legal arguments to the court were conducted outside the presence of the jury, he was in constant attendance throughout the trial. In reviewing press accounts after the trial, I discovered that Texaco had moved for a mistrial after Pennzoil had rested its case. The motion was denied after Kronzer's reply that, "that dog won't hunt."

Texaco was represented by the firm of Miller, Keeton, Bristow & Brown, a small firm of high-powered attorneys. Founder Richard Keeton and Jack "J.C." Nickens were formerly partners at Vinson & Elkins, the biggest law firm in Houston. Robert Brown, formerly the Brown of Liddell, Sapp, Zivley, Brown & Le-Boon, had an impressive track record. A large man with a soft voice, Bob Brown's lengthy, circuitous questioning was sometimes effective on cross-examination.

R. Michael Peterson was the Texaco team's designated document man and was never far from the files containing the hundred of documents introduced into evidence.

But the principal defender of Texaco was lead counsel Richard Miller, a strong advocate whose fire and brimstone was leavened with wry humor. With his erect, military bearing and scornful manner, he continually blasted most of Pennzoil's witnesses and exhibits, even the Pennzoil Company itself, which he called "a big bunch of crybabies."

These are the attorneys who confronted the jury in this case. The jurors who were selected represented a broad cross section of the population of Harris County. We were 10 women and six men, from Tomball to Katy to Clear Lake and all parts in between. Our members included a letter car-

rier, homemakers, a warehouse foreman, nurses, a county employee, an industrial salesman, clerical workers and a city employee; in other words, regular people who work for a living. We were called upon to judge a case involving an army of lawyers and investment bankers in the second largest merger in history.

Who were we to judge? The analogy that kept occurring to me was the Watergate jury that convicted John Mitchell, H. R. Haldeman, and John Ehrlichman. Drawn from the voter pool in Washington, D.C., they were a decidedly blue-collar jury. Under our jury system defendants in criminal and civil cases must answer to juries composed of average people. It would have made a mockery of the system to have the case decided by some anointed panel of experts.

I particularly resented the statement of William Weitzel, senior vice president and general counsel for Texaco, who implied after the trial that the jury was not intelligent or sophisticated enough to rule on the case. We heard 17 weeks of testimony that covered all aspects of this matter, and we rendered our verdict based solely on the evidence that was presented in court. Many jurors felt that Weitzel himself, who testified near the end of the trial, further damaged the Texaco case. His cursory examination of the Pennzoil-Getty dealings apparently convinced him that whatever agreement there was between the parties couldn't stand up under a Texaco assault. By a unanimous vote of 12–0, this jury told Texaco (and the rest of the business community) that a contract is a contract, and if you interfere with someone's contract, you will be held to account.

The Getty family saga was inextricably woven into the fabric of this case. After the death of J. Paul Getty, the majority of the stock of Getty Oil was held by two trusts. Gordon Getty, one of J. Paul's sons, was the sole trustee for the Sarah C. Getty Trust, which controlled 40 percent of the company stock. Sidney Petersen, chairman and chief executive office of Getty Oil, was a bitter foe of Gordon Getty, motivated by the fear that Gordon would find a way to seize control of his late father's company. At a meeting of the Getty Oil board of directors held in Houston on Nov. 11, 1983, Gordon was deceived into leaving the boardroom.

Petersen then brought attorney Barton Winokur of the Dechert, Price & Rhoads firm and Geoffrey Boisi of the investment banking firm of Goldman, Sachs in through the back door. The board then secretly voted to sponsor a lawsuit by one of Gordon's relatives, aimed at having a co-trustee appointed, and to effectively tie a tin can to Gordon's tail. Director Chauncey J. Medberry III was the former chairman of the California-based Bank of America. By a strange coincidence, the proposed co-trustee was the Bank of America.

This scenario, which came out in open court, did nothing to bolster the credibility of Winokur, Boisi and Medberry, all of whom testified for Texaco during the trial. In particular, the arrogance of Bart Winokur was reinforced to me with every hour he spent on the stand (and he was there for about a week). He never seemed to deign to even look at the jury, silently projecting the attitude voiced by Weitzel after the trial. About the only thing Winokur's testimony proved to me was how hard it is to get a lawyer to admit to anything.

That wasn't the case with Martin Lipton of the firm Wachtell, Lipton, Rosen & Katz, who has been involved in over 1,000 corporate takeovers. Under sharp cross-examination by Pennzoil attorney Joe Jamail, Lipton not only admitted he was a fast-talking dealer, he bragged about it: As the attorney for the J. Paul Getty Museum, an 11.8 percent owner of Getty Oil, Lipton participated in drawing up the Getty-Pennzoil agreement, whereby Gordon Getty would become majority owner and chairman of the board of a restructured Getty Oil. Even as he had attorneys from his firm working with Pennzoil lawyers on the final documents, Lipton was on the phone with his former client, Texaco, working to undo the deal.

By the time Lipton appeared as a witness at the trial, the jury had heard hundreds of hours of testimony about what had happened. If the attorneys for Texaco were hoping that Lipton could salvage their case, they were sorely disappointed. By the time Marty Lipton left the stand, he had driven the last nail in Texaco's coffin.

I also take exception to the idea that this case was tried in Pennzoil's backyard. Anybody who lives in Houston is well aware of the large Texaco presence here. For me, the issue of regional bias is a red herring.

As a jury, I believe we were unwilling to accept the idea that different standards of justice apply, depending on what state you're in and how much money you have. We stand by our verdict, and have fulfilled the oath we took as jurors.

There has been a lot of bombast in the press since the verdict, much of it orchestrated, in my opinion, by Texaco public-relations men who were in attendance throughout the trial. But

the truth is in the evidence presented
to the jury. You can find it in the
24,000 pages of the trial transcript.
Free from outside influence and

based solely on this evidence, the jury
has spoken ... and this experiment
called democracy continues.

That was the article as it appeared. Even though I had refused to go
into details of the deliberations, they had still entitled it "Inside the
Jury Room."

The *Chronicle's* Langworthy had made some cosmetic changes that
helped the piece, changing "housewife" to "homemaker," and toning
down some of my characterizations of the key figures of the case. For
example, when I said Winokur "never deigned to look at the jury," they
had softened it to "never seemed to deign," which was a construction I
had never heard of. I had originally described Lipton as "a fast-talking
double dealer," but it appeared in print as "a fast-talking dealer," which
sounded like a description of an auctioneer.

In general, however, I was pleased with the piece. It captured something of the flavor of the trial. After some debate, I elected to include it
here, both as an indication of my immediate post-verdict impressions
and because of the role it would play in the weeks ahead. The *Chronicle*
piece set off reverberations that would trigger an unexpected chain of
events, ultimately including the writing of this book.

36

The reaction to the opinion piece in the *Houston Chronicle* was immediate and intense from every direction. Rather than clearing the air, as I had hoped, the piece had stirred a firestorm of controversy.

Newspapers all over the country printed quotes from the article in their coverage of the case. The *Chronicle* also printed many letters to the editor in response, both favorable and unfavorable.

One Saturday not long after, I sat down with the phone book and called as many of the critical letter writers as I could. Fully one-half admitted (or claimed) to be Texaco employees. Strangely enough, they were almost all friendly and asked me many questions about what had transpired.

Other developments were more worrisome. A *Chronicle* reporter called and told me a Texaco public relations staffer was saying that Joe Jamail had actually written the article that had appeared under my byline.

A three-page handwritten letter from Baine Kerr was among the warmest responses I received and read in part: "Throughout the long weeks of the Pennzoil-Texaco trial, and before, friends and acquaintances asked me what kind of jury would decide the case; were there experts on the jury; and how could an ordinary jury understand what were generally perceived to be complex issues.

"My response was always the same: I have real confidence in the jury system, with fellow citizens chosen at random and clearly not subject to the pressures that could be applied to a single judge or small group of judges. Furthermore, I felt that the issues were basically not all that complicated, even though the factual background was involved. Win or lose, we much preferred to have a jury trial . . .

"All of which is prefatory to my principal reason for writing you, namely, to tell you what a magnificent job you did in your article in the Sunday *Chronicle* in commenting upon the trial and the work of the jury.

"Anyone who reads your article will gain a clear understanding of what the case was all about and, more importantly, how well the jury did its task of listening to the evidence and sifting fact from fiction. The writing is first-rate. While many persons serve on juries and do their very best to reach conclusions based on an objective evaluation of the evidence, there are few who could articulate this process as clearly and accurately as you did."

Kerr's response was particularly gratifying to me during what was a difficult period in the wake of the trial.

This was merely a sideshow, however, for the main attraction, the $10.53 billion verdict itself.

The official reaction from Texaco had been somewhat restrained, at least initially. The day the decision was announced, president Al De-Crane was widely quoted as saying, "The jury's decision will have no impact on the way we run the company."

Aside from some gratuitous insults to the jury from general counsel William Weitzel, Texaco seemed to turn inward, circling the wagons.

Judge Casseb had set a hearing for December 5 and 6, to hear arguments on the Motion to Enter Judgment. As this hearing date approached, Texaco began to mount a massive public relations campaign to bring some pressure to bear on the proceedings—and presumably, Judge Casseb.

DeCrane and Kinnear went on the road and criss-crossed Texas like tent show revivalists, showing up for newspaper and television interviews in a number of cities. At every stop, their message was the same: what's bad for Texaco is bad for you and your neighbors.

The dreaded word "bankruptcy" was bandied about for the first time, as the ultimate threat.

In this highly charged atmosphere, the hearing on the Motion to Enter Judgment convened in front of a packed courtroom. I arrived twenty minutes before it was scheduled to begin and was dismayed to find a huge crowd in the hallway.

The scene made the pandemonium of the last day of the trial pale by comparison. *Pennzoil* v. *Texaco* was now a national story, and the networks all competed with the Texas reporters for a place to stand.

Fortunately, a friendly face materialized out of the crowd. Carl Shaw was still handling courtroom logistics as bailiff of the 151st District Court. He took me down the hall to the front of the courtroom and found me a seat near the clerk's desk. As luck would have it, the seat he found placed me squarely between Al DeCrane and Dahr Jamail, attorney son of Joe Jamail.

When I sat down, DeCrane turned to me and said, "So you're an author now." I smiled and said, "Well, not if you listen to some of the Texaco P.R. people. They've been calling reporters and saying Jamail really wrote the article."

DeCrane looked very serious. "I can assure you it's not any of our people who are doing that," he said. "I've given orders that nothing like that's to occur."

I nodded and changed the subject. "You guys don't know what you've done to us," he offered a few minutes later, a statement to which I had no ready reply; we *did* know what Texaco had done to Pennzoil, but this didn't seem like the time to bring it up.

Rick Lawler had also returned to the courtroom, and Carl managed to find him a seat. The ranks of the attorneys at the Texaco table had swelled precipitously. Dick Miller hovered around the periphery of the scene; Texaco's new lead counsel was David Boies of New York, fresh from a successful defense of CBS in the General William Westmoreland libel suit.

During a break in the hearing, Boies lingered to chat with Dahr and myself. I asked him about the way the national press was handling the story. "Outside of the legal community, nobody in New York was following the trial," he explained, "that's why they're all playing catch-up now."

This fact also explains why Texaco was having such luck getting its version of the facts portrayed. The press abhors a vacuum as much as nature does, and Texaco moved to fill the information vacuum with what I knew to be a distorted view of the facts.

Pennzoil, for its part, had taken the high road, refusing to be drawn into responding to every new attack. Bob Harper, Pennzoil's chief spokesman, came over at one point and I asked him about the Texaco media campaign. He grimaced and said, "We made a conscious decision not to get in a pissing contest with them."

Dahr Jamail gave me a rundown on the new legal muscle Texaco had brought in, including Gibson Gayle and James Sales of Fulbright & Jaworski. "Plaintiff's lawyer got rich from suing General Motors when he was representing them," Dahr said of Sales.

Very much a colorful character in his own right, Dahr Jamail had a full head of hair that was long by the spit-and-polish standards of the corporate world. During the voir dire, he had been present for most of the week. Having no clue as to his actual identity, I speculated that he was a correspondent from the BBC.

When I actually met him, I discovered that he doesn't even have an

English accent. He does have a sharp sense of the absurd as well as the feisty populist instincts that have served his old man so well in taking on the corporate behemoths over the years.

At one juncture, he leaned over and said in a stage whisper, "Those Texaco executives ought to be out pushing brooms." DeCrane stiffened, and I bit a hole in my lip to keep from letting out a frankly horrified laugh.

After all those months on the jury, it was strange to be back in the 151st District Court as a spectator less than three weeks later. Some of the passion and heat of the trial were played out in abbreviated form during two full days of hearings before Judge Casseb.

At the conclusion of this hearing, the judge had the option of entering the jury's verdict, modifying it, or throwing it out entirely.

For Pennzoil, Jamail, Jeffers, and Terrell all took a turn at urging Judge Casseb to enter the verdict. It was also an opportunity to hear Randall Hopkins of Baker & Botts, who had infrequently appeared before the jury. He adopted a forceful (if somewhat academic) approach to the law he was arguing, but Hopkins was a very persuasive advocate. Sitting there listening to him defend the jury's verdict, I felt some degree of vindication as he deflated the arguments raised by the Texaco appeal team.

Similarly, James Kronzer, the senior statesman at the Pennzoil table, cast aspersions at Texaco's complaints about the judge's charge to the jury. He belittled the charge Texaco had submitted on the contract issue, which contained a long list of terms from which the jury was supposed to select those that made up Pennzoil's agreement.

He branded it as an attempt to sow confusion in the jury room and said the jury would have to "play itchy-switchy with these terms" and would have to be "Aristotlean theoreticians" to figure it out. I am convinced that was the first time in the history of the English language that those two phrases have appeared in the same sentence.

When the second day of arguments on the motion drew to a close without any sign of an imminent decision, the hearing was recessed until after the weekend. I was hopeful that Judge Casseb would enter the verdict that we had returned, but speculation was rampant that he would drastically reduce the damages awarded.

One thing that had surprised me since the trial was the number of people who would concede Texaco had in all likelihood interfered with Pennzoil's deal, but continue to rail against the unfairness of the verdict.

As we had known for many months, there was a combination of factors in this case that aroused strong emotions.

*   *   *

In addition to lawyers, reporters, and company officials, the courtroom was crawling with arbitrageurs and their henchmen, straining for a clue about which way Judge Casseb would turn.

When the hearing reconvened on Tuesday, there was a sense that a decision was imminent. Instead, Judge Casseb recessed the hearing and had the attorneys come into his chambers. The rest of the day was a waiting game for the battalion of spectators and reporters, as the attorneys shuttled back and forth between two rooms that had been commandeered.

Brian Wunder, a young Baker & Botts attorney who had participated in the lengthy trial, was not one of the elite circle that was called to the inner sanctum. He later described how he had realized that Judge Casseb was going to enter the judgment when he saw Irv Terrell stick his head out and motion for Clifton Fridge, the senior Pennzoil accountant who had testified at the trial, to come into the back room.

Fridge, Wunder explained, could only have been there to calculate the amount of prejudgment interest, a sure sign that the judgment for Pennzoil would be entered by the court.

Wunder himself got in on the act a little later, when he was dispatched to the Baker & Botts offices to oversee his secretary, Ellen Ferraro, who was typing up a draft of the judgment with changes John Jeffers had called in from the courthouse. When Wunder returned to the courtroom with the typed judgment locked in his brief case, he adopted a totally blank expression to avoid tipping his hand. A friend from a rival law firm came up and asked him where he had been. Wunder didn't reply.

It was nearly dark in Houston, long after the stock market had closed, when Judge Casseb returned to the bench and read the judgment. He upheld the full damage awards totaling $10.53 billion. With prejudgment interest added in, the award to Pennzoil now exceeded $11 billion.

After hearing Judge Casseb affirm the action the jury had taken, I felt like the case had come full circle. Sure, there were many rounds of appeals to come, but the notion of a "runaway jury" had just been negated by the action of Judge Casseb.

37

When Judge Casseb upheld the entire jury award, the attention focused on the Pennzoil-Texaco case increased almost exponentially. The standard lament of "a jury run amok" had to be modified to portray the entire Texas judicial system as a "legal lynch mob" run amok.

The first *Wall Street Journal* editorial about the case was headlined "The Texas Common Law Massacre." Following Judge Casseb's decision, they issued a sequel, cleverly entitled "The Texas Common Law Massacre II."

Those three weeks between the jury verdict and the judge's decision had provided time for the national reporters who had ignored the trial to begin to delve into the intricate case, aided by the partisan viewpoint of aforementioned Texaco media blitz.

Even as the hearing on the Motion to Enter Judgment was unfolding in the second week of December, the mob of reporters gathered at the courthouse scrambled to gain some insight into the case that would give them an advantage over the competition.

The trial transcript was about 25,000 pages long, a voluminous, and detailed document that offered few quick and easy excerpts for those who had not followed the trial. Patricia Manson of the *Houston Post* and Barbara Shook of the *Houston Chronicle* were the only reporters who had covered the trial on a continuing basis. No doubt their clippings were scrutinized by their compatriots in the press, but the difficulties Manson and Shook faced in covering a trial of this length tended to minimize the utility of nineteen weeks of assembled coverage as a true indicator of the trial for the uninitiated.

During those long months, the story had unfolded day by day. With the testimony of many critical witnesses lasting for dozens upon dozens of hours each, it was impossible for Manson and Shook always to extract the relevant passages. Sometimes, the most telling point a witness made would not be brought into focus until it was illuminated by the subsequent testimony of another witness. For most of the trial, even the attorneys could only speculate about how the case would eventually unfold.

The distinct advantage their insight gave to Manson and Shook was amply reflected in the generally excellent accounts they filed with their respective papers after the trial ended, when it became obvious to everyone that this was the biggest business and legal story to come along in many a deadline.

The substance and nuance of their stories naturally found its way into many of the articles written by the horde of reporters who flocked to Houston to cover the latest rounds of the still-unfolding drama.

On the second morning of the hearing on the Motion to Enter Judgment, Rick Lawler and I had appeared on an early-morning program on KHOU-TV, the CBS affiliate in Houston, where we had been peppered with a variety of mostly hostile questions telephoned in by the viewing audience. From my previous experience in political campaigns, I was familiar with the tactic of skewing call-in programs with planted questions for the benefit of your party or candidate. It felt strange to be on the other end of such naked media manipulation.

That many of these questions were planted seemed likely, because they dealt with details of relatively complex issues intrinsic to the case that had only been hinted at in the media coverage of the case.

Rick and I managed to escape relatively unscathed from the encounter, or so we thought at the time.

Later that day during a break at the hearing, Julie Morris of *USA Today* had come over and introduced herself. With my typically dubious charm, I had responded, "Oh, yes, *USA Today*. Jay Leno says that's the newspaper for people who don't know how to read."

Her response will not be recorded here, but she did indicate that she had read my *Chronicle* piece and wanted to talk about the case. In a subsequent telephone interview, Morris asked me a number of questions about my background. Had I ever worked at an oil company? No, I replied, but my wife had worked in the oil industry for a decade. Which companies? Well, until March of that year, Conoco. When pressed, I explained that after DuPont acquired Conoco, there had ultimately been several rounds of early retirements and layoffs. Susie had been caught up in the latter.

Morris seemed surprised by this information. Didn't the attorneys ask you about your wife's employment? In fact, the Juror Information Form provided a space to list the employment of your spouse, where I had duly recorded Susie's contract job at that time with Sohio.

This was the first time any connection between Susie's employment and my service on the jury had even been suggested; it had certainly never crossed my mind during the trial or in the weeks since the verdict. For a resident of Houston to be employed in the oil industry was a thoroughly unremarkable fact, especially in 1985. It struck me as curious that this reporter would be that interested in this fact.

Her article, printed as a sidebar in a column called "INSIDERS: Behind the Scenes in the World of Business," was a bit of *People* magazine–type puffery that seemed to poke a little fun at me, but in a fairly good-natured way. The headline read "Trial Turns Juror into Celebrity," and the photo was sardonically captioned "Shannon: A Star Is Born." I was somewhat embarrassed by the piece, but at least it mentioned that I planned to write a book. It also detailed the fact that my wife had lost her job after DuPont took over Conoco.

Like the previously described television encounter, the *USA Today* piece would have unexpected ramifications.

I didn't have to wait very long. On December 20, 1985, a front page story in *The Wall Street Journal* was bannered "The Case of the Juror's Wife." The lengthy article was written by Matt Moffett and Thomas Petzinger, Jr., two reporters in the *Journal*'s Houston bureau who had been unable to navigate the three blocks from their office to the courthouse during more than four months of wrenching testimony.

I had been aware of the research being done for the article, and had, in fact, granted two lengthy interviews to Moffett. When I chided him about his paper missing the boat on the story, he defensively replied that *The New York Times* had not covered the trial either.

This comment helped me to gain something of a new perspective on the affair: if an event wasn't reported in *The New York Times* or *The Wall Street Journal*, it was relegated to the status of a nonevent.

Clearly, the trial of the *Pennzoil* v. *Texaco* case would have been a significant business story regardless of the ultimate verdict. It was not as if the takeover tactics at question were of little import in the prevailing economic climate. Curiously, the fact that the lawsuit was ignored in New York may have lulled Texaco into a dangerous sense of false complacency.

The December 20 *Journal* piece was a lengthy, in-depth account that was generally accurate. Accuracy and fairness are not equal virtues, however. In my opinion, Texaco operatives managed to use the piece to voice charges against Pennzoil, the judges, and the jury.

In addition to the headline story about my wife's prior employment (which Moffett would later reluctantly admit had been inspired by the *USA Today* piece), the story contained allegations that jury foreman

Rick Lawler had concealed the fact that his estranged father-in-law was a Texaco employee. How Rick would have known such information was not readily apparent, since his wife's father had not been in contact with the family for many years and had never laid eyes on his two grandchildren.

In addition, for the first time allegations of anti-Semitism were leveled at the jury. Shortly before the story went to press, I received a late-afternoon phone call from an obviously reluctant Moffett who said that his editor in New York "wants to know about the anti-Semitism thing."

In my naivete, I asked *"What* anti-Semitism thing?"

Moffett said, "Well, is it true that Terrell said Lipton denied Liman three times?"

I chuckled as I recalled that particular passage of Terrell's excellent final argument, but instantly became alert to the danger in such allegations. Once a charge like that is made, it gains a life of its own.

"That's a cheap shot," I hotly replied. "Anti-Semitism played absolutely no role in this trial, and until you brought it up, I had never even considered it."

Moffett admitted that he, too, had not encountered the issue in his research in Texas, implying that this viewpoint was emanating from editors back East.

The New York Times had carried a somewhat shorter feature story on December 19; the *Journal* story of a day later contained significantly more detail—and controversy.

The pace of the coverage increased; it seemed as if rarely a day passed without some new angle on the case. The flimsiest of stories found space in the page of *The Wall Street Journal.* Speculation about the "Lebanese connection" between Judge Casseb and Joe Jamail (both are of Lebanese descent) seemed to fit nicely with the earlier (and equally groundless) charges of anti-Semitism.

In spite of these red herrings being thrown out by the bargeful, some of the real facts about the case were emerging.

The pro-Texaco sentiments of the early days following the verdict were being tempered somewhat.

Still, it was obvious that six months after I first became involved with the case, the verdict had removed the lawsuit from a Texas courtroom to a new playing field whose size and shape was not yet apparent.

38

This part of the story gets a little confusing but, fortunately, has been widely reported elsewhere. With the bright lights of an ongoing national media event shining on the participants, a series of conflicting legal struggles in courtrooms in Texas, New York, and Washington, D.C. (among others) delineated the new playing field.

The first salvo had been fired by Texaco. When Judge Casseb had upheld the jury verdict, Texaco had immediately gone into federal court in its hometown of White Plains, New York, and obtained an injunction against Pennzoil and the Texas courts.

A provision of the civil law in Texas (and many other states) provides that the loser of a civil court action post a bond to guarantee the award while an appeal was pursued. Since the verdict was for $10.53 billion, a bond of similar size had to be posted or Pennzoil could attach liens to any Texaco property in the state of Texas—and there was a lot of it.

U.S. District Judge Charles Brieant of White Plains, New York, said that such a requirement was absurd and ordered Texaco to post a bond of $1 billion. He didn't stop there, however, and in his order Judge Brieant delivered a blistering rebuke to the jury's decision and hinted that he himself might overturn the verdict.

While Texaco moved to maximize the public impact of what was arguably its first "victory" in the long, bitter history of the case, Pennzoil announced that it would appeal the New York judge's decision to the Second Circuit Court of Appeals.

Texaco was still faced with the task of pressing its appeal in the Texas courts, however. The injunction on the bond issue did buy them some breathing room, but the verdict would have to be appealed on the merits.

The first step was an application to Judge Casseb for a new trial.

Perhaps emboldened by its success in federal court, Texaco filed a huge brief that listed over two hundred reasons why it was entitled to a new trial.

To no one's surprise, the by-now familiar charges against Judge Farris were renewed. Texaco also launched an all-out attack on Judge Casseb, challenging his status as a retired judge, among other things. If there was a judge in the entire state of Texas that Texaco thought was acceptable, it hadn't found him or her.

Rick Lawler and I were rewarded for speaking out in defense of the jury verdict; the Motion for New Trial charged us both with Jury Misconduct. Texaco claimed that Rick wrongfully concealed the fact of his estranged father-in-law's employment with Texaco. The implication was that some sort of family vendetta was then carried out against Texaco.

For my part, Texaco came up with two allegations of Jury Misconduct. The first concerned the by-now famous "Case of the Juror's Wife," regarding Susie's previous employment with Conoco. Specifically, they said I wrongfully concealed this fact and should have volunteered the information in response to a general question asked by Dick Miller during voir dire: "Is there any reason, whether it's religious, philosophical, moral, political, any reason at all, whether I've asked it or not, that you could not be a fair juror in this case? Please raise your hand."

I remembered listening hard to the questions during jury selection. If Miller had asked "Is there anybody here that thinks all the big oil companies are a bunch of crooks?" my hand would have shot up along with many others. Based on my research since the trial (specifically, my reading of *The Seven Sisters* by Anthony Sampson), I would now give the same answer from a much more educated point of view.

The relevance of Susie losing her job in the wake of DuPont's takeover of Conoco still puzzles me, however. If I was prejudiced against corporate takeovers, would that prejudice me against Pennzoil, which had launched the first tender offer for Getty Oil, or Texaco, which had ultimately acquired it? The lack of logic in the Texaco allegation is abundant; clearly, this was a desperation tactic.

The second charge of Jury Misconduct against me was particularly odious, as it was based on an outright lie. The "A.M. Houston" television program on which Rick and I had appeared after the trial had been recorded and transcribed by Texaco attorneys. At one point during the program, it was alleged that $10.53 billion was more than Texaco was worth. I had responded that this was not true, that Texaco's own *Annual Report* would disprove this claim: "When you take Texaco's assets and subtract their liabilities, you're left with a figure over $23 billion."

Yes, I did make that statement, and it happened to be true. The allegation made by Texaco was that because I could cite these figures on a television program on December 10, 1985 (three weeks after the verdict), this somehow proved that *during* the jury deliberations I had obtained a copy of a document not in evidence (the Texaco *Annual Report*) and had used the information to poison the minds of the rest of the jurors against Texaco.

From his conversations with the other jurors, Dick Miller had to know this allegation was not true. In a story printed in the *Houston Post*, I branded the charge as "a despicable lie." Actually, *Post* reporter Patricia Manson was the first one to inform me of this specific allegation. My immediate response to her was "That's a goodamn lie!"

Later, after I had cooled down, I called her back and modified "goddamn" to "despicable."

Had my opinion piece in the *Chronicle* touched a nerve somewhere, or was it just the fact that Rick and I refused to roll over and play dead in the face of what I felt to be Texaco's postverdict disinformation campaign?

The idea of being under personal attack by a powerful entity like Texaco filled me with new resolve. After being exposed to months of their tactics, the only difference here was that the assault was directed at me personally. The fact that they used our families in their attacks on Rick and myself was another example of what Joe Jamail had so aptly described during the trial as "Texaco morality."

When these charges were filed, my resolved increased. I was determined to write this book if I had to print it myself and stand on the street corner and hand out copies.

In retrospect, I have often thought that I should have been thankful that of the more than two hundred reasons Texaco argued it should receive a new trial, I was only implicated in two of them.

The fate of this Motion for a New Trial was sealed when Texaco began its attack on Judge Casseb. Failing to have him removed by a higher court (as it had earlier failed in a its attack on Judge Farris), Texaco was in the position of asking a judge it had personally attacked to rule in its favor.

A constable showed up at my office and slapped a subpoena from Texaco on me; Rich had been served in similar fashion. We had both planned to appear at the hearing anyway; Jamail had said he would call us as witnesses in defense of the false charges Texaco had levied against us. Jamail had also collected affidavits from the other jurors, attesting that the so-called misconduct never took place.

Like Miller's ill-conceived Motion to Recuse Judge Farris, the allegations of Jury Misconduct were doomed from the start.

When the hearing began on the morning of February 20, 1986, Judge Casseb began by asking the attorneys for Texaco if they still challenged his right to preside over the hearing on the motion.

One by one, representatives from each of the law firms representing Texaco in the courtroom that day rose and said that they maintained their challenges. Boies, McMains, Sales, and Miller's junior partner Mike Peterson each affirmed that was their position.

Judge Casseb didn't miss a beat before saying that if that was the case, he would "let the law take it's course." He banged his gavel and adjourned the hearing.

Like the president of the United States, who can legally kill a congressional bill by failing to act before the end of the session (the "pocket" veto), a trial judge in a state district court in Texas can effectively deny a Motion for a New Trial by failing to act on it within a given time frame.

It's not a question of fair or unfair; it's a matter of law. Judge Casseb's unwillingness to preside over a hearing involving attorneys that challenged his right to hear the case was a matter of personal dignity, I think, that was fully supported by constitutional law.

Texaco added the failure of Judge Casseb to hold this hearing to their litany of alleged misdeeds and prepared to take its appeal to a higher court. His action also had the effect of wiping out the charges of Jury Misconduct. Although they could be raised again in a future proceeding, the failure of the trial judge to give credence to such charges almost always results in similar treatment by higher courts.

On February 20, 1986, the same day that Judge Casseb effectively dismissed Texaco's motion for a new trial, the Second Circuit Court of Appeals issued a ruling on Pennzoil's appeal of the injunction Texaco had obtained from Judge Brieant in White Plains.

While affirming the practical aspect of Judge Brieant's ruling (reducing the appeal bond to $1 billion), the Second Circuit opinion also took him to task for the comments he had made attacking the judgment, citing it as an "impermissible appellate review" of issues that were still before a state court in Texas. This had been the thrust of the Pennzoil appeal, that the action of the federal judge in New York had intruded in what was still a state proceeding in Texas.

Shortly after the opinion was handed down, Pennzoil made the decision to carry its appeal of this side issue to the U.S. Supreme Court. Texaco, with its small victory largely intact, still had to press its appeal of the $10.53 billion judgment in the Texas courts.

In the fall of 1986, a three-judge panel from the First District Court of Appeals in Texas heard oral arguments on the extensive briefs filed by Texaco and, in response, by Pennzoil.

Since this case had long since outgrown anything resembling ordinary jurisprudence, it should come as no surprise that the arguments were moved to the large auditorium at the South Texas College of Law in downtown Houston. This court routinely hears arguments in a much smaller room in the same building.

For the case that was now styled *Texaco* v. *Pennzoil*, however, the facilities were stretched to their limits. Over 1,000 lawyers, brokers, reporters, and the genuinely curious public lined up early in the morning to attend the hearing. For the first time in the memory of all present, *tickets* were printed up for a court hearing. The ducats were color coded by group: Texaco, Pennzoil, the press, friends of the court, the college itself, and finally, the rabble—the general public. As a proud member of the last group, I joined the clamor for a seat.

The security for the event was tighter than I've seen for presidential visits. There were three checkpoints at which you had to open your purse or briefcase, walk through metal detectors, and, in some cases, submit to electronic body searchers.

Inside the auditorium, you could see the different classes of ticket holders spread out before you: Pennzoil and Texaco supporters were seated on opposite sides, interestingly behind the distinguished friends of the court, the law school, and the press.

A spillover crowd of perhaps 300 watched the proceedings on closed-circuit television on another floor of the building.

The same principal cast of lawyers who participated in the arguments on the Motion to Enter Judgment carried most of the burden here, Jamail and the Baker & Botts redoubt for Pennzoil; David Boies, Russell McMains, and Richard Keeton, principally, for Texaco.

Arthur Liman attended the hearing, but did not speak on behalf of Pennzoil. "Witness" Liman left that to a senior partner in his firm, Judge Simon Rifkind. The 85-year-old attorney had been appointed to the federal bench by Franklin Roosevelt and had practiced law for longer than Joe Jamail and Dick Miller had been alive.

Judge Rifkind (even the three robed justices addressed him as "Judge") offered a concise, but powerful, response to Texaco's assertion that the jury verdict was contrary to New York law.

After addressing several points raised in Texaco's brief, Rifkind said that "I am a stranger" to their application of New York law and called Texaco's actions in the acquisition of Getty Oil "as brazen an interference with another's contract as I have ever seen."

The assembled masses were, of course, treated to a rhetorical outpouring from Jamail, who said that, left unchecked, Texaco would "bring us back to the laws of the jungle."

Russell McMains, the large, bearded Texaco attorney from Corpus

Christi, thundered and all but roared at the justices, who asked enough questions to keep everyone guessing at their reaction to the arguments.

Having read the briefs filed by both sides, there were no real surprises there for me. Deep into the research and writing of this book, the story just kept rumbling on all around me.

During a break in the day-long hearing, I saw Brian Wunder of Baker & Botts in the lobby. He introduced me to an interested observer, Houston socialite Carolyn Farb. Carolyn herself had attracted some attention when the Texas legal system and her apartment-magnate ex-husband had delivered a $20 million divorce settlement to her a few years back. Since then, she had surprised more than a few people by refusing to be content with tending to charity balls and designer wardrobes. Instead, she became something of a spokesperson on issues of importance to the community.

Drawn to this case by the sheer scope of the drama and the personalities involved, she was chagrined to discover the East Coast press engaging in a furious round of what could only be described as "Texas bashing." The drastic downturn in the oil industry had led to hard times in Texas, especially Houston, where the unlimited horizons of a few years before now seemed sharply defined.

With the Texaco verdict erroneously cited as emblematic of the reasons that the boastful Texans had surely earned their comeuppance, a number of articles in *The New York Times, The Wall Street Journal,* and other national papers and magazines took shots at our wounded economy, culture, and even architecture.

Carolyn, who spent no little amount of time in New York herself, was furious at the one-sidedness of this coverage and was not afraid to say as much (primarily in letters to the editors of the offending publications). Carolyn Farb taking up the gauntlet in this largely regional war was yet another of the bizarre twists and turns generated by this landmark legislation.

The relatively technical nature of the ongoing legal proceedings almost invited this one-dimensional journalistic approach, the sort of reportage eager to come to sweeping conclusions to demonstrate their intimacy with the facts of a complex case.

I have tried (unsuccessfully) to imagine how I would have perceived this case had I not been involved as a juror; the sheer volume of the newspaper and television coverage assures me that it would not have avoided my notice.

While it was definitely not business as usual over at Texaco headquarters, the announcement was made that chairman John McKinley would retire at the end of 1986, more than two years after his scheduled departure.

Rumors about his successor centered on the next two men in the corporation, Vice-Chairman James Kinnear and President Alfred DeCrane. Talk of a "war party" and a "peace faction" within the upper ranks of Texaco persisted. Former naval officer Kinnear was most often mentioned as a leading proponent of the peace faction, while DeCrane was seen as a hard liner on the subject of the ongoing litigation.

Allana Sullivan, the New York–based reporter for *The Wall Street Journal* and resident expert on Texaco's internal politics, documented this new wrinkle in considerable detail. In one article speculating about the order of succession to the autocratic McKinley, Kinnear and DeCrane asserted their independence. When pressed, however, neither man could (or would) be able to recall an occasion when he had disagreed with McKinley.

Those who hoped the appointment of McKinley's successor would signal a new direction in this litigation were disappointed in late 1986, when it was announced that Kinnear and DeCrane would essentially share the prize. DeCrane would replace McKinley as chairman, while Kinnear took DeCrane's position as president. Kinnear, however, was also designated the chief executive officer, which meant that he had in fact attained the highest position in Texaco, the corporation he had served for three decades.

These events were but a few of many that fueled a continuing series of wild fluctuations in the price of both Texaco and Pennzoil stock. From the time of the verdict itself, any rumor that managed to reach Wall Street would trigger a whole new round of speculation.

For most of the last year (1987, year 2 of the epoch), the wheels of justice have turned fitfully. The actions of high courts are hard to predict, especially in high-profile cases. News of any major legal milestone would revive all the attendant recriminations.

In February 1987, the Texas First District Court of Appeals unanimously upheld the jury's verdict. The appeals court ruling reduced the amount of punitive damages from $3 billion to $1 billion. More telling than the reduction itself is the fact that the justices set the punitive damages at $1 billion dollars—said to be more than the combined amount of punitive damages paid in all cases ever. With the accumulated interest, the court had upheld a total damage award still well over $10 billion.

In April 1987, the U.S. Supreme Court overturned Judge Brieant's White Plains ruling limiting Texaco's appeal bond to $1 billion.

Faced with the distinct possibility it would in fact have to post a $10 billion bond, Texaco reportedly offered Pennzoil a last-ditch settlement offer valued at $2 billion. When Pennzoil declined what Liedtke called a "shotgun settlement," Texaco made good on a threat it had first raised over 15 months before—filing bankruptcy. This threat apparently had made little impact on Pennzoil through the many months of fruitless negotiations: having obtained a $10 billion judgment that had already been upheld in the early rounds of appeals gave Pennzoil a strong bargaining position.

Texaco's ultimate threat was to trigger the Doomsday Machine—to take its large, profitable corporation into Chapter 11, filing what would be by far history's largest bankruptcy.

Many financial analysts remained frankly skeptical and continued to discount Texaco's bankruptcy threat, as they had since it was first raised.

For his part, Liedtke responded to this Texaco threat by recalling an anecdote of a more personal nature: "When my daughter Kristie was a little girl, she'd threaten to hold her breath until she died if she couldn't have her way. She'd turn red and scare her mother and me half to death. On our pediatrician's advice, one time we just let her hold her breath till she keeled over. She never did it again. She has a very sweet disposition now."

When Texaco's new CEO Kinnear was told of the Liedtke statement, he fired back, "The truth is he was holding our head under water. You *better* hold your breath."

Given the prevailing atmosphere, it may not seem too surprising in retrospect. But in April 1987, Texaco went into a federal bankruptcy court in White Plains and filed for protection under Chapter 11. Less than a week after the Supreme Court stripped it of the protection of Judge Brieant's ruling, Texaco had again sought shelter in its hometown federal courts.

U.S. Bankruptcy Judge Howard Schwartzberg was, like Judge Brieant, based in White Plains. His rulings throughout the Texaco bankruptcy would be critical to the outcome of the case.

Initially, Judge Schwartzberg said that since Texaco was in Chapter 11, the $1 billion bond it had posted would suffice. Under the provisions of the bankruptcy code, however, Judge Schwartzberg had an obligation to protect Texaco creditors as well as shareholders. The largest Texaco creditor, of course, was Pennzoil with its $10 billion judgment.

Despite the longstanding threats, many people were taken aback by the bankruptcy filing, not least of all the Texaco shareholders whose dividend payments were immediately suspended. These included large numbers of Texaco retirees or their widows who depended on this dividend income. Many large institutional pension plans were also Texaco shareholders. All these interests were subordinated to Texaco management's desire for vindication. The peril that bankruptcy held for management was the very real extent to which they gave up their ability to exercise control over their own corporation.

Liedtke pointed this out in news interviews, predicting that "the euphoria would quickly fade."

This, too, would later seem something of an understatement as new financial players would come to see the bankruptcy as a viable, if unorthodox, business opportunity.

Texaco took its appeal to the Texas Supreme Court while the state legislature was still in session. Simultaneously, a piece of legislation dubbed the "Texaco bail-out bill" wound its way through the Texas House and Senate. Although it was destined to fail, vast resources were expended on lobbying on both sides. Scores of Texaco employees "spontaneously" appeared at the state capitol to lobby lawmakers.

The decision, however, remained with the judiciary. Like the U.S. Supreme Court, the Texas Supreme Court hears only a small percentage of the cases that are submitted for review. In November 1987, the Texas Supreme Court ruled there was "no reversible error" and allowed the decision to stand without a hearing. The ruling came after a thorough review of the record (one justice alone reported spending over 100 hours examining the lower court actions).

Since the Texas high court declined to hear the case, no oral arguments were made. Another round of trashing the Texas judiciary predictably followed, with a new air of desperation.

Texaco's last recourse was an appeal to the U.S. Supreme Court, which only agrees to hear a small number of the cases presented to it.

In December 1987, a bizarre (even by the standards of this case) series of events transpired that made this eventuality seem even more unlikely. Following the "Black Monday" stock market crash in October 1987, a new player had emerged on the scene. Renowned corporate raider Carl Ichan acquired the large (12.5%) block of Texaco stock that had been accumulated by Australian investor Robert Holmes à Court in the wake of the Texaco Chapter 11 filing. Holmes à Court had lost hundreds of millions of dollars in that market fiasco and apparently sold his shares to bolster his financial war chest.

Ichan sensed opportunity in this situation. In addition to his stake in Texaco, he began to purchase Pennzoil shares as well. With the prospect of the legal battle and bankruptcy proceedings dragging on for years, Ichan started to apply pressure. Judge Schwartzberg was not adverse to these new developments; in fact, he issued an order that would allow a settlement of the dispute if approved by Texaco shareholders, Pennzoil, and the committee representing those creditors who were cut off when Texaco filed Chapter 11.

This effectively froze out Texaco's management and board of directors from negotiations on a settlement that would be imposed on them by the same bankruptcy court where they had sought protection. Judge Schwartzberg's decision served as the "reality pill" Joe Jamail had long been prescribing for Texaco management.

Within a matter of days, top-level negotiations between Pennzoil and Texaco produced a $5.5 billion plan to be submitted for approval by the Texaco shareholders. Under this proposed plan (announced on December 19, 1987), Pennzoil would receive a cash settlement of $3 billion, with another $2.5 billion earmarked to pay off the creditors and allow Texaco to emerge from bankruptcy.

Ichan, who had ended up running TWA after an earlier takeover, made it clear that he had no intention of going into the oil business and would support any action that would allow him to take his profits and get out. He railed against stringent antitakeover measures put in place by the Texaco board. Invariably, stock prices soar to new heights when a company is perceived to be a strong takeover target, a situation rendered difficult if not impossible by the action of the Texaco board. Ichan then threatened to offer his own proposal for a shareholder vote. Any such proposal would have a hard time winning the necessary two-thirds approval.

My own experience in electoral politics convinces me that people primarily base their voting decisions on the impact it will have on their pocketbooks. Those Texaco shareholders hadn't received any dividend payments since the April bankruptcy filing, which seemed to indicate that approval of the settlement appeared highly likely.

Pennzoil's willingness to accept a $3 billion settlement of a $10 billion verdict makes sense when you consider how much longer the case could have dragged on in the courts, not to mention the fact that even if the entire award were upheld, it would be paid under the control of the bankruptcy court, which could have diluted the total sum even further.

I certainly don't feel that a settlement for a lesser amount in any way diminishes the jury verdict. In fact, by approving a settlement of this magnitude, Texaco's hometown federal court seems to have vindicated the jury system in general and the much maligned Texas judicial system in particular.

It seems fitting somehow that this landmark legal case would ultimately be resolved in a manner as unconventional as the litigation itself. It is hard to define clear winners and losers in this costly, bitter fight, although Pennzoil certainly proved that you interfere with their contracts at your peril.

Some cynics would say that the only winners are the lawyers on both sides. The astronomical legal fees paid out (the exact amounts were not revealed) lend some justification to this view, but the fact that a rich and powerful corporation like Texaco can be brought to the bar of justice speaks volumes about that most sacred of American doctrines: equal justice under the law.

EPILOGUE

A fresh consideration of the facts of this case is necessarily impossible for me. The figures that leave the most lasting impressions are the human ones, not the zeroes in the verdict.

About a month after the trial, I picked up the phone and called Judge Farris. He had recovered sufficiently to return to his office in the courthouse, but the long-term prognosis was not good. He was formal and restrained in our conversations, but said some interesting things. The first question I asked him was basically whether he agreed with the verdict we had reached after he had stepped down. He said that he concurred with our finding of fact (Pennzoil had a binding agreement with Getty), but still had some questions about the amount of damages. Judge Farris pointedly refused to speculate on what he felt the correct measure of damages was, however, but told me how the case began with him convinced that Pennzoil could not prove its contract.

"By the third day of Mr. Kerr's testimony, as I saw the documents presented in evidence, I began to change my mind," Judge Farris recalled. I thought of the irony of the fact that Texaco had tried so hard to oust this judge who himself tended to support that firm's position.

Judge Farris mentioned that he had read my opinion piece in the *Chronicle*, but was not very forthcoming about his reaction to it. This led me to suspect that he had at least approved of it, for I felt sure he would have expressed any disapproval.

I told Judge Farris that he was the only person in the case with whom I had any previous contact, citing the two times that I had voted for his Democratic opponents. He laughed and said that was all right.

"I hope your name appears on the ballot again, so I'll have a chance to correct the oversight, I told him.

Later, John Jeffers, Julius Glickman, and another attorney would tell me how Judge Farris had recounted to them my story about voting for his opponents and wish to make amends.

The attack on the character and integrity of Judge Farris continued to be one of the featured highlights of the Texaco legal and public relations effort. "I have been savaged," he complained in one interview. This was one tough Marine, however, and he again fought the allegations that had been rejected by Judge Jordan in that special hearing over two years before.

The health of Judge Farris continued to decline. While serving as U.S. attorney, Judge Farris had hired a number of assistants who had gone on to careers of distinction in their own right. Two prominent Houston attorneys in this group, Joe Dooling and Ted Pinson, organized a lunchtime reception that was held at the midtown condominium of Judge Farris and his wife, Wanda.

The judge was not permitting any premature eulogies this day, however. In spite of his weakened condition, he regaled the large gathering with war stories and talk of old times. In addition to the group in attendance, former Farris proteges such as Austin mayor Frank Cooksey and Associate Justice Raul Gonzalez of the Texas Supreme Court sent messages.

Judge Farris had been a stern taskmaster, and Pinson described a fitting toast offered at the reception: "Three cheers for the fairest unfair man in Texas."

"Judge Farris was a Republican man of the people," Pinson would later recall, "and Texas needs more of them."

In September 1986, Judge Anthony J. P. Farris died in Houston. His funeral at the Sacred Heart Co-Cathedral was attended by almost every sitting judge, in silent testimony to the honor of this man who had indeed "been savaged" by some involved in this case.

Presiding Judge Thomas Stovall delivered a moving eulogy to the large, somber crowd of mourners. Joe Jamail was one of the pall bearers. Carl Shaw, the loyal bailiff, attended the judge in death as he had in life. I felt a profound sadness tinged with a genuine regret that Judge Farris had not lived to see the final resolution of the case that had consumed his final year.

The elegant offices of the law firm of Miller, Keeton, Bristow & Brown occupy the top floor of one of the giant glass and steel towers that dominate the Houston skyline.

In the weeks and months after the trial, I had had a number of conversations with Dick Miller about the case. Although he was not heavily involved in the ongoing appeal of the jury verdict, Miller remained the paramount defender of Texaco in my mind. After witnessing his strong advocacy during nearly five months of the trial, it could hardly be any other way.

Since the Pennzoil-Texaco verdict, Miller had not shied away from important cases. His successful defense of a securities case in West Texas in 1987 had been a hard-fought victory in a trial where literally hundreds of millions of dollars were at stake.

My numerous interviews had included two previous face-to-face meetings with Miller, where the conversation had covered a wide range of topics but inevitably returned to the one thing we had in common.

As I neared the deadline for the mansuscript for this book, I wanted to have one last session with Dick Miller.

Although almost two years had passed since the end of the trial, my research had kept the memories fresh. Miller, however, had moved on to other cases and I almost had to prompt him to get a response. It was as if the memory of the case had been filed away in a restricted section of his mind.

After nearly an hour of questions and reminiscences, however, he warmed to the task at hand. What followed was an interview where this formidable trial lawyer revealed some pungent observations about a number of the participants in the drama.

It was immediately clear that he considered his true nemesis in this case to be Hugh Liedtke, not Joe Jamail. He called Liedtke "one of the smartest men in America, and one of the shrewdest and toughest." Referring to what he considered Liedtke's hardball business tactics, Miller added, "If Hugh Liedtke lived in New York, he would make Larry Tisch look like a wimp."

As he had during the trial, he insisted that the case began and ended with the Pennzoil chairman: "Hugh Liedtke will raise hell if he loses, as sure as night follows day. . . . His single-mindedness is what brought this case to trial."

He compared Liedtke to the number two man in the organization, Baine Kerr, a former partner of Miller's during his Baker & Botts days: "The difference between Kerr and Liedtke is the difference between resolve and total resolve. . . . If it had been left to Baine Kerr and [Pennzoil general counsel] Perry Barber, they would have looked for a way to settle this case."

Miller was still rankled about the testimony of Kerr and especially Liedtke: "The judge permitted them to get on the stand and make the same speech over and over, giving patently false evidence, stating that an agreement in principle is binding."

He complained that Liedtke had offered this testimony "in that cornball style like he just got off a drilling rig in Odessa, when in fact he's as sophisticated a man as you'll ever run across."

This detectable element of gruding respect for his adversary did not carry over to some of the other witnesses. It was clear that the way

Miller saw the case was shaped by a view of these witnesses that sharply differed from the reactions of the jury in the courtroom.

Martin Siegel and Geoffrey Boisi, the investment bankers for the Trust and Getty Oil, were cited as two examples that confirmed this view: "When we went up to New York to start working on the case, what struck me was the contrast between Boisi and Siegel, a case of opposites if I ever saw one. Boisi was honest, upright; Siegel was the epitome of crafty, . . . I didn't trust him at all. I knew he would find a way to help Pennzoil."

In fact, the deposition testimony of Siegel had indeed bolstered Pennzoil's case, especially the January 5, 1984, Siegel affidavit filed in the court in California.

Of course, the effect was unintentional. The affidavit had been prepared in the attempt to get the temporary injunction lifted so Gordon Getty could sign the final merger documents with Pennzoil. Within hours, however, Gordon had signed the "I regret" letter declaring his intention to sell the Trust's stock to Texaco.

At the trial, Pennzoil attorneys had confronted nearly every witness with the Siegel affidavit, which seemed to affirm a "done deal" between the Trust and Pennzoil.

Miller said he had learned that Siegel tried to solicit business from Liedtke while the case was pending: "I found out that he had called on Liedtke after the deal was done and the lawsuit was filed, hustling business. [Siegel] went with [head of local Kidder office] to call on Kerr and Liedtke."

In addition, Miller recalled that Siegel had been present at the Pierre when the negotiations between McKinley and Gordon had taken place and that he "sat in silence while deal was done. He didn't volunteer information about that affidavit."

While I can understand Miller's view of Siegel, I don't see the strong contrast with Boisi. The jury had frankly had a hard time buying key parts of Boisi's testimony, although he admittedly did not come across as slick as the urbane Siegel.

One matter that has come into dispute since the trial was the decision to put Martin Lipton on the witness stand for Texaco. A number of sources had Texaco attributing that decision to Miller, but he said, "I didn't want to put Lipton on. I didn't think he could add that much to the case, that it was an unnecessary risk. But I was overruled by my partners and Texaco."

As has been noted, Lipton's testimony had been damaging indeed. I asked Miller about Bruce Wasserstein of First Boston, who had been cited as the source of much of the strategy outlined in the DeCrane notes. Although part of his deposition was read to the jury, Wasserstein had not appeared in the courtroom. Obviously, Texaco could have produced him had they so desired. Miller groaned and shook his head, saying "Wasserstein would have been Lipton in spades."

The meaning of words was another recurrent theme from the trial that came up in this conversation. Miller was incredulous as I pointed to the January 4 Getty press release as an outward sign of a binding agreement between Pennzoil and the Getty entities. He recalled the paragraph that said the transaction was "subject to the execution of a definitive merger agreement."

I had to laugh at Miller's analogy: "If somebody says we're going to have sex tonight subject to how you treat me, that doesn't mean we're going to have sex."

In the end, he was philosophical about the outcome of the case and the fact that some fingers have been unfairly pointed in his direction for blowing the case. "I would have got all the credit for the win. During the trial I never thought about another thing. . . . In the end, we lost. I regret the loss from the standpoint of my client, but I don't brood about it."

He mused aloud about his necessary air of detachment from the cases he tries, observing "I don't get any personal satisfaction from it at all. Any happiness I have is for my client, and maybe relief that I don't have to file for a new trial."

"Sometimes I think I ought to be over at the criminal courthouse defending criminals, but by the end of the first day I'd probably say 'Hell, they all belong in the penitentiary.'"

As always, Dick Miller left me with mixed emotions, but my admiration and respect for him is genuine. In the *Texaco* case, he had played the hand he was dealt and lost. There was no doubt he would fight and win again, and I felt a brief wave of sympathy for those attorneys who will walk into a courtroom and find Dick Miller sitting at the opposing table.

In the wake of the tentative settlement announced in December 1987, the *National Law Journal* published an article that was essentially a review of the impact of the entire case.

Dick Miller was quoted throughout the piece, which ended on this bizarre note:

"Mr. Miller defends not presenting testimony on the damages claim, but said he would have done one thing differently: He would have blocked juror James Shannon from sitting.

"Mr. Shannon, now writing a book tentatively titled 'Texaco and the $10 Billion Jury' was, in Mr. Miller's words, 'responsible for the outcome of the case.'"

I categorically reject this notion as absurd in the extreme. It is not that I am adamant about disavowing personal responsibility for my actions on this jury; I embrace that responsibility and would take the same actions if the trial began tomorrow. But I was only 1 of 12, and the actions of this jury have been thoroughly scrutinized by two appellate courts. Whether the author of the book you now hold was "responsible for the outcome of the case" is a judgment I will leave to the reader.

The time since the trial had made the cast of characters in this drama seem even more remarkable. Rarely a month went by without major headlines involving someone in the case.

Laurence Tisch had bought a major block of stock in CBS, apparently at the network's request, in its battle to resist Ted Turner's take-over bid. As Gordon Getty had discovered in the Getty deal, once Tisch is asked into a situation, all bets are off. In a grand palace coup, the shrewd billionaire had ended up running the network—replacing CBS chairman Thomas Wyman, who had engineered his entrance to the scene.

Martin Siegel left Kidder, Peabody and went with another investment house in New York. While visiting with Martin Lipton at the Wachtell, Lipton offices one afternoon, a federal subpoena was served on Siegel. He had been caught up in the net of the insider trading scandal that rocked Wall Street in 1986. Ivan Boesky had earlier been pinched and had made a deal with the government that apparently included Siegel. It was announced soon after that Siegel, too, would plead guilty to providing inside information to Boesky in, among other deals, Getty-Pennzoil-Texaco.

Arthur Liman became chief Senate counsel for the select committee on the Iran-contra affair and received wide acclaim for his efforts in the televised hearings. It was somewhat disconcerting to see Liman, whom I had experienced as a friendly witness, in his prosecutorial garb. The sharp questioning to which he subjected Hakim, Secord, North, and Poindexter finally brought home the point that Miller had been trying to make with his references to "Witness" Liman. It was readily apparent that Liman is a trial lawyer of great strength and sophistication, a fact that no doubt magnified the damage of his testimony to Texaco's defense.

As Pennzoil's representative outside the two-day Getty board meeting, Liman had been a key player. His testimony at the trial had put the events of the deal making of January 1–3 into clear focus.

*   *   *

With these thoughts and others in mind, I made one last trip to the top of yet another tall building downtown to have a final talk with Joe Jamail. As usual, Jamail was a man in motion. Having instructed his secretary to hold all his calls, he only talked to a dozen people in the two or three hours I was there.

When the phone would ring, Jamail would grab it, and say, "I thought you were going to hold my calls." He'd listen for a second and say, "Well, give him to Denise and let her handle it. . . . What? Oh, that's that Kirk Douglas thing. I'll take it."

When we focused on the case at hand, he asked me who I felt had been his best witness for Pennzoil. I recalled the testimony of Arthur Liman and mentioned the feeling that his subsequent television appearances had given me. Jamail emphasized the fact that Liman was on the scene when the offer-counteroffer-acceptance ritual that resulted in the Pennzoil-Getty agreement was being played out: "Liman was *there*, he looked honest, he is honest, and he never bad-mouthed anybody, in contrast to his friend, Lipton."

During my research I had found material that appeared to challenge that alleged friendship. To Jamail, I recounted a quote from an *American Lawyer* article which had described the relationship between the Liman and Lipton as "conspicuously unfriendly."

Jamail responded, "I don't think they're *friendly* . . . they're just in different worlds, really. Liman is a lawyer, a trial lawyer, an advocate; Lipton is a wheeler and a dealer. I'm not saying he's not a good lawyer, just a different kind. He's not a trial lawyer; he's so busy carving up corporate America and making fortunes at it."

In fact, though, Lipton no longer occupied the central seat in Jamail's shooting gallery. Marty Lipton had been subdued in the public press about the effect his testimony had on the jury in this trial. In all the reading I had done on the case, I hadn't found any vitriolic Lipton responses to the jury verdict. Others had lashed out in his behalf, however, but they all failed to address the central role Lipton had played in the case, first in the Pennzoil deal, then in the Texaco deal . . . with his past and present clients (Texaco and Tisch) who were deeply involved.

For Jamail, however, Lipton had long since ceased to be the principal male factor in the case. "What did they want him to do, come down here and *lie* for them?" Jamail said of Lipton.

Joe Jamail had reliable substitute villains waiting in the wings: the whole class of mergers and acquisitions lawyers and investment bankers. At the beginning of the trial, he had ascribed the blame to "a conspiracy" among this group; the months of the trial and appeals had not diminished his antipathy for this group, whom he saw as "disregarding the goodamn morality and the law, saying 'They don't apply to us, we're sacred.' *Bullshit*. This jury, right here, told them something. And you see what's happened since then. It's part of all the insider business. This case is intrinsically woven into that. Boesky makes money off Siegel off the Getty deal; somebody had to have tipped Boesky."

Jamail's strong words echoed those of Dick Miller at the trial, when he called hostile takeovers "the most maligned tactic in American business today" and predicted they would someday be illegal.

Is that, finally, the lesson to be learned from this case? Maybe it's one of them, but of course it's not that simple.

What could Texaco have done to prevent this fiasco? The obvious answer is that it could have *not* interfered with an agreement that had been announced to the public less than forty-eight hours before.

More practical would have been for Texaco to have followed what it swore throughout the trial was its time-honored corporate policy: to engage only in friendly takeovers. Had Texaco's need for oil reserves been so desperate that it was convinced that its bid for Getty Oil was really friendly? Or was it merely taken in by the assurances of men like Lipton, Siegel, Wasserstein and Perella, operators on both sides of this deal who had met before many times in corporate takeover fights, as both allies and opponents?

In other words, was Texaco itself really a victim of the whole deal-making process? Perhaps, but as Jamail pointed out in the trial, the top executives of Texaco were men "who presumably know how to read and write the English language." The repeated protestations that the Getty entities were not bound to Pennzoil was belied by the language they insisted be inserted in their agreements with Texaco.

That the top men at Texaco chose to ignore these provisions is not really in dispute, much to the sorrow of their shareholders and employees.

The evidence suggests to me that Texaco felt it could continue with its policy if it just changed the definition of the word "friendly." Maybe the real battle for the hearts and minds occurred when DeCrane convinced McKinley that this was really a "friendly" deal, if indeed that occurred. At any rate, these questions do nothing to absolve Texaco from what was clearly its responsibility in this case.

When the dust finally settles and this whole bitter mess starts to fade away, it is my sincere hope that there will be some understanding of what has transpired here and that, ultimately, it will make some little difference in the way we do business in this country.

NOTES

PART I: BLOOD IN THE WATER

Chapter 1

Page 4: The account of J. Paul Getty's life and marriages was taken largely from "The Curse on the House of Getty" by Russell Miller, Los Angeles, July 1983 (adapted from a series in *The Sunday Times*, London, and later expanded in Miller's book, *The House of Getty*, Harold Holt and Co., 1985).

Other details are from *The Great Getty* by Robert Lenzner, Crown, 1985.

Pages 6–8: The early careers of George, Ronald, Paul, and Gordon Getty are described in Miller's *The House of Getty*.

Paul's drug use is mentioned in various Getty family court documents and is confirmed by his own lawyer, Vanni Treeves, in *Oil & Honor* by Thomas Petzinger, Jr., G. P. Putnam's Sons, 1987.

Chapter 2

Page 10: Lansing Hays' behavior was described in an unattributed quotation in *The New York Times* article of December 11, 1983, admitted into evidence during the trial testimony of J. Hugh Liedtke.

Pages 10–21: The struggle within Getty Oil from May 1982 to December 1983 is detailed in "Getty Games" by Stephen Brill, *American Lawyer*, March, 1984.

PART II: THE SIX DAY WAR

Chapter 3

Pages 25–38: The account of the events of January 1–6, 1984 is the author's compilation of the sworn testimony of participants J. Hugh Liedtke, Gordon

P. Getty, Baine P. Kerr, Martin Siegel, Martin Lipton, Harold Williams, Arthur Liman, Geoffrey T. Boisi, Bruce Wasserstein, Alfred C. DeCrane, Laurence Tisch, Barton Winokur, John McKinley, Patricia Vlahakis, James Kinnear, William Weitzel, and Charles (Tim) Cohler.

Pages 34–35: As noted, the account of Texaco's 1940 Nazi entanglement is taken from *The Seven Sisters* by Anthony Sampson, Bantam Books, 1975.

PART III: "THE WHITE KNIGHT GETS SUED BY THE DRAGON"

Chapter 4

Pages 41–43: The January 25, 1984 hearing in Delaware is described in Petzinger's *Oil & Honor* and Brill's "Getty Games," *American Lawyer*, March, 1984.

Chapter 5

Page 44: Joe Jamail described his background in a series of interviews with the author in 1986–1987.

Pages 45–47: The background of the Miller, Keeton firm was described in "Breaking Away: Cracks in the Houston Legal Establishment" by Gary Taylor, *Houston City Magazine*, September, 1985, and in the author's 1986 interviews with Dick Miller.

Chapter 6

Pages 57–69: The account of the October 25, 1984 hearing on the Motion to Recuse was largely derived from the transcript of the proceedings, supplemented by the author's interviews with Dick Miller, Joe Jamail, John Jeffers, and Julius Glickman.

Pages 57–58: The account of the United Gas deal and Pennzoil's involvement with the Nixon campaign are from "The Scrappy Mr. Pennzoil" by James R. Norman with Terri Thompson, Cynthia Green, and Ellen Davis, *Business Week*, January 27, 1987.

PART IV: A SUMMONS FROM THE SHERIFF

Chapter 7

Page 73: The details of how jury candidates are selected in Harris County are from the author's interview with deputy clerk Marion Cleboski.

PART V: "THE MOST IMPORTANT CASE IN THE HISTORY OF AMERICA"

Chapter 8

Pages 86–151: The voir dire (jury selection) is documented in the trial transcript for July 9–16, 1985.

Chapter 9

Page 91: The comment of the prospective juror at the trial of John Wayne Gacy is recorded in *Buried Dreams: Inside the Mind of a Serial Killer* by Tim Cahill based on the reporting of Russ Ewing, Bantam Books, 1986.

Page 92: The comment on the voir dire is contained in "Blood Feud" by Stephen J. Adler, *American Lawyer*, October, 1985.

Page 102: The demographics of the Panel of 100 were compiled by the author from the Juror Information Forms completed by each panel member.

Page 103: Dick Miller's account of Rob Fuller's search for revealing bumperstickers is from a 1987 interview with the author.

Chapter 12

Page 139: The account of Susan Roehm's subpoena service of John McKinley is from the author's interview with Joe Jamail as confirmed by Roehm.

Chapter 14

Pages 155–164: The opening statements of John Jeffers and Richard Keeton are documented in the trial transcript for July 17, 1985.

PART VI: CRITICAL DOCUMENTS I

Pages 165–195: Each document was admitted into evidence during the trial. The author's explanatory notes are derived from trial testimony about the documents.

PART VII: BURDEN OF PROOF

Chapter 15

Pages 200–214: The testimony of Baine Kerr is documented in the trial transcript for July 18–26, 1985.

Page 215: Jamail's comments about the Kerr testimony are from his August 1987 interview with the author.

Chapter 16

Pages 216–220: The video depositions of Sidney Petersen, Peter McNulty, Ralph Copley, Charles Cohler, and Geoffrey Boisi are documented in the trial transcript for July 30–August 5, 1985.

Chapter 17

Pages 221–236: The testimony of Arthur Liman is documented in the trial transcript for August 5–9, 1985.

Chapter 18

Page 237: The Jamail-Miller clash is documented in the trial transcript for August 8, 1985.

Pages 239–243: The video depositions of Patrick Lynch and Alfred DeCrane are documented in the trial transcript for August 12–14, 1985.

Chapter 19

Pages 244–247: The testimony of Dr. Ronald Lewis is documented in the trial transcript for August 14–19, 1985.

Pages 247–248: The video deposition of Martin Siegel is documented in the trial transcript for August 19–20, 1985.

Pages 248–250: The testimony of Clifton Fridge is documented in the trial transcript for August 20–21, 1985.

Chapter 20

Pages 252–264: The testimony of Thomas Barrow is documented in the trial transcript for August 29–September 3, 1985.

Chapter 21

Pages 265–280: The video deposition of Gordon Getty is documented in the trial transcript for August 22–26, 1985.

Pages 270–272: This portion of Gordon Getty's video deposition was not admitted into evidence during the trial, but was transcribed by the author from the unedited videotape.

Chapter 22

Pages 281–307: The testimony of J. Hugh Liedtke is documented in the trial transcript for August 26–28 and September 3–10, 1985.

Page 305: Jamail's claim of pioneering the unzipped fly diversion is contained in "The Taking of Getty Oil," by Steven Coll, Atheneum, 1987.

Chapter 23

Pages 310–331: The testimony of John McKinley is documented in the trial transcript for September 10–18, 1985.

Page 323: The "Godzilla of Big Oil" reference is in Petzinger's *Oil & Honor.*

PART VIII: CRITICAL DOCUMENTS 2

Pages 333–347: Each document was admitted into evidence during the trial. The author's explanatory notes are derived from the testimony at the trial.

PART IX: "THEY NEVER HAD A CONTRACT"

Chapters 24-25

Page 351: The testimony of Barton Winokur is documented in the trial testimony for September 19–20 and September 26–October 8, 1985.

Chapter 25

Pages 360–362: The video deposition of Patricia Vlahakis is documented in the trial transcript for September 24–25, 1985.

Chapter 26

Pages 370–373: The testimony of Geoffrey Boisi is documented in the trial transcript for October 9–10, 1985.

Chapter 27

Pages 374–385: The testimony of Martin Lipton is documented in the trial transcript for October 17–18 and October 29–30, 1985.

Chapter 28

Pages 388–389: The testimony of Henry Wendt is documented in the trial transcript for October 15 and 29, 1985.

Pages 389–391: The testimony of Chauncey J. Medberry III, is documented in the trial transcript for October 31, 1985.

Chapter 29

Pages 393–396: The testimony of Laurence Tisch is documented in the trial transcript for November 4, 1985.

Pages 397–399: The testimony of William Weitzel is documented in the trial transcript for November 6–7, 1985.

Pages 399–402: The testimony of Alfred DeCrane is documented in the trial transcript for November 7–12, 1985.

Page 403: The offering from deposition of Martin Siegel is documented in the trial transcript for November 12, 1985.

PART X: "ARROGANCE, POWER, AND MONEY"

Chapter 30

Pages 407–425: The closing arguments of John Jeffers and Irv Terrell are documented in the trial transcript for November 14, 1985.

Chapter 31

Pages 426–447: The closing argument of Dick Miller is documented in the trial transcript for November 14–15, 1985.

Chapter 32

Pages 448–466: The closing argument of Joe Jamail and the comments of Judge Solomon Casseb are documented in the trial transcript for November 15, 1985.

Chapter 33

Pages 478–480: The jury's questions to Judge Casseb are documented in the trial transcript for November 18, 1985.

Chapter 34

Pages 485–486: The reading of the jury verdict is documented in the trial transcript for November 19, 1985.

Chapter 35

Page 492: The quote from Richard Lawler is reported in Petzinger's *Oil & Honor*.

Pages 494–498: The author's opinion piece appeared in the November 24, 1985 edition of the Houston Chronicle.

Chapter 36

Page 501: The widely-reported comments of William Weitzel were made in a speech to oil industry analysts in New York on November 20, 1985. The meeting had been scheduled well in advance and happened to fall on the day after the jury decision was announced.

Chapter 38

Page 516: The Liedtke-Kinnear exchange is from "The Gambler Who Turned Down $2 Billion," by Stratford P. Sherman, *Fortune*, May 11, 1987.

EPILOGUE

Page 519: The comments of Judge Farris are from telephone interviews with the author in December, 1985 and February, 1986.

Page 520: The description of the reception is from the author's interview with one of the hosts, attorney Ted Pinson.

Pages 521–523: The comments of Dick Miller are from his August, 1987 interview with the author.

Page 524: On pages 332–333 of a deposition given on June 15, 1984, J. Hugh Liedtke confirmed that Martin Siegel called on Pennzoil and Pharr Dusun of the Houston office of Kidder, Peabody after the lawsuit was filed. There was no evidence of any transactions resulting from the meeting.

Page 524: The insider trading investigation of Martin Siegel was reported in "Unhappy Ending: The Wall Street Career of Martin Siegel Was a Dream Gone

Wrong," by James B. Stewart and Daniel Hertzberg, *Wall Street Journal*, February 17, 1987.

Pages 524–525: The comments of Joe Jamail are from his August, 1987 interview with the author.

Page 525: The comment about Lipton and Liman being "conspicuously unfriendly" is from "Getty Games" by Stephen Brill, *American Lawyer*, March, 1984.

INDEX